Image Textures and Gibbs Random Fields

Computational Imaging and Vision

Volume 16

Image Textures and Gibbs Random Fields

by

Georgy L. Gimel'farb

The University of Auckland,
Auckland, New Zealand

SPRINGER SCIENCE+BUSINESS MEDIA, B.V.

A C.I.P. Catalogue record for this book is available from the Library of Congress.

ISBN 978-94-010-5912-1   ISBN 978-94-011-4461-2 (eBook)
DOI 10.1007/978-94-011-4461-2

Printed on acid-free paper

Table of Contents

Preface

Image analysis is one of the most challenging areas in today's computer science, and image technologies are used in a host of applications. This book concentrates on image textures and presents novel techniques for their simulation, retrieval, and segmentation using specific Gibbs random fields with multiple pairwise interaction between signals as probabilistic image models. These models and techniques were developed mainly during the previous five years (in relation to April 1999 when these words were written).

While scanning these pages you may notice that, in spite of long equations, the mathematical background is extremely simple. I have tried to avoid complex abstract constructions and give explicit physical (to be specific, "image-based") explanations to all the mathematical notions involved. *Therefore it is hoped that the book can be easily read both by professionals and graduate students in computer science and electrical engineering who take an interest in image analysis and synthesis.* Perhaps, mathematicians studying applications of random fields may find here some less traditional, and thus controversial, views and techniques.

If you do not like such dreadful things as probabilities, distributions, functions, derivatives, and so on, then you may go directly to the last three chapters containing a large body of experimental results. Experiments have been conducted with very different image textures so that you may readily see when the proposed models do adequately describe particular textures or are completely helpless. If you are still interested in the proposed modelling techniques, then there exist, at least, two options: to change your own attitude to this straightforward mathematics or to simply use its final "easy–to–program" results after skipping the boring derivations. The latter option seems to be a bit more realistic.

Probabilistic image models for image processing and analysis have been the subject of numerous papers, conferences, and workshops over many years, and today you can find many different books on image modelling. So it is desirable to underline in which respects this book differs from the existing literature.

To summarize briefly, **it tailors models and modelling techniques to images, rather than the reverse**. Most of the known probabilistic image models have been developed and successfully used in other scientific

ix

domains such as statistical physics, signal processing, or theory of measurements. An extension of a physical model to images calls for analogies
between the basic notions that describe a physical object (a system of particles, a particle, interactions of particles, an energy of the interaction, a
temperature, and so on) and the notions used to represent an image. Although distant parallels between them do exist, images differ much from
every physical system of interacting particles, and the models borrowed
from physics do not reflect salient features of image textures under consideration. Moreover, some physical notions, for example, a temperature or an
energy, may even be misleading when used in the image modelling context.

Thus, unlike more traditional approaches, this book makes an attempt
to show that modern image analysis is worthy of specific probabilistic image
models and modelling techniques which can be, in some cases, both simpler
and more efficient than their well-known counterparts borrowed from other
scientific domains.

This book covers the main theoretical and practical aspects of image
modelling by Gibbs random fields with multiple pairwise pixel interactions
and demonstrates the ability of these models in texture simulation, retrieval, and segmentation. For ease of reading, it reproduces basic mathematical definitions and theorems concerning Markov and Gibbs random
fields, exponential families of probability distributions, stochastic relaxation
and stochastic approximation techniques and relates these definitions and
theorems directly to texture models.

This book briefly overviews traditional Markov/Gibbs image models
but concentrates most attention on novel Markov and non-Markov Gibbs
models with the arbitrary, rather than pre-defined, structures and the arbitrary strengths of multiple pairwise pixel interactions. The most attractive
feature of our models is that both an interaction structure of a texture
and quantitative interaction strengths specified by Gibbs potentials can be
learnt from a given training sample.

As shown in this book, it is an easy matter to obtain rather close analytic first approximations of Gibbs potentials and use them to recover the
most characteristic interaction structure. The potentials are then refined
by stochastic approximation, and we indicate ways of converting such a
refinement into a controllable simulated annealing technique of generating
images, similar, in a probabilistic sense, to a training sample.

This book discusses a wide range of experiments with various textures
from the well-known album of Brodatz and the digital "VisTex" collection of the MIT Media Laboratory as well as with aerial and space images
of the Earth's surface. Our experiments show that many natural grayscale
textures can be efficiently simulated in the proposed way. Our models allow
the reliable query-by-texture retrieval from large image data bases which

is to some extent scale– and orientation–invariant. Also, considerable attention is given to supervised texture segmentation. It is considered as a simulation of a desired region map using a conditional Gibbs model of region maps corresponding to a grayscale piecewise-homogeneous texture, and the same controllable simulated annealing technique can be used to segment piecewise-homogeneous textures.

Of course, a large number of theoretical and practical problems still remain to be solved, but the current results alone seem to justify our efforts in studying and exploiting these models and techniques.

Acknowledgements

This research was started in the Division of Information Technologies and Systems of the V. M. Glushkov Institute of Cybernetics of the National Academy of Sciences of Ukraine (Kiev, Ukraine)[1]. In June 1994 – December 1995 the investigations were, in the main, conducted at the Computer and Automation Research Institute of the Hungarian Academy of Sciences (Budapest, Hungary). In March 1997 – July 1997 this research was performed at the Institut de Recherche d'Informatique et d'Automatique in Sophia Antipolis, France, under partial financial support of the Ministere de L'Education Nationale, Enseignement Superieur et de la Recherche. I am very much obliged to these institutions.

Since August 1997, this research has continued in the Centre of Image Technology and Robotics of the Department of Computer Science of the University of Auckland in Auckland, New Zealand, under the University of Auckland Research Committee grant XXXXX/9343/3414083.

In addition, I am indebted to Michigan State University, University of Pennsylvania, and University of Washington, USA, and to Deutsche Forschungsgemeinschaft, Germany, which provided financial support for my short visits to the Pattern Recognition and Image Processing Laboratory of Michigan State University in August – September 1994, Intelligent Systems Laboratory of the University of Washington in January – February 1995 and June - July 1996, Medical Image Processing Group of the University of Pennsylvania in August 1996, and Institut für Photogrammetrie of the Universität Bonn in October – December 1996.

It is difficult to overestimate the number of discussions and stimulating collaboration with Dmitry Chetverikov, Wolfgang Förstner, Robert Haralick, Gabor Herman, Anil Jain, Claus Liedtke, Mikhail Schlesinger, Marc Sigelle, Josiane Zerubia, and many other colleagues whom I had the good fortune to meet while strolling across Gibbs fields and stumbling over tangled and intricate textures. These long- and short-term contacts were of inestimable value in obtaining and clarifying results presented in this book. Of course any pitfalls which may exist are of my own doing.

---

[1]In 1997 this Division was transformed into the International Research and Training Centre on Information Technologies and Systems of the National Academy of Sciences and Ministry of Education of Ukraine.

I am very thankful to Dmitry Chetverikov for the digitized fragments of Brodatz textures used in Chapters 5 – 7, to Martin Kirscht for the fragment of a natural SAR image and its training map used in Chapter 7, to Rosalin Pickard and her group at the MIT Media Laboratory (USA) for the public-domain "VisTex" image database used in Chapters 5 and 6, to Leonid Wolfson for the fragment of a space image of the Earth's surface used in Chapter 7, and to Alexey Zalesny for the 5- and 16-region texture collages used in Chapter 7.

I wish to thank my pupil, Alexey Zalesny, for our fruitful joint work for many years. Recently he has extended these models and techniques to more complex color image textures, developed more efficient techniques for estimating characteristic interaction structures, and created software tools for solving various application problems. I expect that he will soon publish these results himself, and I have restricted the book to only my own or joint theoretical results and experiments.

Most of all I give thanks to CROCKISS for encouraging and helping me during more than 22 years.

Instead of introduction

The book considers spatially homogeneous and piecewise-homogeneous image textures and presents novel probabilistic techniques for their modelling and processing. Let us postpone for a while a very attractive but never-ending discussion about which images should be referred to as textures and whether such a texture means the same for human and computer eyes, as these topics will be touched upon in Chapter 1.

We treat image textures as samples of specific Gibbs random fields with multiple pairwise interactions between gray levels or other signals in the pixels. In later chapters you will encounter both positive and negative results of simulating, retrieving, and segmenting natural image textures using these Gibbs models.

Our eyes solve image retrieval and segmentation problems so easily that you may be puzzled why even the most successful computational models rank far below, both in quality and processing rate. For me, human vision is a miracle of Nature, and more than three dozen years spent in computer vision and image processing domains have only strengthened this belief. It is difficult to imagine that our vision (and, generally, all plant and animal life) could be created in line with the conventional theory of evolution. Thus I do not expect that the foreseeable future will bridge wide gap between human and computer visual skills. But, it is still a very attractive challenge to fit our visual abilities to a Procrustean bed of mathematical models and algorithms and to emulate, somehow, one or another side of human vision with these fundamentally "inhuman" tools. Fortunately, there are many natural image textures that can be closely approximated by our models and modelling techniques. So these models and techniques are worthy of investigation.

What you do and do not find here ...

If you, dear reader, have no time to go into detail but wish to find more information than in the Preface, here is a brief review of the book.

You will find in this book novel types of Gibbs random field image models that differ from or generalize the known ones. In many cases the proposed models are more convenient for describing spatially homogeneous and piecewise-homogeneous textures than traditional models.

Image homogeneity is quantitatively defined in this book in terms of conditional probability distributions of spatial signal combinations. Mostly the distributions are assumed to be translation invariant, that is, in different parts of a homogeneous image that can be superimposed by translation, the sample relative frequencies of signal combinations are expected to be almost the same. Examples of such translation-invariant homogeneous textures are shown below (the upper row contains four natural image textures, and the bottom one shows similar textures simulated as described in this book).

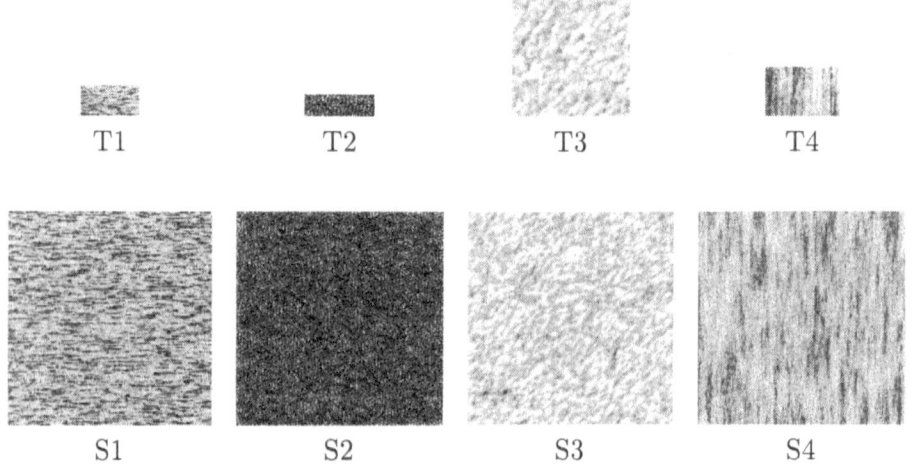

Training (T) and simulated (S) samples of the textures "Fur" (1), "Blood Cells" (2), "Pressed Cork" (3), and "Wood Grain" (4).

The translation-invariant textures constitute only a tiny part of all possible image textures but the described techniques of texture simulation and retrieval can use more general transformations of the homogeneous parts including not only translation but also limited rotation and scale invariance of the corresponding distributions.

Gibbs image models are specified by a finite set of quantitative parameters describing the geometric structure and quantitative strengths of pixel interactions. The interaction structure is formed by characteristic subsets of interacting pixels which are included in a model. The interaction strengths are given by so-called Gibbs *potentials*, or potential functions, that control the probability distributions of all the possible gray level combinations in these subsets.

It should be particularly emphasized that an interaction between the pixels, or equivalently, between the gray levels or other signals in the pixels, has no physical meaning. It reflects only the fact that some spatial signal combinations in a particular texture are more frequent than others: the less uniform the probability distribution of signal combinations in the pixels that have a particular spatial arrangement, the stronger the interaction between these pixels.

As in the majority of more traditional image models, our models are restricted to no more than pairwise pixel interactions. However, unlike the traditional models, we presume that every model involves an arbitrary structure of multiple pairwise pixel interactions, and the structure may vary for different textures. Also, we presume that every model may have arbitrary interaction strengths.

Strange as it may seem, our models with multiple pairwise pixel interactions turn out to be both more general and computationally simpler than the well-known particular cases such as auto-binomial or Gauss-Markov models, which are widely used in statistical physics and theory of measurements. These traditional models have mostly pre-defined structures and potential functions, contrary to our models which place almost no restriction on the interaction structure and potential values.

As distinct from traditional models, our models permit us to learn, or estimate, from a given training sample not only the potentials but also the structure of multiple pairwise pixel interactions.

You will find that such a generalization of the Gibbs models simplifies notably the computational techniques for learning the model parameters. The learning techniques introduced in this book are based firstly on analytic and subsequently on stochastic approximations of the unconditional or conditional maximum likelihood estimates (MLE) of the potentials. After the learning stage, every spatially homogeneous or piecewise-homogeneous image texture is described quantitatively by a particular set of the most characteristic families of translation invariant pixel pairs that specifies the interaction structure and by particular potentials for each family that specify the interaction strengths.

You will find that the proposed models give a more penetrating insight into the physical meaning of the parameters to be learnt with respect to image textures. The model parameters have a natural interpretation in terms of spatial self-similarity between the image patches, namely, in terms of statistical repeatability of signal combinations. In particular, a grayscale image texture is described as a whole by a set of particular gray level co-occurrence or difference histograms collected over the characteristic families of pixel pairs. The histograms are sufficient statistics for our models, and the book shows how the potentials depend explicitly on them.

The signal histograms allow the formation of a *model-based interaction map* for recovering the interaction structure. The interaction map shows the contributions of various families of pixel pairs to the overall probability of a given training sample, and even a simple thresholding can recover the characteristic families.

Also, **you will find** in the book a new modelling scenario of simulating samples in the vicinity of a given training sample as regards their probabilities in the context of a particular Gibbs model. The proposed scenario, called Controllable Simulated Annealing (CSA), is simpler, but more efficient in texture simulation or segmentation than the traditional modelling scenarios. These traditional scenarios are based on a prior estimation of the model parameters and subsequent image generation by stochastic relaxation or on a conventional "blind" stochastic gradient search by simulated annealing.

You will find in Chapters 5 – 7 many experimental results of simulation (generation), retrieval, and segmentation of different image textures. These results outline pros and cons of the proposed Gibbs models. We restrict our consideration to only supervised texture simulation and segmentation when each texture is represented by an adequate training sample (e.g., in the case of a piecewise-homogeneous texture this sample contains both the grayscale textured image and the corresponding map of homogeneous regions in this image). Under the alternative modelling scenario, both the simulation of spatially homogeneous or piecewise-homogeneous textures and the segmentation of piecewise-homogeneous textures are achieved in almost the same way. Also, the proposed models allow for computationally simple scale and orientation adaptation for texture simulation or retrieval.

You will not find in this book a thorough review of other known texture models although some limitations of more traditional Gibbs models, being overcome by the proposed approach, are outlined. Such comprehensive reviews can be easily found elsewhere.

Also, **you will not find** the experimental comparisons with other models because this book does not intend to provide evidence that the proposed models are much better. In fact, sometimes they do outperform, but in other cases may be inferior to, the known counterparts. The proposed models and modelling frameworks simply differ from and possess very attractive learning capabilities in comparison to the more traditional ones.

Generally, each mathematical model has its own place in describing images, and all of them are too simple to represent even a tiny part of such a natural miracle as (human) vision. So, when a model does emulate a small part of natural images, it deserves to be known and used in practice.

Now, after this brief review, let us refer to Chapter 1 and enter the exciting world of image textures, structures, and multiple pairwise pixel interactions...

Texture, Structure, and Pairwise Interactions

In this chapter we will consider at some length what an image texture means for human and computer vision. Also, this chapter presents the basic theoretical background of probabilistic image modelling by Gibbs and Markov random fields. Modern theory of random fields is extremely deep and intricated, but we will explore in this book only rather simple features of Gibbs random fields that are necessary to study the image models and modelling techniques under consideration.

1.1. Human and computational views

Such terms as spatial image structure and image texture are widely used in the present-day image analysis and computer vision. Nonetheless, both the terms, as is stated, for instance, by Haralick and Shapiro (1992), lack formal and precise definitions. Before any attempt to quantitatively define them, it is interesting to compare informal human views that can be found in most standard lexicons, say, The Oxford Dictionary (1971, 1989), Webster's Dictionary (1959, 1986), The World Book Dictionary (1990), and so on. Some instances given in Tables 1.1 and 1.2 and the similar ones enable the following separation.

A **structure** (from the Latin *structura* which means "to build, arrange") is defined in a broad sense as a mutual relation of the elements of the whole object or a fabric or framework of putting them together.

A **texture** (from the Latin *textura* which means "weaving", "web", or "structure") relates to a specific structure of visual or tactile surface characteristics of particular objects such as natural woven ones (fabrics, tissues, weaves, webs, paintings). In a broad sense, the texture defines also the structure or composition of an object with regard to its components.

Probably, most extensively both terms are elaborated in modern petrography. The *texture* describes smaller features of a rock depending upon a size, shape, arrangement, orientation, and distribution of the components but the *structure* relates to the larger features such as foldings, faults, crackings, etc. In other areas the texture is also referred mostly to small-scale surface features or fine structures of the objects and only in rare cases directly replaces the structure.

TABLE 1.1. Informal definitions (The Oxford Dictionary, 1971; 1989)

Texture	Structure
1. The process or art of weaving; the fabricating or composing of schemes, writings, etc.	*1.* Manner of building or construction; the way in which an edifice, machine, implement, etc. is made or put together.
2. A woven fabric, or any natural *structure* having an appearance or consistence as if woven.	*2.* The mutual relations of the constituent parts or elements of a whole as determining its peculiar nasture or character. of a whole.
3. The character of a textile fabric as to its being fine, coarse, close, loose, etc., resulting from the way in which it is woven.	*3.* The coexistence in a whole of distinct parts having a definite manner of arrangement.
4. In extended use: The constitution, *structure*, or substance of anything with regard to its constituents or formative elements.	*4. In a wider sense*: A *fabric* or framework of material parts put together.

TABLE 1.2. Informal definitions (Webster's Dictionary, 1959; 1986)

Texture	Structure
1. Something composed of closely interwoven elements.	*1.* The action of building; something constructed or built; something made up of more or less interdependent elements or parts; something having a definite or fixed pattern of organization.
2. The essential part of something, an identifying quality.	*2.* The manner of construction; the way in which parts of something are put together or organized.
3. The size and organization of small constituent part of a body or substance; the visual or tactile surface characteristics and appearance of something.	*3.* The arrangement of particles or parts in a substance or body; the interrelation of parts as dominated by the general character of the whole.
4. A basic scheme or *structure*; the overall *structure* of something incorporating all or most of parts.	*4.* The elements or parts of an entity or the position of such elements or parts in their external relationships to each other.

Although both the notions - the structure and the texture - are rather close in the meaning, the former is much more universal. One may consider an internal structure of a given texture, that is, may study its particular elements and their spatial arrangement. But, it seems rather odd to consider "a texture of a structure" although this involves no contradiction.

Thus, for a human, the texture relates mostly to specific, spatially repetitive, to some extent, (micro)structures which are obtained by "weaving" of the object elements so that their arrangement matches, in a broad sense, visual (or tactile) features of a woven fabric. Such a textured surface has a rough translational invariance in that it is formed by repeating a particular patch or a few patches in different spatial positions (generally, with some local variations of scale, orientation, or other geometric and photometric features of the patches).

Image textures can be specified as (*i*) grayscale, color, or multiband images of natural textured surfaces and (*ii*) simulated patterns which approach, within certain limits, these natural images. Image sensors involve additional geometric transformations as well as specific linear or non-linear photometric transformations of the perceived surfaces. Spatially uniform photometric transformations and different overall changes of the image orientation or scale do not usually affect a particular type, or class of a texture. But, the translational invariance of a spatially homogeneous image texture can be lost after non-uniform geometric and photometric transformations.

The texture is somewhat more restricted than the structure itself, and each human definition involves various informal qualitative structural features such as fineness, coarseness, smoothness, granulation, lineation, directionality, roughness, regularity, randomness, *etc.* These features, defining a spatial arrangement of specific texture constituents, help to single out the desired texture types, for instance, fine or coarse, close or loose, plain or twilled or ribbed textile fabrics. The human classifications are too fuzzy to form a basis for formal definitions of image textures, because it is still unclear how to associate these features, more or less easily perceived by a human, with computational models that describe the image textures.

As a result, present computational views on image textures are conditioned by particular natural examples generally recognized as the *de facto* prototypes, such as the well-known photoalbum of Brodatz (1966) and more recent digital "VisTex" collection created in the MIT Media Laboratory by Pickard et al. (1995)[1]. To simulate or analyze the image textures, we need computational models which are adequate to describe these natural images.

---

[1] Although in past years, there were several critical comments concerning the abuse of the Brodatz textures and the need of more diverse natural prototypes, the two abovementioned collections are most extensively used in today's studies on texture analysis and synthesis, and this book is no exception.

1.2. Spatial homogeneity, or self-similarity of textures

Each computational model of a texture involves a particular set of basic constituents, or elements and rules of their spatial arrangement. The texture elements are called *textons* by Julesz (1981) and Julesz and Bergen (1983) or *texels* by Haralick and Shapiro (1992). In a finite lattice, each texel is supported by a specific spatial combination of the pixels. The texel is represented by gray levels in these pixels so that a particular support produces a set of possible texels. In this book we restrict our consideration to only the grayscale textures. But the Gibbs models which will be studied in the subsequent chapters can be easily extended to color or multi-band textured images.

Generally, there exists a hierarchy of texels starting from the most primitive ones. Complex elements are built of one or several primitives arranged spatially in a particular way. In terms of the texels, the basic property of an image to be a homogeneous texture is its *spatial self-similarity*, that is, (statistical) repetitiveness of the texels, supported by the same or geometrically similar pixel combination, over the image. Of course, this definition is still informal but, at least, it offers a way of building the formal one which has to specify quantitatively what and how must be repeated in a texture.

For human, the spatial self-similarity, or repetitiveness, is quite definitive. But it is very hard or impossible to give it a general constructive formal definition. An early self-similarity concept of Chetverikov (1987) referred to a subimage (patch) of the minimum size that can still be considered as a texture and to a binary similarity relation between the patches. But, no ways were proposed how to find them in practice. The recent and more elaborated approach of Zalesny (1994) explores the distances between conditional probability distributions of a particular function of gray level combinations in the patches to quantitatively measure the self-similarity. The patches are certain connected subsets of pixels which are equivalent under a given group of geometric transformations. But, in the general case, it is unclear how to find these patches and functions for testing the spatial self-similarity of a given image.

To avoid impracticable generalizations, the self-similarity concept is applied in this book to only the *translation invariant* image textures. Translational invariance of a texture is the simplest type of its spatial homogeneity. It is typical for many human-made textured objects such as fabrics, nettings, weavings, etc. In the idealized case of a spatially regular texture, the translational invariance is produced by the identical texels. These latter differ only by their positions in the image, and the positions are governed by the fixed rules of the relative arrangement of the texels.

Most natural image textures have only the approximate translational self-similarity. It permits the texels of a single type to geometrically and photometrically vary in the different positions preserving the continuity of an image. For example, there could be local scale – orientation or offset – contrast variations of the texels shown, for example, in Figure 1.1. In addition, images of a natural 3D surface covered by a translation-invariant texture may demonstrate certain affine or projective transformations that describe the viewing conditions and surface geometry.

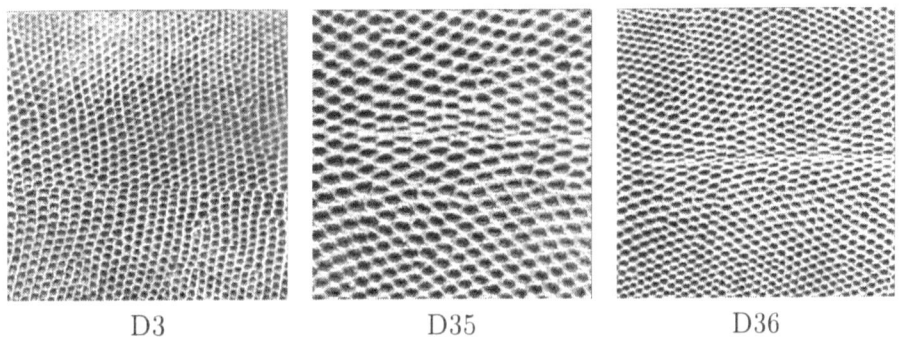

D3   D35   D36

Figure 1.1.   Local scale–orientation variations of the texels in the 512 × 512 fragments of the natural textures D3 (Reptile skin), D35 (Lizard skin), and D36 (Lizard skin) from the Brodatz collection.

As a result, the self-similarity of a natural image texture has to take account of the admissible local geometric changes over the image that preserve the image continuity and do not effect the "textural" visual appearance of the surface. The admissible photometric transformations of images representing the same natural object preserve such an appearance, too. In the simplest case, the photometric transformations are linear (changes in the total offset and contrast of an image). But many sensors or preprocessing techniques result in certain non-linear changes of the gray scale. Therefore, the texture models should be invariant with respect to all these changes.

This book considers only the simplest grayscale texels supported either by a single pixel or by a pixel pair. To underline the translational invariance of a texture, the pixel pairs are stratified into the "translation-invariant" families. Each family contains all the pairs in the lattice that have the same relative spatial arrangement of the two pixels and differ only by their absolute positions. We assume a *probabilistic self-similarity* such that all the possible signal pairs in a particular family are considered as the texels but with very different probabilities of being met in a texture; only a few texels have the probabilities which are significantly different from zero.

Such restrictions allow to precisely define the spatial self-similarity in terms of marginal probability distributions of the signals and signal pairs in a translation-invariant image texture.

Definition 1.1 *Two grayscale images represent the same translation-invariant texture if they have the same or, at least, closely similar marginal probability distributions of gray levels in the pixels and of gray level co-occurrences in the pixel pairs that belong to a particular characteristic subset of the families of translation-invariant pixel pairs.*

This definition suggests that the self-similarity can be quantitatively measured by using one or another statistical goodness-of-fit test for the sample relative frequency distributions of gray levels and of particular gray level co-occurrences in the images. As will be shown below, Gibbs random fields as image models directly exploit this scheme.

1.3. Basic notation and notions

This section presents the basic notions and designations used in the subsequent chapters. Main abbreviations and special notation are listed in Tables 1.3 and 1.4, respectively.

TABLE 1.3. Abbreviations

CSA	Controllable Simulated Annealing
DEM	Digital Elevation Model
GLCH	Gray Level Co-occurrence Histogram
GLDH	Gray Level Difference Histogram
GLH	Gray Level Histogram
GL/RLH	Gray Level and Region Label Histogram
GLC/RLCH	Gray Level and Region Label Co-occurrence Histogram
GLD/RLCH	Gray Level Difference and Region Label Coincidence Histogram
GPD	Gibbs Probability Distribution
GRF	Gibbs Random Field
IDB	Image Data Base
IRF	Independent Random Field
MLE	Maximum Likelihood Estimate
MRF	Markov Random Field

Probabilistic image modelling is based on assumption that each digital image is a sample of a particular random field and can be met in experiments with a particular probability. An image model relates the image signals (e.g., gray levels) to the probability. Also if the model is genera-

TABLE 1.4. Special notation.

$\{\dots\}$	Set of values.
$[\dots]$	Vector or array
$[s_i : i \in \mathbf{c}]$	Vector of scalars s_i with indices i passing through a support $\mathbf{c}$.
$\|s\|$	Absolute value of a scalar s.
$\mathbf{N}_i$	Neighborhood of a site i in a lattice.
$\mathbf{c}^i$	All components of a set $\mathbf{c}$ except the component i.
$\|\mathbf{s}\|$	Cardinality of a set $\mathbf{s}$.
$s : \mathbf{c} \to \mathbf{U}$	Scalar function of the argument $i \in \mathbf{c}$ which takes values from a finite set $\mathbf{U}$. The finite set $\mathbf{c}$ of argument values is a *support*, or supporting set of the function.
$\mathbf{s} = [s_i : i \in \mathbf{c}]$	Graph of a function s, or an array of values s_i taken by the function for each element i of a support $\mathbf{c}$.
$\mathbf{x} \bullet \mathbf{y}$	Dot product of vectors $\mathbf{x}$ and $\mathbf{y}$.
$\mathcal{R}$	Set of real numbers.

tive then various image samples can be simulated in such a way that their sample relative frequencies tend to the probabilities specified by the model.

For completeness, we start with general definitions of the images and random fields on finite lattices. Let $\mathbf{R} = [(m, n) : m = 0, \dots, M - 1; n = 0, \dots, N - 1]$ be a finite 2D arithmetic lattice containing MN sites, or pixels, with integer Cartesian coordinates (m, n). With respect to a continuous optical image, the rectangular lattice $\mathbf{R}$ has M horizontal rows and N vertical columns of pixels. Mostly, we will use the shorthand notation i to designate a pixel $(m, n) \in \mathbf{R}$.

Definition 1.2 (Ritter et al., 1990) *Digital image* $\mathbf{s} = [s_i : i \in \mathbf{R}]$ *on a finite 2D arithmetic lattice* $\mathbf{R}$ *is the graph of a function* $s : \mathbf{R} \to \mathbf{U}$ *that maps the supporting lattice* $\mathbf{R}$ *to a finite set* $\mathbf{U}$ *of integer signal values.*

To put it another way, a digital image is an array $\mathbf{s} = [s_i : i \in \mathbf{R}]$ of values s_i taken from the same finite signal set $\mathbf{U}$.

Definition 1.3 *(Cramer and Leadbetter, 1967; Feller, 1970) Random field* $\mathbf{S}$ *on a finite 2D arithmetic lattice* $\mathbf{R}$ *is an array of random variables* $\mathbf{S} = [S_i : i \in \mathbf{R}]$ *supported by the lattice sites* $i \in \mathbf{R}$.

What this means is that any random field $\mathbf{S}$ with the components S_i taking values s_i from the same finite signal set $\mathbf{U}$ can be considered as a probabilistic model of images (as well as of other finite 2D arrays of single-type measurements). But to be of practical use the model should provide high probabilities of the desired images and zeroth or almost zeroth probabilities of other images. Usually the desired images to be modelled constitute only a tiny part of all the possible images.

Let $\mathcal{S}$ denote a *parent population*, that is, the set of all the possible samples $\mathbf{s} = [s_i : i \in \mathbf{R}]$ with the signals from a given finite set: $s_i \in \mathbf{U}$. Each random field $\mathbf{S}$ is specified by a particular joint probability distribution $\Pr : \mathcal{S} \to [0, 1]$ giving the probabilities $\Pr(\mathbf{S} = \mathbf{s})$ of the samples $\mathbf{s} \in \mathcal{S}$ over the parent population. The joint probability $\Pr(\mathbf{S} = \mathbf{s})$ depends on the signals in each sample $\mathbf{s} = [s_i : i \in \mathbf{R}]$ and has the ordinary features:

$$\forall \mathbf{s} \in \mathcal{S} \;\; \Pr(\mathbf{S} = \mathbf{s}) \geq 0; \;\; \sum_{\mathbf{s} \in \mathcal{S}} \Pr(\mathbf{S} = \mathbf{s}) = 1.$$

The parent population of the images is of size $|\mathcal{S}| = |\mathbf{U}|^{|\mathbf{R}|}$, so that there is usually a combinatorial number of the possible images, and the probability $\Pr(\mathbf{S} = \mathbf{s})$ of each image is extremely small by value. So the models are expected to simply provide essentially higher probabilities of the desired images relative to the other images in the parent population.

This book considers the following three types of grayscale textures: spatially homogeneous images, region maps, and piecewise-homogeneous images. Examples of more or less homogeneous image textures are given in Figure 1.1, and examples of piecewise-homogeneous textures and their region maps are shown in Figure 1.2. The images and region maps differ only by a physical meaning of the signals s_i in the pixels.

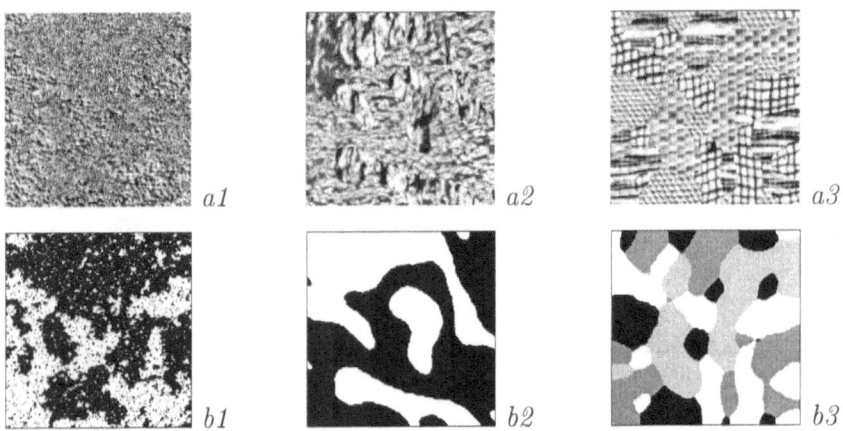

Figure 1.2. Two-region (*a1*, *a2*) and four-region (*a3*) piecewise-homogeneous textures and corresponding region maps (*b1 – b3*).

Digital homogeneous grayscale image $\mathbf{g} = [g_i : i \in \mathbf{R}; \; g_i \in \mathbf{Q}]$ takes values from a finite set $\mathbf{Q} = \{0, \ldots, Q\}$ of gray levels. Gray levels represent optical signals that correspond to brightness of the object points within the field-of-view (FOV) of an image sensor and are measured in metric scale. Usually, the set $\mathbf{Q}$ contains 256 gray levels ($Q = 255$). Our texture simulation, segmentation, and retrieval experiments in Chapters 5 – 7 are conducted with the lesser number of gray levels, namely, with 16 gray levels ($Q = 15$), for reducing the number of model parameters to be estimated.

A *gray range* $[q_{\min}(\mathbf{g}), q_{\max}(\mathbf{g})]$ of a particular grayscale image $\mathbf{g}$ is bounded by the minimum, $q_{\min}(\mathbf{g}) = \min_{i \in \mathbf{R}} \{g_i\}$, and the maximum gray value, $q_{\max}(\mathbf{g}) = \max_{i \in \mathbf{R}} \{g_i\}$, in the image.

Digital region map $\mathbf{l} = [l_i : i \in \mathbf{R}; l_i \in \mathbf{K}]$ takes values from a finite set $\mathbf{K} = \{0, \ldots, K\}$ of integer region labels. Region labels, measured in nominal scale, indicate parts of the sensor's FOV occupied by different objects. Region map $\mathbf{l}$ splits the lattice $\mathbf{R}$ into $|\mathbf{K}|$ non-overlapping regions $\mathbf{R}_k = [i : i \in \mathbf{R}; l_i = k] \subset \mathbf{R}$ such that $\bigcup_{k \in \mathbf{K}} \mathbf{R}_k = \mathbf{R}$. The gray-coded region labels are shown in Figure 1.2, *b1–b3*.

Digital piecewise-homogeneous image $(\mathbf{g}, \mathbf{l})$ is represented by the pair containing a grayscale image $\mathbf{g}$ and its region map $\mathbf{l}$. In the case, the objects are different homogeneous textures, and the region map represents these homogeneous objects so that each pixel i supports the two scalar signals, (g_i, l_i), giving the gray level and the region label, respectively. It is evident that each homogeneous image $\mathbf{g}$ has a singular region map $\mathbf{l}$ that contains only a single region coinciding with the whole lattice $\mathbf{R}$.

We denote $\mathcal{G}$ and $\mathcal{L}$ the parent populations of grayscale images $\mathbf{g}$ and region maps $\mathbf{l}$, respectively, supported by the same lattice $\mathbf{R}$.

The above notation is summarized in the following table:

Image $\mathbf{s}$	Signal set ($\mathbf{U}$) of	Parent population ($\mathcal{S}$)
Grayscale homogeneous image $\mathbf{g}$	Gray levels $\mathbf{Q}$	$\mathcal{G}$
Region map $\mathbf{l}$	Region labels $\mathbf{K}$	$\mathcal{L}$
Piecewise-homogeneous image $(\mathbf{g}, \mathbf{l})$	Gray level and region label pairs $\mathbf{Q} \times \mathbf{K}$	$\mathcal{G} \times \mathcal{L}$

Thereafter, we use the relative, conditional, and marginal probabilities of image signals related to the joint probability $\Pr(\mathbf{S} = \mathbf{s})$. These probabilities have the following definitions.

Let a sample $\mathbf{s}'$ have the non-zero probability, $\Pr(\mathbf{S} = \mathbf{s}') > 0$. Then the ratio $\Pr(\mathbf{S} = \mathbf{s}) / \Pr(\mathbf{S} = \mathbf{s}')$ is referred to as the *relative* probability of the sample $\mathbf{s}$ with respect to the sample $\mathbf{s}'$.

Let $\mathbf{R}^i$ denote the lattice $\mathbf{R}$ with exception of a pixel i. Let $\mathbf{S}^i$ and $\mathbf{s}^i$ denote all the components of the random field $\mathbf{S}$ and sample $\mathbf{s}$ with the exception of the component S_i and s_i, respectively, in the pixel i. The *conditional* probability $\Pr(S_i = u|\mathbf{S}^i = \mathbf{s}^i)$ of a signal value u in the pixel i, given the fixed signals $\mathbf{s}^i$ in all the other pixels, is obtained from the joint probability $\Pr(\mathbf{S} = \mathbf{s}) \equiv \Pr(S_i = s_i, \mathbf{S}^i = \mathbf{s}^i)$ as follows (Feller, 1970):

$$\Pr(S_i = u|\mathbf{S}^i = \mathbf{s}^i) = \frac{\Pr(S_i = u, \mathbf{S}^i = \mathbf{s}^i)}{\Pr(\mathbf{S}^i = \mathbf{s}^i)};$$

$$\sum_{u \in \mathbf{U}} \Pr(S_i = u|\mathbf{S}^i = \mathbf{s}^i) = 1.$$

Here, $\Pr(\mathbf{S}^i = \mathbf{s}^i)$ is the *marginal* probability of the sample components $\mathbf{s}^i$ supported by the sublattice $\mathbf{R}^i$:

$$\Pr(\mathbf{S}^i = \mathbf{s}^i) = \sum_{u \in \mathbf{U}} \Pr(S_i = u, \mathbf{S}^i = \mathbf{s}^i).$$

Let $\mathbf{R}^\mathbf{J}$ denote the lattice $\mathbf{R}$ with exception of the subset $\mathbf{J}$. Let $\mathbf{S}^\mathbf{J} = [S_i : i \in \mathbf{R}^\mathbf{J}]$ and $\mathbf{s}^\mathbf{J} = [s_i : i \in \mathbf{R}^\mathbf{J}]$ be the components of the random field $\mathbf{S}$ and the sample $\mathbf{s}$ supported by the subset $\mathbf{R}^\mathbf{J}$, respectively. Let $\mathbf{S}_\mathbf{J} = [S_j : j \in \mathbf{J}]$ and $\mathbf{s}_\mathbf{J} = [s_j : j \in \mathbf{J}]$ be the components of the random field $\mathbf{S}$ and the sample $\mathbf{s}$ supported by the subset $\mathbf{J}$, respectively. Then the conditional probabilities are generally defined for the signal combinations over any subset $\mathbf{J} \subset \mathbf{R}$ of the pixels, given the fixed signals in all the other pixels $\mathbf{R}^\mathbf{J}$ and the joint distribution $\Pr(\mathbf{S} = \mathbf{s})$, as follows:

$$\Pr(\mathbf{S}_\mathbf{J} = \mathbf{s}_\mathbf{J}|\mathbf{S}^\mathbf{J} = \mathbf{s}^\mathbf{J}) = \frac{\Pr(\mathbf{S}_\mathbf{J} = \mathbf{s}_\mathbf{J}, \mathbf{S}^\mathbf{J} = \mathbf{s}^\mathbf{J})}{\Pr(\mathbf{S}^\mathbf{J} = \mathbf{s}^\mathbf{J})}$$

where $\Pr(\mathbf{S}^\mathbf{J} = \mathbf{s}^\mathbf{J})$ is the marginal probability of the sample components supported by the sublattice $\mathbf{R}^\mathbf{J}$:

$$\Pr(\mathbf{S}^\mathbf{J} = \mathbf{s}^\mathbf{J}) = \sum_{j \in \mathbf{J}} \sum_{u_j \in \mathbf{U}} \Pr(\mathbf{S}_\mathbf{J} = [u_j : j \in \mathbf{J}], \mathbf{S}^\mathbf{J} = \mathbf{s}^\mathbf{J}).$$

From here on, for brevity, in most cases we omit the explicit specifications of the random field components and denote the above joint, marginal, and conditional probabilities $\Pr(\mathbf{s})$, $\Pr(\mathbf{s}^i)$, and $\Pr(s_i|\mathbf{s}^i)$, respectively.

1.4. Random fields and probabilistic image modelling

Probabilistic signal modelling pursue the goal of simulating, or generating random samples described by a given signal model $\Pr(\mathbf{s})$. Generally, by this is meant that the sample relative frequencies of generated samples should approximate the probability distribution $\Pr(\mathbf{s})$. But the image modelling goal must be specified in a less general way because the parent populations are of combinatorial size, and each particular simulation experiment can result in only a negligibly small part of the whole population.

This book is concerned with only image modelling by *supervised learning*. In this case the desired types of image textures are specified by one or several examples called *training samples*. The image model is given in a parametric form, $\Pr(\mathbf{s}, \theta)$, and has to be "focussed" on a desired small subset of images by estimating, or learning, the quantitative parameters θ from the training samples. Then the simulation has to produce the images with probabilities in a close vicinity of the top-rank probabilities in the chosen model.

We shall restrict our consideration exclusively to the maximum likelihood estimates (MLE) of the model parameters describing the geometric structure and quantitative strengths of multiple pairwise pixel interactions over the lattice. Under certain assumptions, the MLE θ^* ensures the maximum probability of the training sample $\mathbf{s}^\circ$ in the model $\Pr(\mathbf{s}^\circ, \theta^*)$ comparing to all other variants $\Pr(\mathbf{s}^\circ, \theta)$, $\theta \neq \theta^*$:

$$\theta^* = \arg \max_\theta \Pr(\mathbf{s}^\circ, \theta).$$

The model is assumed to be adequate for describing the desired image type if the relative probability of the training sample $\mathbf{s}^\circ$ (with respect to the maximum probability $\max_{\mathbf{s} \in \mathcal{S}} \Pr(\mathbf{s}, \theta^*)$, which can be provided by this model) is sufficiently close to one.

Computational complexity of the random field image modelling can be roughly evaluated by a number of signals per pixel that take part in generating a sample. In the general case, the joint probability distribution $\Pr(\mathbf{s})$ is an intractable function of $|\mathbf{R}|$ variables $[s_i : i \in \mathbf{R}]$. Therefore it is practicable only if it can be decomposed into the tractable components that each depend (at least, in the average) on a relatively small number of variables. This feature is offered by particular Markov and Gibbs random fields which can thus serve as image models.

Below we overview the basic properties of these fields with respect to modelling the homogeneous and piecewise-homogeneous image textures.

1.4.1. NEIGHBORHOODS AND MARKOV RANDOM FIELDS

Probabilistic relations between the signals in different pixels are usually called *pixel interactions*, and random field image models are conveniently described in terms of interacting pixel pairs called *neighbors*.

Definition 1.4 (Besag, 1974; Moussouris, 1974) *A pixel $j \in \mathbf{R}$ is called the neighbor of a pixel $i \in \mathbf{R}$ if the conditional probability $\Pr(s_i|\mathbf{s}^i)$ of the signal value s_i depends on the signal value s_j.*

Collection of all the neighbors of the pixel i in the lattice forms the *neighborhood* $\mathbf{N}_i$ of the pixel i.

The conditional probability of a signal in the pixel i depends only on the signals in the neighborhood $\mathbf{N}_i$, so that the obvious relation $\Pr(s_i|\mathbf{s}^i) \equiv \Pr(s_i|[s_j : j \in \mathbf{N}_i])$ holds. In other words, this conditional probability is actually supported by only the subset $i \cup \mathbf{N}_i$ of the pixels.

Definition 1.5 (Besag, 1974; Hassner and Sklansky, 1980) *A set* $\mathbf{N_S} = \{\mathbf{N}_i : i \in \mathbf{R}\}$ *of all the neighborhoods in the lattice comprises a neighborhood system of a random field* $\mathbf{S}$.

A neighborhood system is represented by a *neighborhood graph*. Vertices of the neighborhood graph are associated with the pixels $i \in \mathbf{R}$, and arcs join all the pairs of neighbors. The neighborhood graph displays the pairwise interactions in the images modelled by a particular random field.

Definition 1.6 (Cramer and Leadbetter, 1967; Feller, 1970; Dobrushin and Pigorov, 1975) *A random field* $\mathbf{S}$ *with a given neighborhood system* $\mathbf{N_S}$ *is called a Markov random field (MRF) if all the supports* $i \cup \mathbf{N}_i$ *of the conditional probabilities* $\Pr(s_i|[s_j : j \in \mathbf{N}_i])$ *are the proper subsets of the lattice* $\mathbf{R}$, *that is, have cardinalities that are less than the cardinality* $|\mathbf{R}|$ *of the lattice:*

$$|i \cup \mathbf{N}_i| < |\mathbf{R}|; \ \forall i \in \mathbf{R}.$$

The order of a MRF is specified by the maximum cardinality of the neighborhoods, $\max_{i \in \mathbf{R}}|\mathbf{N}_i|$. Markovian property means that each component of a MRF is independent of some other components. Therefore, the MRFs are of order between 1 and $|\mathbf{R}| - 2$, and the field of the maximum order $|\mathbf{R}| - 1$ should be referred to as the *non-Markov* random field.

The computational complexity of probabilistic image modelling depends mostly on computing the conditional probabilities $\Pr(s_i|\mathbf{s}^i)$, $i \in \mathbf{R}$. That is why only the MRFs of small order have become popular as image models (Chellappa and Jain, 1993; Li, 1995). In actual fact, the total computational complexity per image is dictated by all $|\mathbf{R}|$ conditional probabilities, that is, by the sum, rather than the maximum, of the cardinalities $|\mathbf{N}_i|$ of the pixel neighborhoods. Therefore the following assertion holds:

Lemma 1.1 *The computational complexity of random field image modelling depends on the average order of the field specified by the average cardinality of the pixel neighborhoods per pixel,* $\frac{1}{|\mathbf{R}|}\sum_{i \in \mathbf{R}}|\mathbf{N}_i|$.

The average order is equal to the ordinary one if all the pixel neighborhoods are of the same cardinality. But the average order and hence the computational complexity can be relatively small, as will be shown later, even for some simple non-Markov random fields. Therefore it is more convenient to use the average order rather than the ordinary order for ranking the random field models. Moreover, the number of the signals that take part in computing each conditional probability may be even lesser than the neighborhood cardinality if the current signal value $s_i \in \mathbf{U}$ influences the actual neighborhood.

1.4.2. GIBBS PROBABILITY DISTRIBUTIONS

A Gibbs random field (GRF) is specified by a Gibbs probability distribution (GPD) that defines the explicit geometric structure and quantitative strengths of pixel interactions. Let a subset of the pixels $\mathbf{c} = [i, j, k, \ldots] \subset \mathbf{R}$ supporting a function $f_\mathbf{c} : \mathbf{U}^{|\mathbf{c}|} \to \mathcal{R}$ of the signals in these pixels be referred to as *clan* (Moussouris, 1974). Signals in the pixels over a clan $\mathbf{c}$ in a particular image $\mathbf{s}$ form a *signal combination* $(s_i : i \in \mathbf{c})$ on the clan.

The *interaction structure* is given by a particular set $\mathbf{C}$ of clans $\mathbf{c}$. The *interaction strength* for each clan $\mathbf{c}$ is given by a non-negative and non-constant function $f_\mathbf{c}(s_i : i \in \mathbf{c})$ of the signal combinations supported by the clan. The GPD is represented by the normalized product of these functions:

$$\Pr(\mathbf{s}) = \frac{1}{Z} \prod_{\mathbf{c} \in \mathbf{C}} f_\mathbf{c}(s_i : i \in \mathbf{c}). \tag{1.1}$$

Here, Z is the normalizing factor, called the partition function, or statistical sum in statistical physics (Isihara, 1971):

$$Z = \sum_{\mathbf{s} \in \mathcal{S}} \prod_{\mathbf{c} \in \mathbf{C}} f_\mathbf{c}(s_i : i \in \mathbf{c}).$$

Loosely speaking, the GRF can be considered as a natural extension of the independent random field (IRF). The IRF is the simplest random field model, and the joint probability distribution of the IRF is given by Eq. (1.1) with the single-pixel clans $\mathbf{c} = [i]$; $\mathbf{C} = \mathbf{R}$. In this case the functions $f_i(s_i)$ represent the marginal pixelwise probabilities of the signals $s_i \in \mathbf{U}$. In the general case, the functions $f_\mathbf{c}(s_i : i \in \mathbf{c})$ of Eq. (1.1) have no such straightforward association with the marginal probabilities. But the higher the value of a function $f_\mathbf{c}(s_i : i \in \mathbf{c})$, the higher the probability of a signal combination $[s_i : i \in \mathbf{c}]$.

We will consider only a particular case of the GPDs with the strictly positive factors $f_\mathbf{c}(s_i : i \in \mathbf{c}) > 0$; $\mathbf{c} \in \mathbf{C}$. Because the factors are assumed to be strictly positive, the GPDs satisfy the so-called *positivity condition* $\Pr(\mathbf{s}) > 0$; $\forall \mathbf{s} \in \mathcal{S}$ of Hammersley and Clifford discussed in (Besag, 1974). Equivalently, one may specify the strictly positive marginal probabilities of the signals s_i: $\Pr(s_i) > 0$ $\forall i \in \mathbf{R}$ (Besag, 1974; Liggett, 1985). The positivity condition means that there are no constraints on the local spatial signal combinations, so that each combination has a non-zero probability to appear in a clan. In other words, whatever the Gibbs model, each sample $\mathbf{s}$ of the parent population $\mathcal{S}$ can be met with non-zero probability.

Usually, the factors of the GPD in Eq. (1.1) are represented in the exponential form:

$$f_\mathbf{c}(s_i : i \in \mathbf{c}) \equiv \exp\left(V_\mathbf{c}(s_i : i \in \mathbf{c})\right) \qquad (1.2)$$

where $V_\mathbf{c} : \mathbf{U}^{|\mathbf{c}|} \to \mathcal{R}$ is a real-valued function of the signal combinations $[s_i : i \in \mathbf{c}]$ in the clan $\mathbf{c}$. This function has the bounded above absolute value $|V_\mathbf{c}(s_i : i \in \mathbf{c})| < \infty$. The set $\mathbf{V} = (V_\mathbf{c} : \mathbf{c} \in \mathbf{C})$ of the functions in Eq. (1.2) is called the (*Gibbs*) *potential* (Moussouris, 1974; Dobrushin and Pigorov, 1975). The components $V_\mathbf{c} : \mathbf{c} \in \mathbf{C})$ are also referred to as Gibbs potentials. The constraints $|V_\mathbf{c}(s_i : i \in \mathbf{c})| < \infty$ on the potential values follow from the positivity condition.

The GPD of Eq. (1.1) is specified by the Gibbs potential as follows:

$$\Pr(\mathbf{s}) = \frac{1}{Z} \exp \sum_{\mathbf{c} \in \mathbf{C}} V_\mathbf{c}(s_i : i \in \mathbf{c}). \qquad (1.3)$$

To use the GPD of Eqs. (1.1) or (1.3) for describing the translation invariant textures, it is convenient to stratify the set $\mathbf{C}$ into several families $\mathbf{C} = [\mathbf{C}_a; \ a \in \mathbf{A}]$, of the translation invariant clans $\mathbf{c}_a \subset \mathbf{R}$. Here and in the following, a denotes the index of a clan family and $\mathbf{A}$ is a fixed index set.

The clans of the same family $\mathbf{C}_a$ have the same fixed relative arrangement of the pixels and differ only by their absolute positions in the lattice. To ensure the translation invariant interaction strength over a clan family, each family possesses its own potential function $V_a : \mathbf{U}^{|c_a|} \to \mathcal{R}$ of the signal combinations in the clans.

The following GPD specifies a GRF with the translation invariant interaction structure $\mathbf{C} = [\mathbf{C}_a; \ a \in \mathbf{A}]$:

$$\Pr(\mathbf{s}) = \frac{1}{Z} \exp \sum_{a \in \mathbf{A}} \sum_{c_a \in \mathbf{C}_a} V_a(s_i : i \in \mathbf{c}_a). \qquad (1.4)$$

Usually, the exponent is called the total (Gibbs) *energy* of pixel interactions under a given interaction structire $\mathbf{C}$ and potential $\mathbf{V}$. We denote $e(\mathbf{s}|\mathbf{V})$ the total energy in Eq. (1.4):

$$e(\mathbf{s}|\mathbf{V}) = \sum_{a \in \mathbf{A}} e_a(\mathbf{s}|\mathbf{V}_a) \qquad (1.5)$$

where $e_a(\mathbf{s}|\mathbf{V}_a)$ is the *partial energy* for each family $\mathbf{C}_a$:

$$e_a(\mathbf{s}|\mathbf{V}_a) = \sum_{c_a \in \mathbf{C}_a} V_a(s_i : i \in \mathbf{c}_a). \qquad (1.6)$$

Gibbs models similar to the model in Eq. (1.4) will be used below, in Chapter 2, to describe various homogeneous and piecewise-homogeneous grayscale textures and region maps.

It should be noted that the notion "energy" is preserved in our image models only due to tradition and must not be linked to the physical energy of a system of particles.

1.4.3. TRADEOFFS BETWEEN GIBBS AND MARKOV FIELDS

In practice our concern is only with those GRFs that yield a low computational complexity of image modelling. In other words, the factors of a GPD in Eq. (1.1) or, what is the same, the potentials in Eq. (1.3) are expected to depend, at least, in the average, on a relatively small number of variables.

Traditionally, the GRFs are considered as a subclass of the MRFs due to the well-known theorem of Hammersley and Clifford (Grimmett, 1973; Besag, 1974) about the equivalence between the two fields. To formulate the theorem, one needs the following

Definition 1.7 (Besag, 1974) *A clique is either a single vertex or a subset of vertices forming a complete subgraph. The number of the pixels in a clique is referred to as the order of the clique.*

Theorem 1.1 (Hammersley–Clifford) *Under the positivity condition, the joint probability distribution of a MRF* **S** *over the parent population S can be represented as the GPD of Eq. (1.3) where the potential functions $V_c(\ldots)$ of random variables S_i are supported by cliques* **c** *of the neighborhood graph describing the pixel interactions in the MRF.*

The most general form of this theorem is presented by Winkler (1995). The Hammersley–Clifford theorem states that the clans of a Markov/Gibbs model are the *cliques* of a neighborhood graph for the MRF under consideration. Proof of this theorem is given, in particular, by Besag (1974). Also, it can be proved using the theorem of Brook (1964) about an arbitrary function f with several expansions into factors. According to this latter theorem, there always exists an expansion of f that contains no factors depending simultaneously on variables x and y if each such factor is absent in at least one of the possible expansions.

In the general case, the equivalence between Markov and Gibbs models, not necessarily supported by an arithmetic lattice, has been considered by Moussouris (1974). It was proved that (i) every GRF is a MRF and (ii) every MRF, under the positivity condition, is a GRF with clan potentials uniquely determined by explicitly specified ratios of the factors in Eq. (1.2). The proof does not exclude the limiting cases of the non-Markov fields of order $|\mathbf{R}| - 1$. Therefore, as is shown, for instance, in (Liggett, 1985; Winkler, 1995), every random field on the finite lattice **R** that satisfies the positivity condition (that is, has strictly positive marginal probabilities of the signals s_i in each site i) can be represented as a GRF with a particular potential.

If the positivity condition does not hold, then some MRFs cannot be represented by GPDs. Examples of such MRFs with the local constraints on signal combinations are given in (Moussouris, 1974). For texture modelling, the positivity condition seems to be not too restrictive because of the unavoidable output noise of image sensors. This noise prevents to *a priori* exclude any signal combination from the images under consideration.

On the other hand, it is only the computational complexity that is of chief interest in practical image modelling and has stimulated a search for computationally feasible Markov/Gibbs models of low order. As discussed earlier and will be shown in Chapter 2, not only the MRFs but also some non-Markov GRFs may also be considered as image models due to their small average order. In these models the actual neigborhood of each pixel coincides, in principle, with the total lattice but only the clans of cardinality 2, representing the most significant part of the lattice-wide interactions, take part in the majority of computations. Then, on the average, only a relatively small number of variables per pixel take part in image modelling. Therefore, this book is principally concentrated on the various GRFs of low average order.

In the subsequent text, for brevity, all the supports of the factors or potentials in the GPDs are referred to by the same terms "cliques" and "clique families" even when we consider non-Markov GRFs where, in the strict sense of the word, these supports may not be cliques of a neighborhood graph.

1.5. Physics and image modelling: what an interaction means...

GPDs, in a general way, were introduced first in statistical physics to describe equilibrium probability distributions of the states of a large statistical system of interacting particles under different physical conditions (Isihara, 1971). Physical interactions describe a particular natural phenomenon, say, gravitational, electrical, magnetic, or other one, that follows the well-known physical laws. Therefore, the interaction structures and strengths in a GPD have obvious and constructive physical parallels.

Generally, a GPD describes properties of a statistical system **s** composed of a very big but finite number N of similar particles, such as atoms or molecules, in terms of generalized coordinates ξ_i and pulses η_i of each particle i and of an overall temperature T of the system. Not counting details unrelated to our case, the coordinates and pulses of all the particles are combined in a Hamiltonian function, or *Hamiltonian*, giving a total physical energy of a system:

$$\mathcal{H}(\mathbf{s}) = \frac{1}{kT} \sum_{i=1}^{N} \xi_i \cdot \eta_i \tag{1.7}$$

where k is a specific physical constant. The probability that a particular system $\mathbf{s}^\circ$ is observed depends on the total energy of this system, and the most probable dependence is represented by the GPD:

$$\Pr(\mathbf{s}^\circ) = \frac{\exp\left(-\mathcal{H}(\mathbf{s}^\circ)\right)}{\displaystyle\sum_{\mathbf{s}\in\mathcal{S}} \exp\left(-\mathcal{H}(\mathbf{s})\right)} \tag{1.8}$$

where $\mathcal{S}$ is the parent population of all the possible systems $\mathbf{s}$ of N particles. The GPD in Eq. (1.8) is expanded into factors that express the global behavior of a physical system in terms of a quantitative structure and strengths of local particle interactions represented by the signal products in the Hamiltonian of Eq. (1.7).

In the image modelling context, each straightforward analogy with physical interactions between particles, with energies of particle systems, with a temperature of a system, and so on, is mostly misleading[2]. Generally, the *pixel interaction* means almost no more than gray levels or region labels in the pixels within a particular clique are interdependent. The signal interdependence in the cliques of a particular clique family (and hence the interaction strength for the clique family) is closely related to marginal probabilities of signal combinations in the cliques; the larger the probability distribution of signal combinations deviates from the equiprobable, or uniform distribution, the stronger the interaction between the supporting pixels[3]. Therefore both the geometric structure of pixel interactions and Gibbs potentials of a Gibbs image model should adequately reflect the most characteristic, that is, the least uniform probability distributions of signal combinations in the lattice in a particular image texture to be modelled. Such types of interactions have little or nothing to do with the typical physical ones described by the signal products.

But good use made of traditional Gibbs models on finite multidimensional lattices, in particular, auto-binomial and Gauss–Markov models, in modern statistical physics and measurement theory, gave rise to a common opinion that every such GPD can be easily adapted to describe two-

[2]The Hamiltonian appears negative in Eq. (1.8) since the physical energy can only take non-negative values and the probability to observe a physical system depends inversely on the system's energy. In our image models the Hamiltonian appears positive as energies of pixel interactions have no physical meaning and can be with alternating signs.

[3]All the signal combinations are equiprobable for the IRF described by the GPD of Eq. (1.8) with the zero-valued Hamiltonian.

dimensional image textures. The adaptation is done by learning, or statistical estimation of the control parameters that specify the interaction structure and strength from a given training sample; see, for instance, (Kashyap, 1980; Kashyap and Chellappa, 1983; Chellappa, 1985; Besag, 1986; Derin and Elliot, 1987; Younes, 1988; Jacobsen, 1989; Cohen et al., 1991; Cohen and Patel, 1991; Comets, 1992; Picard and Elfadel, 1992; Besag, 1993; Devijver, 1993; Gidas, 1993). It should be noted that parameter learning is not common for physics and has been developed for certain GPDs only in image modelling. As a result, the potentials and interaction structures well-justified in physics result sometimes in notable difficulties of the parameter estimation.

Physical Gibbs models on finite lattices which are most interesting in image modelling have been much investigated by Dobrushin (1968), Averintsev (1970, 1972), Besag (1974) and others; see also the surveys of Dobrushin and Pigorov (1975) and Dubes and Jain (1989). These investigations have resulted in a general description of a GRF by a Gibbs potential $\mathbf{V}$ under the assumption that an explicit structure of interactions between the lattice sites is known and represented by a set $\mathbf{C}$ of interacting cliques $\mathbf{c}$ of the lattice sites i: $\mathbf{C} = \{\mathbf{c} \subset \mathbf{R}\}$. The potential

$$\mathbf{V} = [V_{\mathbf{c}}(u_i : i \in \mathbf{c}; u_i \in \mathbf{U}) : \mathbf{c} \in \mathbf{C}]$$

describes the interaction energies in the cliques in terms of certain signals s_i measured in the interacting sites by specifying the Hamiltonian of Eq. (1.3) in a general way:

$$\mathcal{H}(\mathbf{s}) = -\sum_{\mathbf{c} \in \mathbf{C}} V_{\mathbf{c}}(s_i : i \in \mathbf{c}). \tag{1.9}$$

Such a representation makes it possible to define the interconsistent conditional probabilities of signal combinations in the lattice (these probabilities are of most interest in analyzing and simulating samples of a particular GRF). Let $\mathbf{U}$ be a finite set of the signal values measured in each lattice site. Let $\mathbf{I} \subset \mathbf{R}$ and $\mathbf{J} = \mathbf{R}\backslash\mathbf{I}$ denote an arbitrary subset of the lattice sites and its complement, that is, the lattice $\mathbf{R}$ except the subset $\mathbf{I}$, respectively ($\mathbf{R} = \mathbf{I} \bigcup \mathbf{J}$). Let $\mathbf{u}_{\mathbf{I}}^{\circ} = [u_i^{\circ} : i \in \mathbf{I}]$ denote an arbitrary signal combination in the sites of the subset $\mathbf{I}$. Let $\mathbf{s}_{\mathbf{J}}^{\circ} = [s_j^{\circ} : j \in \mathbf{J}]$ be a fixed signal combination in the sites of the complementary subset $\mathbf{J}$. All the conditional probability distributions $\Pr(\mathbf{u}_{\mathbf{I}}^{\circ}|\mathbf{s}_{\mathbf{J}}^{\circ})$ of the signal combinations over every subset $\mathbf{I}$, provided that the signals over its complement $\mathbf{J}$ have the fixed values, are defined by the joint GPD in Eq. (1.3) as follows:

$$\Pr(\mathbf{u}_{\mathbf{I}}^{\circ}|\mathbf{s}_{\mathbf{J}}^{\circ}) = \frac{\exp\left(-\mathcal{H}(\mathbf{u}_{\mathbf{I}}^{\circ}, \mathbf{s}_{\mathbf{J}}^{\circ})\right)}{\sum_{i \in \mathbf{I}} \sum_{u_i \in \mathbf{U}} \exp\left(-\mathcal{H}(\mathbf{u}_{\mathbf{I}}, \mathbf{s}_{\mathbf{J}}^{\circ})\right)}.$$

Both the joint GPD and all the conditional GPDs are not uniquely defined by the potential of Eq. (1.2) because the probabilities are invariant to any constant offset b of the Hamiltonian. The probability $\Pr(\mathbf{s})$ is not changed when the Hamiltonian $\mathcal{H}(\mathbf{s})$ is replaced by $\mathcal{H}(\mathbf{s}) + b$ because every constant offset $V_{\mathbf{c}}(\ldots) + \beta$ of the potential for each clique $\mathbf{c} \in \mathbf{C}$ can be simply reduced from the nominator and denominator of the GPD in Eq. (1.8).

There exist many ways to obtain a unique representation of a GPD.

Theorem 1.2 (Dobrushin and Pigorov, 1975; Liggett, 1985) *Let $S^* = S\backslash\mathbf{s}^*$ be the parent population of the samples $\mathbf{s}$ with exception of the sample $\mathbf{s}^*$ and $\mathbf{s}_{\mathbf{c}} = [s_i : i \in \mathbf{c}]$ denotes the signal combination in the clique $\mathbf{c}$ for a particular sample $\mathbf{s}$. The GPD with a potential $\mathbf{V}$ is uniquely represented with a relative Hamiltonian $\mathcal{H}(\mathbf{s}) - \mathcal{H}(\mathbf{s}^*)$ obtained by reducing an arbitrary chosen sample $\mathbf{s}^*$ from the nominator and denominator of Eq. (1.3):*

$$\Pr(\mathbf{s}^\circ) = \frac{\exp\left(\sum\limits_{\mathbf{c} \in \mathbf{C}} (V_{\mathbf{c}}(\mathbf{s}_{\mathbf{c}}^\circ) - V_{\mathbf{c}}(\mathbf{s}_{\mathbf{c}}^*))\right)}{1 + \sum\limits_{\mathbf{s} \in S^*} \exp\left(\sum\limits_{\mathbf{c} \in \mathbf{C}} (V_{\mathbf{c}}(\mathbf{s}_{\mathbf{c}}) - V_{\mathbf{c}}(\mathbf{s}_{\mathbf{c}}^*))\right)}. \tag{1.10}$$

Generally, the representation of Eq. (1.10) offers a combinatorial number $|S|$ of the equivalent relative Hamiltonians and corresponding normalized potentials $\mathbf{V}^*$ resulting in the same GPD:

$$V_{\mathbf{c}}^*(\mathbf{s}_{\mathbf{c}}) = V_{\mathbf{c}}(\mathbf{s}_{\mathbf{c}}) - V_{\mathbf{c}}(\mathbf{s}_{\mathbf{c}}^*); \quad \mathbf{c} \in \mathbf{C}.$$

In practice, we need a more definite method of normalizing the potentials, and this point will be discussed at greater length in the next section.

Potential centering.   In the general case, the unique representation of a GPD can be obtained by normalizing the potentials with respect to a specific "vacuum" sample. Below we briefly review this technique which has been detailed in (Winkler, 1995, Theorems 3.3.1 and 3.3.3).

Let $\mathbf{s}^*$ denote the vacuum sample. Let $V_{\mathbf{c}}(\mathbf{s})$ be an arbitrary potential for a particular clique $\mathbf{c}$. For every sample $\mathbf{s}$, this potential depends only on signals $\mathbf{s}_{\mathbf{c}} = [s_i : i \in \mathbf{c}]$ in the clique $\mathbf{c}$. The normalized potential must be equal to zero for every clique $\mathbf{c}$ in the vacuum sample: $V_{\mathbf{c}}^*(\mathbf{s}_{\mathbf{c}}^*) = 0$. Then the normalization of an arbitrary potential $V_{\mathbf{c}}(\mathbf{s})$ is obtained as follows:

$$V_{\mathbf{c}}^*(\mathbf{s}) = \sum_{\sigma \subset \mathbf{c}} (-1)^{|\mathbf{c} - \sigma|} V_{\mathbf{c}}(\hat{\mathbf{s}}^\sigma)$$

where $|\mathbf{c} - \sigma|$ is the cardinality of the subset complementing the subset σ to the whole clique $\mathbf{s}$ and $\hat{\mathbf{s}}^\sigma$ denotes the sample that coincides with the sample $\mathbf{s}$ on the subset σ and with the vacuum sample $\mathbf{s}^*$ off the subset σ.

Let $\mathbf{c} = [i, j]$ be a second-order clique. It has the following four subsets $\sigma \in \{[i, j], [i], [j], [\emptyset]\}$ where $\emptyset$ denotes the empty subset so that the normalized potential is as follows (Lebedev et al., 1983):

$$V^*_{[i,j]}(s_i, s_j) = V_{[i,j]}(s_i, s_j) - V_{[i,j]}(s_i, s_j^*) - V_{[i,j]}(s_i^*, s_j) + V_{[i,j]}(s_i^*, s_j^*).$$

Here, we introduce another, more flexible normalizing scheme for the potentials of Gibbs models with multiple pairwise pixel interactions. It is based on a *potential centering* constraint, proposed first in (Gimel'farb, 1996a). The potential centering excludes the possible constant offsets of the potential values for all the signal combinations in every clique by reducing these offsets from the nominator and denominator of the GPD in Eq. (1.3). The potential centering simplifies the specification of a "vacuum" image.

Definition 1.8 *The centered potential* $\mathbf{V}$ *has the zeroth total sum of the potential values for all the signal combinations* $\mathbf{u_c} \in \mathbf{U}^{|\mathbf{c}|}$ *in every clique* $\mathbf{c}$:

$$\forall \mathbf{c} \in \mathbf{C} \quad \sum_{\mathbf{u_c} \in \mathbf{U}^{|\mathbf{c}|}} V_\mathbf{c}(\mathbf{u_c}) \equiv \sum_{i \in \mathbf{c}} \sum_{u_i \in \mathbf{U}} V_\mathbf{c}(u_i : i \in \mathbf{c}) = 0. \qquad (1.11)$$

Therefore, the potential centering replaces each potential value $V_\mathbf{c}(\mathbf{s_c})$ of the initial arbitrary potential $\mathbf{V} = [V_\mathbf{c}(\mathbf{u_c}) : \mathbf{c} \in \mathbf{C}; \mathbf{u_c^\circ} \in \mathbf{U}^{|\mathbf{c}|}]$ with the following centered one:

$$V^*_\mathbf{c}(\mathbf{s_c}) = V_\mathbf{c}(\mathbf{s_c}) - \frac{1}{|\mathbf{U}|^{|\mathbf{c}|}} \sum_{\mathbf{u_c} \in \mathbf{U}^{|\mathbf{c}|}} V_\mathbf{c}(\mathbf{u_c}).$$

The GPDs with the centered potentials have the following obvious feature.

Lemma 1.2 *Let the parent population* S *contain an "average" sample* $\mathbf{s}^*$ *so that the Hamiltonian of Eq. (1.9) for the average sample, computed with the centered potential* $\mathbf{V}$, *is equal to zero,* $\mathcal{H}(\mathbf{s}^*) = 0$. *Then the potential centering alone results in the unique representation of the GPD in Eq. (1.10), and the average sample* $\mathbf{s}^*$ *represents the vacuum one.*

We will consider in the next section how to choose the average sample $\mathbf{s}^*$ that constitutes the natural vacuum one in our case.

1.6. GPDs and exponential families of distributions

There is a finite number $|\mathbf{U}^{|\mathbf{c}|}|$ of the possible signal combinations $\mathbf{s_c}$ in each clique $\mathbf{c}$ of a family $\mathbf{C}_a$ in the GPD of Eq. 1.4. Therefore the partial energies of a sample $\mathbf{s}$ in Eq. (1.6) can be easily rewritten as the dot product

$$e_a(\mathbf{s}|\mathbf{V}_a) = \mathbf{V}_a \bullet \mathbf{H}_a(\mathbf{s}); \qquad (1.12)$$

of the potential vector $\mathbf{V}_a$ for the clique family $\mathbf{C}_a$ and the vector $\mathbf{H}_a(\mathbf{s})$ representing the sample histogram of the signal combinations collected over this family.

Each component $H_a(\mathbf{u}|\mathbf{s})$ of a sample histogram $\mathbf{H}_a(\mathbf{s}) = [H_a(\mathbf{u}|\mathbf{s}) : \mathbf{u} \in \mathbf{U}^{|\mathbf{c}|}]$ specifies how many times a particular signal combination $\mathbf{u} = [u_i : i \in \mathbf{c}] \in \mathbf{U}^{|\mathbf{c}|}$ is encountered among the actual combinations $\mathbf{s}_{\mathbf{c}} = [s_i : i \in \mathbf{c}]$ in the cliques $\mathbf{c} \in \mathbf{C}_a$ in the sample $\mathbf{s}$ (Lloyd, 1984):

$$H(\mathbf{u}|\mathbf{s}) = \sum_{\mathbf{c} \in \mathbf{C}_a} \delta(\mathbf{u} - \mathbf{s}_{\mathbf{c}}). \tag{1.13}$$

Here, $\delta(\ldots)$ denotes the generalized Kronecker function equal to 1 if $\mathbf{u} = \mathbf{s}_{\mathbf{c}}$ and 0 otherwise.

A component $H_a(\mathbf{u}|\mathbf{s})$ of a histogram and its relative value $F_a(\mathbf{u}|\mathbf{s}) = \frac{H_a(\mathbf{u}|\mathbf{s})}{|\mathbf{C}_a|}$ with respect to the cardinality of the clique family $|\mathbf{C}_a|$ are referred to as a *sample frequency* and *sample relative frequency* of the signal combination $\mathbf{u}$, respectively. For every clique family $\mathbf{C}_a$; $a \in \mathbf{A}$, the sample frequency and relative frequency distributions have the following features:

$$\begin{aligned} \sum_{\mathbf{u} \in \mathbf{U}^{|\mathbf{c}|}} H_a(\mathbf{u}|\mathbf{s}) &= |\mathbf{C}_a|; \\ \sum_{\mathbf{u} \in \mathbf{U}^{|\mathbf{c}|}} F_a(\mathbf{u}|\mathbf{s}) &= 1. \end{aligned} \tag{1.14}$$

As is apparent from Eq. 1.12, the total energy of Eq. (1.5) has the similar dot-product form:

$$e(\mathbf{s}|\mathbf{V}_a) = \mathbf{V} \bullet \mathbf{H}(\mathbf{s}). \tag{1.15}$$

where the total histogram vector $\mathbf{H}(\mathbf{s}) = [\mathbf{H}_a(\mathbf{s}) : a \in \mathbf{A}]$ consists of all the sample frequency vectors for the clique families.

The dot-product representation of the total energy in Eq. (1.15) shows that the GPD of Eq. (1.4) has the general form of an exponential family of probability distributions[4]:

$$\Pr(\mathbf{s}^{\circ}|\mathbf{V}) = \frac{1}{Z_{\mathbf{V}}} \exp(\mathbf{V} \bullet \mathbf{H}(\mathbf{s}^{\circ})) \tag{1.16}$$

where the normalizing factor $Z_{\mathbf{V}}$ is as follows:

$$Z_{\mathbf{V}} = \sum_{\mathbf{s} \in \mathcal{S}} \exp(\mathbf{V} \bullet \mathbf{H}(\mathbf{s})).$$

[4]An in-depth analysis of the exponential families of distributions is given, for example, by Barndorff-Nielsen (1978).

It follows from Eq. 1.16 that the sample frequency distributions of signal combinations are *sufficient statistics* to completely determine the model. By this is meant that every sample of the Gibbs model in Eqs. 1.4 or 1.16 is fully specified by the histograms $\mathbf{H}(\mathbf{s}) = [\mathbf{H}_a(\mathbf{s}) : a \in \mathbf{A}]$ of signal combinations in a given subset $\mathbf{C} = [\mathbf{C}_a : a \in \mathbf{A}]$ of the clique families.

As will be shown later, in Chapters 3 and 4, the basic feature of the exponential families, namely, the unimodality of the log-likelihood function which is briefly reviewed in the next section, simplifies considerably the maximum likelihood estimates (MLE) of parameters specifying the GPDs with multiple pairwise pixel interactions.

Unimodality of the log-likelihood function.   Let $G = \sum_{a \in \mathbf{A}} (|\mathbf{U}|^{|c_a|-1})$ be the overall number of the independent scalar components of the total histogram vector $\mathbf{H}(\mathbf{s})$. The total number of all the possible signal combinations in the clique families $\mathbf{C} = [\mathbf{C}_a : a \in \mathbf{A}]$ is equal to $\sum_{a \in \mathbf{A}} (|\mathbf{U}|^{|c_a|}) = G + |\mathbf{A}|$.

According to Eqs. (1.11) and (1.14), both the centered potential vector $\mathbf{V}$ and the histogram vector $\mathbf{H}(\mathbf{s})$ in Eq. (1.16) lie in the same Euclidean G-dimensional subspace of the vector space $\mathcal{R}^{(G+|\mathbf{A}|)}$ of signal combinations.

The log-likelihood function $L(\mathbf{V}|\mathbf{s}^\circ) = \frac{1}{|\mathbf{R}|} \log \Pr(\mathbf{s}^\circ|\mathbf{V})$ for the GPD in Eq. (1.16) is as follows:

$$L(\mathbf{V}|\mathbf{s}^\circ) = \mathbf{V} \bullet \mathbf{F}(\mathbf{s}^\circ) - \tfrac{1}{|\mathbf{R}|} \log \sum_{\mathbf{s} \in \mathcal{S}} \exp |\mathbf{R}| (\mathbf{V} \bullet \mathbf{F}(\mathbf{s})) \qquad (1.17)$$

where $\mathbf{F}(\mathbf{s})$ is the vector of sample relative frequencies: $\mathbf{F}(\mathbf{s}) = \frac{\mathbf{H}(\mathbf{s})}{|\mathbf{R}|}$. It is straightforward to show that this log-likelihood function has the following first and second partial derivatives:

$$\begin{aligned}
\tfrac{\partial}{\partial \mathbf{V}} L(\mathbf{V}|\mathbf{s}^\circ) &= \mathbf{F}(\mathbf{s}^\circ) - \mathcal{E}\{\mathbf{F}(\mathbf{s})|\mathbf{V}\}; \\
\tfrac{\partial^2}{\partial \mathbf{V}^2} L(\mathbf{V}|\mathbf{s}^\circ) &= -\mathbf{Cov}\{\mathbf{F}(\mathbf{s})|\mathbf{V}\}.
\end{aligned} \qquad (1.18)$$

Here, $\mathcal{E}\{\ldots\}$ denotes the mathematical expectation:

$$\mathcal{E}\{\mathbf{F}(\mathbf{s})|\mathbf{V}\} = \sum_{\mathbf{s} \in \mathcal{S}} \mathbf{F}(\mathbf{s}) \cdot \Pr(\mathbf{s}|\mathbf{V})$$

and $\mathbf{Cov}\{\mathbf{F}(\mathbf{s})|\mathbf{V}\}$ is the covariance matrix of the realtive sample frequencies:

$$\mathbf{Cov}\{\mathbf{F}(\mathbf{s})|\mathbf{V}\} = \mathcal{E}\{\mathbf{F}^2(\mathbf{s})|\mathbf{V}\} - (\mathcal{E}\{\mathbf{F}(\mathbf{s})|\mathbf{V}\})^2.$$

The covariance matrix is at least non-negative definite, so that the matrix of the second partial derivatives, or Hessian of the log-likelihood function in Eq. (1.18) is at least non-positive definite.

If the Hessian is strictly negative definite for all the possible vectors $\mathbf{V}$, then the log-likelihood function of Eq. (1.17) has the single maximum and the potential $\mathbf{V}^\circ$ which yields this maximum is implicitly given by the obvious relationship:

$$\mathbf{F}(\mathbf{s}^\circ) = \mathcal{E}\{\mathbf{F}(\mathbf{s})|\mathbf{V}^\circ\}. \tag{1.19}$$

As is shown by Barndorff-Nielsen (1978), the GPD in Eq. (1.16) is the regular exponential family distribution with the minimal canonical parameter $\mathbf{V}$ and minimal sufficient statistic $\mathbf{H}(\mathbf{s})$ if and only if the potential vectors $\mathbf{V}$ are affinely independent and the histogram vectors $\mathbf{H}(\mathbf{s})$ are affinely independent[5]. These conditions simply ensure that the Hessian is strictly negatively defined so that the regular exponential family distribution is unimodal with respect to the potential values.

The following theorem rephrases the corresponding theorem 9.13 in (Barndorff-Nielsen, 1978) for the GPD of Eq. (1.16).

Theorem 1.3 *The log-likelihood function of Eq. (1.17) has a maximum if and only if the histogram vector $\mathbf{H}(\mathbf{s})$ is lying in the interior of its vector space, that is, if all the sample relative frequencies of signal combinations in the clique families do not reach their limiting values, that is, are as follows:*

$$0 < F_a(\mathbf{u}|\mathbf{s}^\circ) < 1; \quad \mathbf{u} \in \mathbf{U}^{|c|}; \quad a \in \mathbf{A}.$$

The MLE $\mathbf{V}^\circ$ of the potential vector is given by Eq. (1.19).

In our case the potential vectors are affinely independent because they have no restrictions, other than the centering conditions of Eq. (1.11). The affine independence of the histogram vectors has to be proved for each particular GPD. For several Gibbs image models introduced in Chapter 2 such proofs are given in Chapter 3.

Potential centering - one more remark... Generally, if the potentials are unconstrained then the GPD of Eq. (1.16) is a non-unique representation of the GRF. The unique representation is obtained by using the relative Hamiltonians of Eq. (1.10), that is, the relative total energies $e(\mathbf{s}|\mathbf{V}) - e(\mathbf{s}^*|\mathbf{V})$ as the exponents. But the derivatives in Eq. (1.18) depend on the differences between the relative frequencies so that the fixed energy $e(\mathbf{s}^*|\mathbf{V})$ as well as the fixed relative frequency distribution $\mathbf{F}(\mathbf{s}^*)$ of the vacuum, or "base" sample $\mathbf{s}^*$ in the relative Hamiltonian are simply excluded from the consideration.

It is easy to verify that the relative total energy does not depend on a constant offset of the potential values for every clique family. Therefore

---

[5]By definition, the *affinely independent* vectors span over the whole subspace $\mathcal{R}^G$ and do not lie in any proper subspace of $\mathcal{R}^G$ (Barndorff-Nielsen, 1978).

the centering conditions of Eq. (1.11) are the only general constraints on the potentials that can be deduced from a relative Hamiltonian if we take no explicit account of the base sample $\mathbf{s}^*$. Although we do not necessarily need a unique representation of the GPD for studying its basic differential features, in particular, for computing the MLEs of the potentials, these centering conditions allow the specification of the adequate potentials for a Gibbs model to be considerably simplified.

Moreover, the centering conditions result in a natural choice of a vacuum sample for every "non-physical" Gibbs model with multiple pairwise pixel interactions. Let us consider an independent random field (IRF) which is supported by the lattice $\mathbf{R}$ and has the equiprobable random signals S_i; $\mathrm{Pr}_i(S_i = u) = \frac{1}{|\mathbf{U}|}$; $u \in \mathbf{U}$, in the lattice sites. The IRF is specified by the GPD of Eq. (1.16) when all the potential values for the signal combinations $\mathbf{u} = [u_i : i \in \mathbf{c}] \in \mathbf{U}^{|\mathbf{c}|}$ in the cliques are equal to zero, $V_{\mathbf{c}}(\mathbf{u}) = 0$.

Sample relative frequencies of every signal combination $\mathbf{u} = [u_i : i \in \mathbf{c}]$ in the IRF samples have the same mathematical expectation, or marginal probability $\mathrm{Pr}_{\mathrm{irf}}(\mathbf{u}) = \frac{1}{|\mathbf{U}|^{|\mathbf{c}|}}$, and the same variance $\mathrm{Pr}_{\mathrm{irf}}(\mathbf{u})\left(1 - \mathrm{Pr}_{\mathrm{irf}}(\mathbf{u})\right)$. It follows that an imaginary "average" IRF sample having the uniformly distributed relative frequencies $F(\mathbf{u}) = \mathrm{Pr}_{\mathrm{irf}}(\mathbf{u})$ of all the signal combinations can be considered, for every GRF, as a natural vacuum, or base sample of Lemma 1.2 which ensures the potential centering conditions. By this is meant in practice that the centering conditions refer to an implicit choice of the vacuum sample $\mathbf{s}^* \in \mathcal{S}$ that yields the minimum deviation between the relative frequencies of signal combinations in the cliques and the above marginal probabilities $\mathrm{Pr}_{\mathrm{irf}}(\mathbf{u})$ as compared with all the other samples $\mathbf{s}$ of the parent population $\mathcal{S}$.

Therefore, it is the deviations of an actual probability distribution of signal combinations in a clique family with respect to the equiprobable one that govern the potential values for these combinations around zero: the higher the positive deviation, the larger the positive potential value.

1.7. Stochastic relaxation and stochastic approximation

Gibbs models permit the simulation (generation) of samples by using a computationally feasible pixelwise stochastic relaxation. The relaxation process creates a Markov chain of the samples that have a desired GPD when the chain reaches an equilibrium. This section gives a brief overview of the stochastic relaxation.

1.7.1. GENERAL FEATURES OF STOCHASTIC RELAXATION

Definition 1.9 (Bharucha-Reid, 1960; Kemeney and Snell, 1960) *Stochastic relaxation is a process of generating a Markov chain* $(\mathbf{s}_t : t = 0, 1, 2, \ldots)$, *of samples that has a given distribution* $\Pr(\mathbf{s})$ *in an equilibrium.*

A Markov chain of samples is obtained by randomly choosing each next sample with a conditional probability that depends only on a current sample. The chain is described by an initial distribution $\Pr_0(\mathbf{s})$ of samples and a family of conditional probability distributions:

$$\mathbf{P}_t = [p_t(\mathbf{s}_{t+1}|\mathbf{s}_t) : \mathbf{s}_t, \mathbf{s}_{t+1} \in \mathcal{S}]; \; t = 0, 1, 2, \ldots;$$
$$\sum_{\mathbf{s} \in \mathcal{S}} p_t(\mathbf{s}|\mathbf{s}') = 1 \; \forall \mathbf{s}' \in \mathcal{S},$$

to be used at each step t of generation. Here, $p_t(\mathbf{s}_{t+1}|\mathbf{s}_t)$ is a probability of choosing the sample $\mathbf{s}_{t+1}$ in the next position $t + 1$ of a chain if the sample $\mathbf{s}_t$ is in the current position t. The probability $p_t(\mathbf{s}_{t+1}|\mathbf{s}_t)$ and the $|\mathcal{S}| \times |\mathcal{S}|$ matrix $\mathbf{P}_t$ are called the *transition probability* and *transition matrix*, respectively (Feller, 1970). If the transition matrix does not depend on the step t then the corresponding Markov chain is called *homogeneous*, otherwise it is *inhomogeneous*.

Theorem 1.4 (Bharucha-Reid, 1960; Kemeney and Snell, 1960) *Let* $\Pr_t(\mathbf{s})$ *be a probability distribution of the samples in a position t in the inhomogeneous Markov chains generated with the transition matrices* $\mathbf{P}_t$. *Let* $\Pr_0(\mathbf{s})$ *be a known distribution of the samples* $\mathbf{s}$ *in the starting position $t = 0$. Then each generation step, $t = 1, 2, \ldots$, transforms the distribution* $\Pr_t(\mathbf{s})$ *into the distribution* $\Pr_{t+1}(\mathbf{s})$ *as follows:*

$$\forall \mathbf{s} \in \mathcal{S} \quad \Pr_{t+1}(\mathbf{s}) = \sum_{\mathbf{s}' \in \mathcal{S}} p_t(\mathbf{s}|\mathbf{s}') \cdot \Pr_t(\mathbf{s}').$$

Definition 1.10 (Feller, 1970) *A finite Markov chain is irreducible if for all pairs of samples* $\mathbf{s}, \mathbf{s}' \in \mathcal{S}$ *there is a positive probability* $\Pr^{[\tau]}(\mathbf{s}|\mathbf{s}')$ *of reaching the sample* $\mathbf{s}'$ *from the sample* $\mathbf{s}$ *in a finite number τ of transitions:* $1 \leq \tau < \infty$.

Definition 1.11 (Feller, 1970) *A finite Markov chain is aperiodic if for all samples* $\mathbf{s} \in \mathcal{S}$ *the greatest common divisor of all integers $\tau \geq 1$ that yield a positive probability of reaching the sample* $\mathbf{s}$ *from the same sample* $\mathbf{s}$ *in the τ transitions,* $\Pr^{[\tau]}(\mathbf{s}|\mathbf{s}) > 0$, *is equal to 1.*

If generated Markov chains are irreducible then every desired sample $\mathbf{s} \in \mathcal{S}$ can be, in principle, reached from an arbitrary starting sample. Aperiodicity of a chain ensures the generation can converge to every single sample.

Let $\Pr^\tau(\mathbf{s}|\mathbf{s}')$ be a τ-step transition probability, or the conditional probability of generating the image $\mathbf{s}$ by τ successive steps of stochasic relaxation that starts from the image $\mathbf{s}' \in S$. The following theorem paraphrases the like theorems in (Derman, 1954; Bharucha-Reid, 1960; Kemeney and Snell, 1960; Feller, 1970) to our particular case:

Theorem 1.5 *Let the τ-step conditional probabilities $\Pr^{[\tau]}(\mathbf{s}|\mathbf{s}')$ of a transition in τ steps from every sample $\mathbf{s}' \in S$ to any other $\mathbf{s} \in S$ be strictly positive for a fixed finite number $\tau < \infty$. When t tends to the infinity, the generated irreducible and aperiodic chains $[\mathbf{s}_t : t = 0, 1, 2, \ldots]$ reach an equilibrium having a unique final distribution $\Pr(\mathbf{s}) = \Pr_\infty(\mathbf{s})$ specified by the transition probabilities. For a homogeneous Markov chain with the transition probabilities $(p(\mathbf{s}|\mathbf{s}'); \mathbf{s}, \mathbf{s}' \in S)$, the limiting, or stationary, distribution is uniquely determined by the equilibrium equations:*

$$\forall \mathbf{s} \in S \quad \Pr(\mathbf{s}) = \sum_{\mathbf{s}' \in S} p(\mathbf{s}|\mathbf{s}') \cdot \Pr(\mathbf{s}'). \tag{1.20}$$

Because there are only $|S|$ restrictions, the $|S|^2$ transition probabilities resulting in the same limiting distribution $\Pr(\mathbf{s})$ can be chosen in a variety of ways. The following sufficient but not necessary conditions that yield the relations of Eq. (1.20) are widely used to simulate images:

$$\forall \mathbf{s}, \mathbf{s}' \in S \quad p(\mathbf{s}|\mathbf{s}') \cdot \Pr(\mathbf{s}') = p(\mathbf{s}'|\mathbf{s}) \cdot \Pr(\mathbf{s}). \tag{1.21}$$

Convergence to an equilibrium. It is easily shown that the Markov chains generated by stochastic relaxation tend to equilibrium. The proof below was given by Creutz (1983). Let $\Delta(\Pr_t(\mathbf{s}), \Pr(\mathbf{s}))$ to denote the distance between the distributions $\Pr_t(\mathbf{s})$ and $\Pr(\mathbf{s})$ defined as the total sum of the absolute deviations:

$$\Delta(\Pr_t(\mathbf{s}), \Pr(\mathbf{s})) = \sum_{\mathbf{s} \in S} |\Pr_t(\mathbf{s}) - \Pr(\mathbf{s})|.$$

Then the following inequality holds:

$$
\begin{aligned}
\Delta(\Pr_{t+1}(\mathbf{s}), \Pr(\mathbf{s})) &= \sum_{\mathbf{s} \in S} \left| \sum_{\mathbf{s}' \in S} p(\mathbf{s}|\mathbf{s}') \cdot \Pr_t(\mathbf{s}') - \Pr(\mathbf{s}') \right| \\
&\leq \sum_{\mathbf{s} \in S} \sum_{\mathbf{s}' \in S} p(\mathbf{s}|\mathbf{s}') \, |\Pr_t(\mathbf{s}') - \Pr(\mathbf{s}')| \\
&= \Delta(\Pr_t(\mathbf{s}), \Pr(\mathbf{s})).
\end{aligned}
$$

If there are positive transition probabilities $p(\mathbf{s}|\mathbf{s}') > 0$ for all the sample pairs $\mathbf{s}, \mathbf{s}' \in S$, then while $\Pr_t(\mathbf{s}) \neq \Pr(\mathbf{s})$ there always exist both positive and negative deviations $\Pr_t(\mathbf{s}) - \Pr(\mathbf{s})$ of the probabilities[6]. Therefore the above inequality is strict unless the chain is already in equilibrium, so that the generated Markov chains do approach the equilibrium.

But, it is only a purely theoretical possibility because the number of relaxation steps to reach the equilibrium is usually too big to be practicable. As shown in (van Laarhoven, 1988, Comments to Theorem 2.5), to closely approximate a stationary distribution, the number of transitions along a Markov chain formed by stochastic relaxation should be at least quadratic in the number of all the possible samples, that is, quadratic in the cardinality of the parent population, $|\mathcal{S}|^2$. Since this number depends exponentially on the lattice, $|\mathcal{S}| = |\mathbf{U}^{\mathbf{R}}|$, the time for reaching the equilibrium typically grows exponentially with the lattice size. Therefore, in image modelling we cannot expect that a "pure" stochastic relaxation process will reach an equilibrium or the generated samples will have the desired GPD.

In Chapter 3 we will introduce a more suitable image simulating scheme called Controllable Simulated Annealing (CSA) that is based on solving the equation system (1.19) by stochastic approximation (Younes, 1988); see also (Wasan, 1969; Nevel'son and Has'minskiĭ, 1973). This solution uses in turn a stochastic relaxation technique to generate samples of a GRF under a particular Gibbs potential $\mathbf{V}$. But in this case the simulation produces an inhomogeneous Markov chain of samples that converges directly to a desired result. Next section outlines in brief the stochastic approximation technique to be used in our case.

1.7.2. STOCHASTIC APPROXIMATION

The discrete Robbins–Monro procedure of stochastic approximation solves Eq. (1.19) by a stochastic gradient algorithm. The algorithm is defined by the recursive relation:

$$\mathbf{V}_{t+1} = \mathbf{V}_t + \lambda_{t+1}\left(\mathbf{F}(\mathbf{s}^\circ) - \mathbf{F}(\mathbf{s}_{t+1})\right) \qquad (1.22)$$

where t denotes the approximation step ($t = 0, 1, 2, \ldots$) and $\mathbf{F}(\mathbf{s}_t)$ is the vector representing the sample relative frequency distributions of signal combinations in the clique families. The distributions are collected over a sample $\mathbf{s}_t$ generated at step t under the GPD $\Pr(\mathbf{s}|\mathbf{V}_{t-1})$ with the previous potential $\mathbf{V}_{t-1}$. When $t \to \infty$, this procedure converges almost surely[7] to a desired solution $\mathbf{V}^\circ$ of Eq. (1.19) under rather general conditions; to study

---

[6]Here and in Eq. (1.20) one can equally consider the τ-step transition probabilities. Under the positivity condition, all the $|\mathbf{R}|$-step probabilities of transitions between any two samples $\mathbf{s}$ and $\mathbf{s}'$ so that each step changes only a single pixel $i \in \mathbf{R}$ without repetition are strictly positive.

[7]That is, the probability of convergence is equal to 1.

them at length see, for instance, (Wasan, 1969, Chapter 5, Theorem 1) and (Nevel'son and Has'minskiĭ, 1973, Theorem 4.4 and Corollary 4.1).

Firstly, these conditions restrict the rate of growth of the variance of vectors $\mathbf{F}(\mathbf{s})$ about the fixed goal vector $\mathbf{F}(\mathbf{s}^\circ)$ when the length of a potential vector tends to the infinity, $|\mathbf{V}| \to \infty$. Roughly speaking, the variance should not grow faster than $|\mathbf{V}|^2$, and in our case this always holds because the variance of sample relative frequencies is bounded.

Secondly, the factors λ must comply with the requirements:

$$\sum_{t=0}^{\infty} \lambda_t = \infty; \quad \sum_{t=0}^{\infty} \lambda_t^2 < \infty.$$

As shown by Younes (1988), a particular choice of these factors permits us to estimate the convergence rate for a particular choice of these factors. The following theorem restates the theorem 4.1 in (Younes, 1988).

Theorem 1.6 (Younes, 1988) *Let us consider the algorithm in Eq. (1.22) with $\lambda_t = \frac{1}{\gamma(1+t)}$ where $\mathbf{s}_{t+1}$ is obtained from $\mathbf{s}_t$ by stochastic relaxation under the GPD $\Pr(\mathbf{s}|\mathbf{V}_t$[8]. Let $\mathbf{V}^\circ$ be the solution of Eq. (1.19). Let d° be the maximum Euclidean distance between all the vectors $\mathbf{F}(\mathbf{s})$ and the vector $\mathbf{F}(\mathbf{s}^\circ)$:*

$$d^\circ = \max_{\mathbf{s} \in \mathcal{S}} |\mathbf{F}(\mathbf{s}) - \mathbf{F}(\mathbf{s}^\circ)| .$$

Let d_{step} be the maximum Euclidean distance between the vectors $\mathbf{F}(\mathbf{s})$ and $\mathbf{F}(\mathbf{s}')$ so that the sample $\mathbf{s}$ is obtained from $\mathbf{s}'$ by a single step of pixelwise stochastic relaxation when the signal at only one site is changed. If $\gamma > 2d^\circ d_{\text{step}}|\mathbf{R}|$ then the potential vector $\mathbf{V}_t$ converges almost surely to $\mathbf{V}^\circ$ and there exists a constant $\varepsilon_0 > 0$ so that almost surely the Euclidean distance $d_t = |\mathbf{V}_t - \mathbf{V}^\circ|$ decreases faster than $t^{-\varepsilon}$ for $\varepsilon < \varepsilon_0$.

One can readily see that the above constants d° and d_{step} have the following approximate values for the Gibbs model in Eq. (1.4):

$$d_{\text{step}} \cong \frac{\sqrt{2|\mathbf{A}|}}{|\mathbf{R}|}; \quad d^\circ \cong \sqrt{2\sum_{a \in \mathbf{A}} \rho_a^2} \cong \sqrt{2|\mathbf{A}|}$$

where $\rho_a = \frac{|\mathbf{C}_a|}{|\mathbf{R}|}$ is the relative size of the clique family $\mathbf{C}_a$. Therefore, in our case the stochastic approximation procedure in Eq. (1.22) should converge almost surely to the MLE $\mathbf{V}^\circ$ if the above constant $\gamma > 4|\mathbf{A}|$, that is, the factors $\lambda_n < \frac{1}{4|\mathbf{A}|(1+t)}$.

As was indicated by Younes (1988), this theoretically derived value of γ is too large to expect that the procedure will converge in a reasonable

---

[8]To be more specific, by using the pixelwise stochastic relaxation with the Gibbs sampler that will be discussed in the next section.

time so that a heuristic choice of $\lambda_t = \frac{1}{c_1 + c_2 t}$ with $c_1 = 1,000$ and $c_2 = 1$ or 10 gave the best results in experiments. In this case, the probability of non-convergence of the procedure is of order $\frac{1}{c_1}$.

This stochastic approximation procedure allows us to introduce an explicit stopping rule. Two such rules were proposed in (Younes, 1988):

(i) to iteratively compute the distance between the mean vector $\frac{1}{T}\sum_{t=0}^{T}\mathbf{F}(\mathbf{s}_t)$

and the goal one $\mathbf{F}(\mathbf{s}^\circ) = \mathcal{E}\{\mathbf{F}(\mathbf{s})|\mathbf{V}^\circ\}$ and stop when the distance is less than a given threshold, and

(ii) to stop when the distance between the potential vectors separated by a fixed number $\tau \geq 1$ of steps is less than a given floating threshold:

$$|\mathbf{V}_{t+\tau} - \mathbf{V}_t| \leq \frac{\delta\tau}{c_1 + c_2 t}$$

where δ is a constant that specifies the desired accuracy of an approximation: $|\mathbf{F}(\mathbf{s}^\circ) - \mathcal{E}\{\mathbf{F}(\mathbf{s})|\mathbf{V}^\circ\}|\, le\delta$.

1.7.3. PIXELWISE STOCHASTIC RELAXATION

Generally, a *pixelwise stochastic relaxation* is computationally most feasible for generating the Markov chains of samples under a given GPD $\mathrm{Pr}(\mathbf{s})$ assuming that the GPD conforms to the positivity condition. In this particular case, only transitions that update a single pixel i in the lattice have the non-zero values, and the transition probabilities $p(\mathbf{s}|\mathbf{s}') \equiv 0$ for all the image pairs $\mathbf{s}$ and $\mathbf{s}'$ that differ in two or more pixels. Therefore, every two successive images $\mathbf{s}_t$ and $\mathbf{s}_{t+1}$ along a Markov chain may differ at most by a single signal value.

Let $\mathbf{s} = (s_i = u, \mathbf{s}^i)$ and $\mathbf{s}' = (s_i = u', \mathbf{s}^i)$ be the two images which differ at most in a single pixel i. For such image pairs, the relations in Eq. (1.21) can be rewritten as follows:

$$\begin{aligned} p(s_i = u|s_i = u', \mathbf{s}^i) \cdot \mathrm{Pr}(s_i = u'|\mathbf{s}^i) &= \\ p(s_i = u'|s_i = u, \mathbf{s}^i) \cdot \mathrm{Pr}(s_i = u|\mathbf{s}^i) \end{aligned} \tag{1.23}$$

where $\mathrm{Pr}(s_i = u|\mathbf{s}^i)$ denotes a conditional probability of a signal $u \in \mathbf{U}$ in a pixel i, given the fixed signals $\mathbf{s}^i$ in all other pixels.

The GPDs are of prime interest for generating images by the pixelwise stochastic relaxation because of the ability to specify the interconsistent conditional probabilities in Eq. (1.23). These latter depend only on the neighboring signals:

$$\Pr(s_i = u | \mathbf{s}^i) \equiv \Pr(s_i = u | s_j : j \in \mathbf{N}_i)$$

$$= \frac{\exp \sum_{\mathbf{c} \in \omega_i} V_{\mathbf{c}}(s_j : j \in \mathbf{c}; s_i = u)}{\sum_{u' \in \mathbf{U}} \exp \sum_{\mathbf{c} \in \omega_i} V_{\mathbf{c}}(s_j : j \in \mathbf{c}; s_i = u')}. \tag{1.24}$$

Here, $\omega_i = \{\mathbf{c} : \mathbf{c} \in \mathbf{C}; i \in \mathbf{c}\}$ is the subset of the cliques that contain the pixel i. These probabilities, which can be easily computed, allow us to generate a Markov chain of the images

$$\mathbf{s}_0 \to \mathbf{s}_1 \to \ldots \to \mathbf{s}_t \to \ldots$$

so that each next image $\mathbf{s}_{t+1}$ is obtained from the current one $\mathbf{s}_t$ by choosing the signal in a single pixel i_t in accord with the conditional probabilities of Eq. (1.24). Here, i_t, $t = 0, 1, \ldots$, is a random sequence of pixels tracing the lattice $\mathbf{R}$.

Let $\mathbf{s}_{\mathbf{N}}^i = [s_j : j \in \mathbf{N}_i]$ be a particular signal combination over the neighborhood $\mathbf{N}_i$ of a pixel i in a sample $\mathbf{s}$. Under the positivity constraint, the pixelwise stochastic relaxation involves only the non-zero transition probabilities $p(s_i = u | s_i = u', \mathbf{s}_{\mathbf{N}}^i)$ for all the signals $u, u' \in \mathbf{U}$ in the pixel i and for every signal combination $\mathbf{s}_{\mathbf{N}}^i$ in each neighborhood $\mathbf{N}_i$. In such a case the τ-step transition probabilities $\Pr^{[\tau]}(\mathbf{s}|\mathbf{s}')$ become first non-zero for all the image pairs $\mathbf{s}, \mathbf{s}' \in \mathcal{S}$ after a sequence of $|\mathbf{R}|$ successive relaxation steps that trace all the pixels $i \in \mathbf{R}$ in an arbitrary order but without repetition. This permits to introduce the following

Definition 1.12 *A single pass round all the $|\mathbf{R}|$ pixels in the lattice, without repetition, under an equiprobable random choice of each next pixel is referred to as a* macrostep *of the pixelwise stochastic relaxation.*

The Markov chain of images generated at each macrostep has in the limit $(t \to \infty)$ the equilibrium joint probability distribution of Eq. (1.3) that specifies the transition probabilities in Eq. (1.24). But it should be noted once more that there are no practical statistical criteria to verify that a particular Markov chain of images has already approached the equilibrium during the relaxation, and the theoretical estimates result in the combinatorial numbers of the macrosteps (van Laarhoven, 1988; Winkler, 1995).

Implementation.   The pixelwise stochastic relaxation generates a desired Markov chain of images by repeating iteratively, at each macrostep t, the following two operations of tracing the lattice and forming the next image from a current one.

1. First, it chooses a current arbitrary route of visiting all the pixels in the lattice without repetitions. The route is called a lattice *coloring* (Besag, 1974) or *visiting scheme* (Winkler, 1995).
2. Second, in each pixel i along a chosen route, the current signal, $s_i = u'$, is replaced by a signal u picked up randomly from the known set $\mathbf{U}$ according to the pixelwise transition probabilities

$$p(s_i = u | s_i = u', \mathbf{s}_N^i); \quad u \in \mathbf{U}.$$

The conditional probabilities of Eq. (1.24) allow us to specify the pixelwise transition probabilities in Eq. (1.23) which depend in this case only on the neighboring signals $\mathbf{s}_N^i = [s_j : j \in \mathbf{N}_i]$.

The two most known stochastic relaxation techniques are the heat-bath, or *Gibbs sampler* (Geman and Geman, 1984) and the computationally simpler *Metropolis sampler* (Metropolis et al., 1953). Both the samplers were introduced originally in statistical physics and differ by the pixelwise transition probabilities that are used for getting each next signal. An in-depth analysis of their statistical properties can be found, for instance, in (Hammersley and Handscomb, 1964; Cross and Jain, 1983; Lebedev et al., 1983; Geman and Geman, 1984; Gelfand and Mitter, 1993).

1. The Gibbs sampler chooses a new signal u according to the probabilities of Eq. (1.23):

$$p(s_i = u | s_i = u', \mathbf{s}_N^i) = \Pr(s_i = u | \mathbf{s}^i); \quad u \in \mathbf{U},$$

that is, it uses the conditional probabilities of Eq. (1.24) as the transition ones. This sampler calls for computing $|\mathbf{U}|$ probabilities of all possible signals $u \in \mathbf{U}$ in a pixel. The new signal, chosen randomly in accord with these probabilities, replaces the current signal in each current pixel i.

2. The Metropolis sampler chooses first equiprobably a signal $u \in \mathbf{U}$ and then substitutes it for $s_i = u'$ in line with the ratio $p_i(u|u') = \frac{\Pr(s_i=u|\mathbf{s}^i)}{\Pr(s_i=u'|\mathbf{s}^i)}$ of the two conditional probabilities:

$$p(s_i = u | s_i = u', \mathbf{s}_N^i) = \begin{cases} 1 & \text{if } p_i(u|u') > 1; \\ p_i(u|u') & \text{otherwise.} \end{cases}$$

Therefore, the Metropolis sampler reduces the computations to only the two transition probabilities per pixel.

Let $e_i(s_i = u | \mathbf{s}_N^i, \mathbf{V})$ denote a conditional pixelwise energy of the interaction of a signal $s_i = u$ in the pixel i with the fixed neighboring signals $\mathbf{s}_N^i$:

$$e_i(s_i = u | \mathbf{s}_N^i, \mathbf{V}) = \sum_{\mathbf{c} \in \omega_i} V_\mathbf{c}(s_i = u, \mathbf{s}_N^i).$$

The Metropolis sampler determines a new signal u for replacing the current one u' in the pixel i in the two following stages. First, a candidate u for the new signal is selected by a random equiprobable choice. If

$$\Pr(s_i = u|\mathbf{s}^i) \geq \Pr(s_i = u'|\mathbf{s}^i)$$

or, similarly, if $e_i(s_i = u|\mathbf{s}_N^i, \mathbf{V}) - e_i(s_i = u'|\mathbf{s}_N^i, \mathbf{V}) \geq 0$, then the new signal u replaces the current signal u'. Otherwise the new signal u is accepted only with the probability

$$p_i(u|u') = \exp\left(e(s_i = u|\mathbf{s}_N^i, \mathbf{V}) - e_i(s_i = u'|\mathbf{s}_N^i, \mathbf{V}) \right).$$

The latter probability decreases exponentially as the difference between the current and new conditional pixelwise energies increases.

The second stage is equivalent to but is less computationally complex than a random choice between the signals u and u' according to their relative probabilities:

$$\frac{\Pr(s_i = u|\mathbf{s}^i)}{\Pr(s_i = u|\mathbf{s}^i) + \Pr(s_i = u'|\mathbf{s}^i)} \quad \text{and} \quad \frac{\Pr(s_i = u'|\mathbf{s}^i)}{\Pr(s_i = u|\mathbf{s}^i) + \Pr(s_i = u'|\mathbf{s}^i)}.$$

Visiting scheme. To form an arbitrary visiting scheme, it is not necessary to randomly choose each pixel. As shown in (Besag, 1974), if the structure of pixel interaction is known then the lattice $\mathbf{R}$ can be split into several conditionally independent sublattices $\mathbf{R}_b$; $b \in \mathbf{B}$. Here, $\mathbf{B}$ denotes an index set for the disjoint sublattices that cover the whole lattice: $\mathbf{R} = \bigcup_{b \in \mathbf{B}} \mathbf{R}_b$.

Each pixel in a sublattice $\mathbf{R}_\beta$ has the neighbors only in other sublattices $\mathbf{R}_b$; $b \in \mathbf{B}^\beta$, where $\mathbf{B}^\beta$ is a subset of the indices, other than β. The splitting is uniquely defined by a given neighborhood graph.

Because each sublattice contains no neighboring (or interacting) pixels, the pixels of a single sublattice may be visited in any order with no impact on the results of generation. Therefore an arbitrary visiting scheme can be formed by only randomizing the choice of the successive sublattices and implementing a few regular routes within each sublattice to be randomly chosen for a current macrostep.

CHAPTER 2

Markov and Non-Markov Gibbs Image Models

We start this chapter with a brief overview of traditional Markov/Gibbs image models and then present new Markov and non-Markov models with multiple pairwise pixel interactions. These latter models proposed first in (Gimel'farb, 1996a)–(Gimel'farb, 1997a) generalize the traditional models in the following directions.

- The models are invariant to simple gray range transformations of images, namely, to an arbitrary gray range offset or scaling that does not affect the visual appearance of image textures.
- Every model has an arbitrary interaction structure specified by a characteristic set $\mathbf{C} = [\mathbf{C}_a : a \in \mathbf{A}]$ of clique families containing translation invariant cliques of second order.
- An interaction strength for every clique family is specified by an arbitrary Gibbs potential depending generally on signal co-occurrences in a clique.

Therefore they seem to be most suitable for describing, simulating, retrieving, and segmenting spatially homogeneous and piecewise-homogeneous image textures and their region maps.

We restrict our consideration to texture homogeneity specified by only translation invariance of an interaction structure. But, as will be shown later, more complex geometric transformations such as rotation or limited rescaling can be taken into account in texture simulation or retrieval experiments. In such a case we assume that a geometric transformation acts only upon the interaction structure of a model and changes it in a particular way in different parts of a supporting lattice. But, it is still unclear how to take into account the varying interaction structure in estimating parameters of a model or segmenting a piecewise-homogeneous image.

2.1. Traditional Markov/Gibbs image models

In the general case, the GPD of Eq. (1.3) restricted to only pairwise pixel interactions was introduced, for instance, in (Besag, 1974; Besag, 1986). Restriction to the pairwise interactions is common both to physics and image modelling because it simplifies the theoretical analysis and results in the lesser computational complexity of modelling (Dobrushin, 1968;

Averintsev, 1970; Averintsev, 1972; Dobrushin and Pigorov, 1975; Chellappa and Jain, 1993; Li, 1995; Winkler, 1995).

For the most part, pixel interactions in (Besag, 1974; Besag, 1986) are not subclassified into families of the translation-invariant cliques:

$$\Pr(\mathbf{s}) \propto \exp\left(\sum_{i \in \mathbf{R}} V_i(s_i) + \sum_{[i,j] \in \mathbf{C}} V_{i,j}(s_i, s_j)\right).$$

As a special case of the equally treated pixels and interacting pairs, the above GPD was represented in (Besag, 1986) using the signal histogram,

$$\mathbf{H}_{\text{pix}}(\mathbf{s}) = [H_{\text{pix}}(u|\mathbf{s}) : u \in \mathbf{U}],$$

and the signal co-occurrence histogram,

$$\mathbf{H}_{\text{pair}}(\mathbf{s}) = [H_{\text{pair}}(u, u'|\mathbf{s}) : [u, u'] \in \mathbf{U}^2],$$

that describes all the combined pairwise interactions:

$$\Pr(\mathbf{s}) \propto \exp\left(\sum_{u \in \mathbf{U}} V_{\text{pix}}(u) H_{\text{pix}}(u|\mathbf{s}) + \sum_{[u,u'] \in \mathbf{U}^2} V_{\text{pair}}(u, u') H_{\text{pair}}(u, u'|\mathbf{s})\right)$$

but without an explicit reference to the exponential families of distributions.

Although Besag (1986) had mentioned that it seems desirable generally to adopt large pixel neighborhoods with translation invariant potentials, most works in image modelling (and the cited papers are no exception) exploited only the two types of GPDs describing MRFs with pixelwise and pairwise pixel interactions, namely, the *autobinomial* and *autonormal*, or Gauss–Markov, models. These Markov/Gibbs models, called *automodels* by Besag (1974), have interaction structures which are mostly pre-defined and Gibbs potentials of very specific functional forms.

2.1.1. AUTO-BINOMIAL AND GAUSS/MARKOV MODELS

History in brief. Extensive investigations of the autobinomial model with respect to image textures were initiated by Hassner and Sklansky (1978, 1980), Cross and Jain (1983), Lebedev et al. (1983). The Gauss–Markov image model was studied in (Kashyap, 1980; Kashyap, 1981; Kashyap and Chellappa, 1983; Chellappa, 1985). These pioneering results were amplified and directed to more practical problems of image restoration and segmentation in (Geman and Geman, 1984; Chellappa, 1985; Besag, 1986; Derin and Cole, 1986). Then this area was explored in (Cohen and Cooper, 1987; Derin and Elliot, 1987; Marroquin, 1987; Marroquin et al., 1987;

Lebedev, 1988; Gimel'farb and Zalesny, 1989; Geman et al., 1990; Karssmei-jer, 1990; Gimel'farb, 1991; Gimel'farb and Zalesny, 1991; Chatterjee, 1993; Geman et al., 1993; Gimel'farb and Zalesny, 1993a; Gimel'farb and Zalesny, 1993b; Marroquin, 1993) and many other works.; see also the comprehensive surveys (Kashyap, 1986; Dubes and Jain, 1989; Chellappa et al., 1993; Tuceryan and Jain, 1993). It is simply impossible to give here an exhaustive list of these investigations. They attract a widespread attention even at present (Elfadel and Picard, 1994; Levitan et al., 1995; Li, 1995; Winkler, 1995) although the automodels, borrowed from physics, are not completely adequate for describing image textures because they ignore the fact that pixel interactions have no physical meaning.

It should be noted that we do not consider in this book the once very popular causal Markov models such as Markov mesh random fields (Derin et al., 1984; Devijver, 1993). These models have simply extended ordinary 1D Markov processes onto a 2D lattice by assuming specific "oriented" pixel neighborhoods which result in asymmetric descriptions of inherently symmetric pixel interactions. Gibbs models discussed in this book permit us to avoid such asymmetry, but generally at the expense of more complicated and intensive computations.

Autobinomial model. It exploits only the nearest 4-neighborhood of a pixel, and the GPD contains the four cliques families with the following intra-clique shifts between the pixels:

$$\iota = (\mu, \nu) \in \mathbf{N} = \{(-1, 1), (0, 1), (1, 1), (1, 0)\}.$$

The potentials for each family are proportional to the product of the signals in the interacting pixels: $V_{\mu,\nu}(s, s') = \theta_{\mu,\nu} \cdot s \cdot s'$. The four coefficients, $\theta_{-1,1}$, $\theta_{0,1}$, $\theta_{1,1}$, $\theta_{1,0}$, are the independent model parameters to be learnt.

In physics, a direct proportionality between the interaction strength and signal product is readily justified. In image modelling, the interaction strength specifies the probabilities of co-occurrences of the particular signals (s, s') in the cliques. Therefore, the autobinomial model leads to rather exotic probabilities. In particular, it is hard to explain why the neighboring pair of the equal gray levels (q, q) should have the same interaction strength as the different signals $(1, q^2)$ or $(\frac{q}{2}, 2q)$ but four times weaker strength than the equal gray levels $(2q, 2q)$. For the majority of natural and synthetic image textures, the product of gray levels or region labels does not relate to their frequencies in the images. Also, the pre-defined interaction structure of the model in principle cannot take account of the distant neighbors typical to most textures.

Parameter estimation for autobinomial models. The model meets computational difficulties in learning the model parameters, so that its flexibility in adapting to a given training sample is strongly limited.

Because both the nominator and denominator (the partition function Z) of the model depend on the control parameters $\theta_{...}$, there are no ways to analytically compute the maximum likelihood estimate (MLE) of the parameters. To avoid these difficulties, several *ad hoc* approximations have been proposed, in particular, the pseudo-maximum-likelihood estimate (PMLE) in (Besag, 1986) and the marginal frequency estimate (MFE) in (Derin and Elliot, 1987).

The PMLE can be used for an arbitrary GPD. It replaces the GPD by a product of several conditional probabilities which each depend on the desired parameters to be estimated. The lattice $\mathbf{R}$ is split into the same conditionally independent sublattices $\mathbf{R}_b$; $b \in \mathbf{B}$; $\mathbf{R} = \bigcup_{b \in \mathbf{B}} \mathbf{R}_b$, where $\mathbf{B}$ denotes an index set, which are used for a visiting scheme of stochastic relaxation discussed in Chapter 1. The conditional probability of a combination $\mathbf{s}_\beta$ of signals within a sublattice $\mathbf{R}_\beta$, under the fixed other signals and parameters $\boldsymbol{\theta}$, is as follows:

$$\Pr(\mathbf{s}_\beta | \mathbf{s}_b : b \in \mathbf{B}^\beta, \boldsymbol{\theta}) = \prod_{i \in \mathbf{R}_\beta} \Pr\left(s_i | s_j : j \in \mathbf{N}_i, \boldsymbol{\theta}\right). \tag{2.1}$$

Here, $\Pr\left(s_i | s_j : j \in \mathbf{N}_i, \boldsymbol{\theta}\right)$ is the conditional probability of the signal s_i in the pixel $i \in \mathbf{R}$, given the signals $[s_j : j \in \mathbf{N}_i]$ in its neighborhood $\mathbf{N}_i$.

It is easily seen that the partition function Z of the GPD is not used to compute the conditional probabilities of Eq. (2.1) and obtain the PMLE. For the autobinomial GPD, the conditional probability is as follows:

$$\Pr\left(s_i | s_j : j \in \mathbf{N}_i, \boldsymbol{\theta}\right) = \frac{\exp\left(\sum_{\iota \in \mathbf{N}} \theta_\iota \cdot s_i \cdot (s_{i-\iota} + s_{i+\iota})\right)}{\sum_{u \in \mathbf{U}} \exp\left(\sum_{\iota \in \mathbf{N}} \theta_\iota \cdot u \cdot (s_{i-\iota} + s_{i+\iota})\right)}. \tag{2.2}$$

The PMLE is computed by maximizing the product of the conditional probabilities of Eq. (2.1) for all the sublattices $\mathbf{R}_\beta : \beta \in \mathbf{B}$:

$$\prod_{\beta \in \mathbf{B}} \Pr(\mathbf{s}_\beta | \mathbf{s}_b : b \in \mathbf{B}^\beta, \boldsymbol{\theta})$$

with respect to the desired parameters $\boldsymbol{\theta}$.

Both the PMLE and MLE are asymptotically consistent for estimating the potentials of finite range, and the PMLE results in fairly good approximations of the true parameters for the autobinomial model (Comets, 1992;

Winkler, 1995). Thus, the PMLE is usually considered as a practicable alternative to the conventional MLE although the PMLE is obviously less efficient than the MLE[1].

For the MFE, the situation is much worse because this estimate can only be used in practice if all the neighborhoods N_i and the set of signal values U are of very small cardinalities. Here, the conditional probabilities of Eq. (2.2) are approximated by the conditional sample frequencies for a given training sample. In general, this gives a system of $|U|^{1+|N_i|}$ non-linear equations for the probabilities and sample relative frequencies that contain $|\theta|$ unknown parameters.

This system is usually highly overdetermined, and usually the logarithmic ratios of probabilities and frequencies are used to simplify the problem. But the ratio of frequencies is a very poor statistical estimate for the probability ratio. In addition, a good part of the frequencies cannot be estimated at all because the corresponding signal combinations are simply absent in a given training sample. So, generally the MFE results in a very poor approximation of the true parameter values.

Autonormal, or Gauss-Markov model. This model is much more flexible than the autobinomial one because it assumes that all the pixels within a given rectangular, square, or circular window w form a clique, that is, they are the neighbors of each other (Kashyap and Chellappa, 1983; Kashyap, 1986; Cohen and Patel, 1991; Chellappa et al., 1993). In most cases, w is a rectangular window

$$w = ((\mu, \nu) : \mu = -m_{\text{win}}, \dots, 0, \dots, m_{\text{win}}; \nu = -n_{\text{win}}, \dots, 0, \dots, n_{\text{win}}).$$

The model contains a single family of the cliques of order $|w|$ that include both the close-range and long-range pixel interactions.

The potential is proportional to a squared error of the autoregressive prediction, or approximation of the signal s_i in the central pixel i of the window w from the neighboring signals $s_{i+\iota}$; $\iota \in w$; $\iota \neq (0,0)$ in all other pixels of the window:

$$V\left(s_{i+\iota} : \iota \in w\right) = \left(\sum_{\iota \in w} \theta_\iota \cdot s(i + \iota)\right)^2$$

where the regression coefficient $\theta_{(0,0)} = -1$.

---

[1]As will be shown in Chapters 3 and 4, the generalized Gibbs models with pairwise pixel interactions permit us to compute directly the conventional MLEs by using first analytic and then stochastic approximation of the potentials. Therefore, in this case we have, in principle, no need to replace the MLE by the PMLE.

This model can also be represented as the model with multiple short- and long-range pairwise interactions within the window (Besag, 1974) having particular signal products with interdependent coefficients as Gibbs potentials. But such repesentation is merely formal.

As regarding image modelling, the Gauss–Markov model has some drawbacks. First, it assumes the infinite continuous range of the signal values and, in the strict sense, does not represent the digital images that have a finite set $\mathbf{U}$ of the signal values with rather small cardinality. Secondly, the model is based on the initial assumption that the residual errors of the autoregressive prediction are statistically independent and have the same normal distribution for each possible set of the regression coefficients $\boldsymbol{\theta} = (\theta_\iota : \iota \in \mathbf{w}; \theta_{0,0} = -1)$. In fact, this assumption is very restrictive, but it is necessary to exclude the dependence of the partition function Z from the model parameters—regression coefficients to simplify their estimation.

To avoid prohibitively large volumes of computations, an exotic toroidal lattice has to be assumed to derive an analytical scheme of estimating the coefficients $\boldsymbol{\theta}$ (Kashyap, 1980; Kashyap and Chellappa, 1983). Then the MLEs for the coefficients can be obtained from a given learning sample by solving a system of $(2m_{\text{win}} + 1) \cdot (2n_{\text{win}} + 1)$ linear equations with the same number of the variables $\boldsymbol{\theta}$.

The size of the system of equations is given by the squared number of the variables or, what is almost the same, of the window size $|\mathbf{w}|$. Therefore the large windows cannot be used in practice. Usually, the Gauss–Markov image models exploit the windows which are not greater than $15 \times 15...17 \times 17$ (Kashyap, 1980; Kashyap and Chellappa, 1983; Kashyap, 1986; Chellappa et al., 1993). Compared to the auto-binomial model, the Gauss–Markov model introduces much more diverse texels and hence has been successfully used to simulate, segment, and classify different image textures with a limited spatial range of pixel interactions (Kashyap, 1986; Chellappa et al., 1993).

But it is difficult to explicitly adapt the interaction structure to a particular texture, say, by choosing the most characteristic window size for a given type of texture or by excluding less significant interactions from a window $\mathbf{w}$. To find a characteristic window one has to exhaust all the possible variants of subwindows (Kashyap and Chellappa, 1983), and generally such a process has combinatorial complexity.

An alternative way to estimate the parameters of a Gauss–Markov model is based on representing the model in the spatial frequency domain. Spectral components of a Gauss–Markov random field form an IRF, and the probabilities of spectral components are functions of the regression coefficients $\boldsymbol{\theta}$ (Cohen and Patel, 1991). The spectral representation escapes the large systems of equations, but comes across another difficulty. Spectral components have rather poor statistical properties (Kendall and Stuart, 1966) that may lead to inadequate parameter estimates of the model. In particular, the standard deviation of each spectral component has the same order as the component itself, so that the sample spectrum usually contains very large fluctuations and cannot be considered as a reliable representative of a true theoretical one. These fluctuations can be decreased by local smoothing of a spectrum, but this results in the considerably biased estimates of the components (Kendall and Stuart, 1966).

2.1.2. MODELS WITH MULTIPLE PAIRWISE INTERACTIONS

For describing spatially homogeneous and piecewise-homogeneous textures, the interaction structure seems to be more important that the potential. As is evident from the foregoing, traditional automodels cannot reflect in full measure the characteristic features of image textures mainly because of the pre-defined interaction structure. But, they are widely used in practice due to a common opinion that there are no practicable alternatives.

Contrary to this opinion, it is an easy matter to generalize the Gibbs models as to involve an arbitrary structure of multiple translation invariant pairwise pixel interactions. In a generalized model each pixel appears in several characheristic second-order cliques, and the interaction structure $\mathbf{C} = \bigcup_{a \in \mathbf{A}} \mathbf{C}_a$ is formed by uniting the corresponding second-order clique families $\mathbf{C}_a$.

Each second-order clique family

$$\mathbf{C}_a \equiv \mathbf{C}_{\mu_a, \nu_a} = [(i, j) : i, j \in \mathbf{R}; \; i - j = (\mu_a, \nu_a)]$$

contains all the translation invariant pixel pairs (i, j) that have a fixed inter-pixel, or intra-clique shift (μ_a, ν_a); if $i = m, n$ then $j = (m - \mu_a, n - \nu_a)$. The shift (μ_a, ν_a) may also be described by the orientation angle $\varphi_a = \arctan(\mu_a / \nu_a)$ and Cartesian distance $\rho_a = (\mu_a^2 + \nu_a^2)^{\frac{1}{2}}$ between the pixels.

The inter-pixel shifts (μ_a, ν_a); $a \in \mathbf{A}$, that specify the interaction structure may be arbitrary, depending on a texture to be modelled. The "star-like" texels, represented by gray level combinations in the cliques containing the same pixel, are considered as most characteristic for this texture as a whole (see, for instance, Figure 2.1).

Basically, such simplification leaves aside most of the natural textures that usually have larger and more diverse texels and greater geometric differences between the self-similar patches. Nevertheless, there are many natural and artificial image textures that can be modelled adequately by simple Gibbs models with multiple pairwise pixel interactions. This specific class of spatially homogeneous and piecewise homogeneous image textures is called *stochastic textures* in (Gimel'farb, 1996a). A stochastic texture has pixels and pixel pairs as supports for (primitive) texels and is completely specified by the structure and strengths of the translation invariant pairwise pixel interactions.

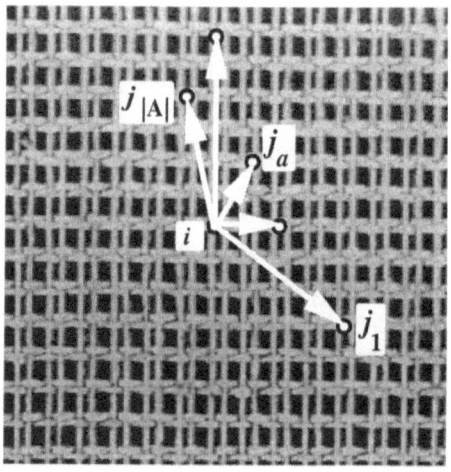

Figure 2.1. Pairwise pixel interactions.

There were only a few attempts to broaden the class of Gibbs image models by introducing potential functions and/or clique families that differ from the autobinomial or autonormal models. In particular, one of the Gibbs models in (Moussouris, 1974) as well as the model of region maps for a piecewise-homogeneous grayscale texture in (Derin and Cole, 1986) used the nearest-neighbor cliques of orders 2–4 shown in Figure 2.2. The associated potentials for the 9 clique families have only two values: $V_a(s_j : j \in \mathbf{c}_a) = \theta_a$ if all signals s_j in the clique $\mathbf{c}_a$ are equal and $-\theta_a$ otherwise.

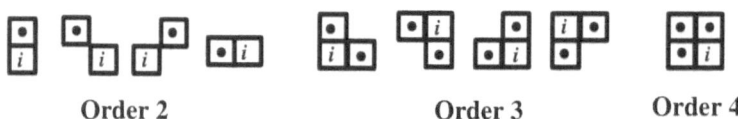

Order 2   Order 3   Order 4

Figure 2.2. Nearest-neighbor cliques of order 2–4.

The region map models in (Gimel'farb and Zalesny, 1993a) generalize the latter model by using more diverse potential functions and introducing the second-order cliques with both the short-range (nearest-neighbor) and long-range pixel interactions. Also, the model of region maps for a piecewise-homogeneous grayscale texture in (Gimel'farb and Zalesny, 1991) exploited the conditional sample histograms of region labels for every fixed gray level as potentials. As will be shown in Chapters 3 and 4, such a heuristic choice is in fact very close to a theoretically justified one.

The grayscale texture model of Gimel'farb and Zalesny (1993b) used both arbitrary multiple pairwise interactions and partly pre-defined multimodal potential functions for each clique family:

$$V_a(q, q_a) = \theta_0 \cdot (\theta_1 + \min\{|q - q_a - \theta_2|, \theta_3, |q - q_a - \theta_2|\}).$$

These multimodal potentials permit us to emphasize certain signal differences which are typical for the "wavy" image patterns. As was indicated in (Gimel'farb and Zalesny, 1991), the likelihood function for this GPD is convex with respect to the scaling factors θ_0, so that these factors can be estimated for the different clique families by stochastic approximation. Unfortunately, all other parameters of the above potentials cannot be estimated in a similar way and have to be given using heuristic considerations.

The scaled sample gray-level difference histograms (GLDH) collected over the multiple second-order clique families were proposed by Gimel'farb and Zalesny (1993b) as heuristic estimates of the potential functions for each family a:

$$V_a (d = q - q_a) = \theta_a \cdot H_a (d|\mathbf{g}^\circ).$$

Here, $H_a(d|\mathbf{g}^\circ)$ denotes a component of the GLDH for a given training sample $\mathbf{g}^\circ$ of grayscale texture, and θ_a is a scaling factor. Such a potential emphasizes the most frequent gray level differences (the larger the number of occurrences of a grey level difference d in the cliques of a family $\mathbf{C}_a$ in the training sample, the larger the potential value). Similar heuristic choice of the potentials, but only for the cliques of the nearest neighborhood, was proposed also by Lohmann (1994). As mentioned above, these GLDH-based heuristics are surprisingly close to the theoretically justified MLEs of potentials for the generalized Gibbs models with multiple pairwise pixel interactions, described in the subsequent sections.

2.2. Generalized Gibbs models of homogeneous textures

2.2.1. NON-MARKOV GIBBS IMAGE MODEL

Gray range normalization.   Usually, the gray range $[q_{min}(\mathbf{g}), q_{max}(\mathbf{g})]$ of a grayscale image $\mathbf{g}$ may arbitrarily change without affecting the visual perception of a texture. As regards the probabilistic image modelling, it follows that the probability of each image sample should be invariant to the admissible scale changes of its gray range. In other words, the images $\mathbf{g} \in \mathcal{G}$ that differ with only the gray range scale have to be equivalent by probability to a particular reference image, $\mathbf{g}^{rf}$. The reference image is obtained by normalizing the initial gray range $[q_{\min}(\mathbf{g}), q_{\max}(\mathbf{g})]$. The normalization transforms the initial range into a reference one $[0, Q]$ specified by a given set $\mathbf{Q}$ of gray values as follows:

$$\forall i \in \mathbf{R} \quad g_i^{\mathrm{rf}} = \frac{g_i - q_{\min}(\mathbf{g})}{q_{\max}(\mathbf{g}) - q_{\min}(\mathbf{g})} \, Q.$$

In the case when only the offset changes are admissible the normalization is much simpler:

$$g_i^{\mathrm{rf}} = g_i - q_{\min}(\mathbf{g}).$$

Image modelling. Non-Markov Gibbs models of grayscale images under the admissible gray range changes are obtained by embedding the above normalization directly into the potentials. The resulting GPD is as follows:

$$\Pr(\mathbf{g}|\mathbf{V}) = \frac{1}{Z_{\mathbf{V}}} \cdot \exp\left(e(\mathbf{g}^{\mathrm{rf}}|\mathbf{V}_{\mathrm{pix}}) + \sum_{a \in \mathbf{A}} e_a(\mathbf{g}^{\mathrm{rf}}|\mathbf{V}_a)\right) \qquad (2.3)$$

where $e(\mathbf{g}^{\mathrm{rf}}|\mathbf{V}_{\mathrm{pix}})$ is the partial energy of the pixelwise interactions:

$$e(\mathbf{g}^{\mathrm{rf}}|\mathbf{V}_{\mathrm{pix}}) = \sum_{i \in \mathbf{R}} V_{\mathrm{pix}}(g_i^{\mathrm{rf}}) \qquad (2.4)$$

and $e_a(\mathbf{g}^{\mathrm{rf}}|\mathbf{V}_a)$ denotes the partial energy of the pairwise pixel interactions in the clique family $\mathbf{C}_a$:

$$e_a(\mathbf{g}^{\mathrm{rf}}|\mathbf{V}_a) = \sum_{(i,j) \in \mathbf{C}_a} V_a(g_i^{\mathrm{rf}}, g_j^{\mathrm{rf}}). \qquad (2.5)$$

Here, $\mathbf{V} = [\mathbf{V}_{\mathrm{pix}}, \mathbf{V}_a : a \in \mathbf{A}]$ denotes the vector of the centered potential values in the cliques, $\mathbf{V}_{\mathrm{pix}} = [V_{\mathrm{pix}}(q) : q \in \mathbf{G}]$ is the potential vector for the gray levels in the pixels, and $\mathbf{V}_a = [V_a(q, q') : (q, q') \in \mathbf{G}^2]$ is the potential vector for gray level co-occurrences in the cliques of the family $\mathbf{C}_a$.

Potential centering and other features. As follows from Eq. (1.11), the partial energies in Eq. (2.3) are invariant to the potential centering:

$$\sum_{q \in \mathbf{Q}} V_{\mathrm{pix}}(q) = 0; \quad \forall a \in \mathbf{A} \quad \sum_{q,q' \in \mathbf{Q}^2} V_a(q, q') = 0. \qquad (2.6)$$

The model in Eq. (2.3) specifies a non-Markov GRF because each pixel $i \in \mathbf{R}$ depends after the normalization on all the other pixels $j \in \mathbf{R}^i$. More fully, the local pairwise pixel interactions, represented by the second-order cliques (j_a, i), $(i, j'_a) \in \mathbf{C}_a$; $|i - j_a| = |j'_a| = \text{const}_a$; $a \in \mathbf{A}$, containing a particular pixel i, are supplemented with a lattice-wide interaction. This latter one supplies the minimum $q_{\min}(\mathbf{g})$ and the maximum $q_{\max}(\mathbf{g})$ gray levels for normalizing the initial image $\mathbf{g}$.

It should be pointed out that the above non-Markov model is "almost" the Markov/Gibbs one with regard to the average volume of computations per pixel for generating a Markov chain of the reference samples by stochastic relaxation. The lattice-wide interaction manifests itself in the conditional probability $\Pr(g_i^{\mathrm{rf}} | [g_j^{\mathrm{rf}} : j \in \mathbf{N}_i], \mathbf{V})$ of the gray levels for a single pixel i only if the pixel i supports the solitary minimum (0) or maximum gray level (Q) in the lattice. Because such a solitary extremum can be affected by pixelwise relaxation, the actual neighborhood of the pixel coincides with the whole lattice: $\mathbf{N}_i = \mathbf{R}^i$. In this case the current reference sample $\mathbf{g}^{\mathrm{rf}}$ has to be renormalized for computing the conditional probability of each gray level $q \in \mathbf{Q}$ in the pixel i. The renormalization is based on the actual gray range for the image obtained by substituting q for the current signal g_i^{rf}. Otherwise, the minimum and maximum gray levels do not depend on the updated gray level g_i^{rf} in the pixel i, and only the local neighborhood $\mathbf{N}_i$, specified by characteristic second-order clique families $[\mathbf{C}_a : a \in \mathbf{A}]$ of the model in Eq. (2.3), has to be taken into account.

2.2.2. MARKOV/GIBBS MODEL

Invariance to gray range offsets. Let the admissible signal transformations be restricted to only the offset changes. Then the exact Markov/Gibbs model that needs no image normalization is specified by assuming the potentials are invariant to an arbitrary gray range offset. Under such an assumption, the potential for the pixelwise interactions has to be set to zero, and potentials for the pairwise interactions depend only on the gray level differences $d_{i,j} = g_i - g_j$. These latter are independent of the uniform offset of the image gray range.

Image modelling. The resulting Markov/Gibbs model is as follows:

$$\Pr(\mathbf{g}|\mathbf{V}) = \frac{1}{Z_\mathbf{V}} \cdot \exp\left(\sum_{a \in \mathbf{A}} e_a(\mathbf{g}|\mathbf{V}_a)\right) \tag{2.7}$$

where $e_a(\mathbf{g}|\mathbf{V}_a)$ is the partial energy of the pairwise pixel interactions:

$$e_a(\mathbf{g}|\mathbf{V}_a) = \sum_{(i,j) \in \mathbf{C}_a} V_a(g_i - g_j). \tag{2.8}$$

Here, $\mathbf{V} = [\mathbf{V}_a : a \in \mathbf{A}]$ is the vector of the centered potentials for the second-order clique families $\mathbf{C} = [\mathbf{C}_a : a \in \mathbf{A}]$, and $\mathbf{V}_a = [V_a(d) : d \in \mathbf{D}]$

is the centered potential vector for the clique family $\mathbf{C}_a$ with components that depend on the gray level differences d in the cliques. We denote $\mathbf{D} = \{-Q, \ldots, 0, \ldots, Q\}$ a finite set of the gray level difference values. The partial energy in Eq. (2.8) only differs by the potential dependence of the gray level differences rather than co-occurrences from the similar energy in Eq. (2.5).

Potential centering and other features.   The model of Eq. (2.7) is invariant to the potential centering:

$$\forall a \in \mathbf{A} \quad \sum_{d \in \mathbf{D}} V_a(d) = 0. \tag{2.9}$$

Basic features of the non-Markov model in Eq. (2.3) are similar or obtained with minor changes from the like features of the Markov/Gibbs model of Eq. (2.7). The major distinction between them is in the stochastic relaxation for generating the image samples. In the Markov/Gibbs case, each relaxation step involves a local summation of the potentials over the "star–like" neighbourhood $\mathbf{N}_i$ formed by the second-order cliques containing the current pixel i. The maximum size of such a neighborhood is $|\mathbf{N}_i| = 2 \cdot |\mathbf{A}|$. The non-Markov model has to directly produce the normalized reference samples having the gray range $[0, Q]$. As indicated earlier, in the non-Markov model the local summation holds for all the pixels of the generated reference sample, except for the pixels supporting the current solitary maximum (Q) or minimum (0) signal. Only in this rare case of the solitary extremum signal the actual neighbourhood of the pixel becomes lattice–wide and the potentials are summed up over the total lattice. Thus, the computational complexity of the relaxation does not increase substantially in the non-Markov case relative to the Markov/Gibbs one. Even if each macrostep involves the two lattice-wide neighborhoods, the average neighborhood size per pixel is growing from $2 \cdot |\mathbf{A}|$ to roughly $2 \cdot |\mathbf{A}| + 2$.

2.2.3. SIMPLIFIED NON-MARKOV MODEL

The non-Markov model of Eq. (2.3) can also be simplified by assuming that the potentials for the second-order cliques depend only on the signal differences $d_{i,j} = g_i^{\mathrm{rf}} - g_j^{\mathrm{rf}}$. This particular case is quite similar to the general-case model in Eq. (2.3):

$$\Pr(\mathbf{g}|\mathbf{V}) \quad = \quad \frac{1}{Z_\mathbf{V}} \cdot \exp\left(e(\mathbf{g}^{\mathrm{rf}}|\mathbf{V}_{\mathrm{pix}}) + \sum_{a \in \mathbf{A}} e_a(\mathbf{g}^{\mathrm{rf}}|\mathbf{V}_a) \right), \tag{2.10}$$

except that the partial energies of the pairwise pixel interactions depend on the gray level differences rather than on the gray level co-occurrences.

The energies are quite similar to the energies in Eq. (2.8) except that the signals are taken from the normalized reference image, $\mathbf{g}^{\mathrm{rf}}$, instead of the initial one, $\mathbf{g}$:

$$e_a(\mathbf{g}^{\mathrm{rf}}|\mathbf{V}_a) = \sum_{(i,j)\in\mathbf{C}_a} V_a(g_i^{\mathrm{rf}} - g_j^{\mathrm{rf}}). \tag{2.11}$$

The potential vectors $\mathbf{V}_a = [V_a(d) : d \in \mathbf{D}]$ contain the potential values for the gray level differences in the cliques of a family $\mathbf{C}_a$ that satisfy the centering conditions of Eq. (2.9).

2.2.4. EXPONENTIAL FAMILY REPRESENTATIONS

The partial and total energies in Eqs. (2.3), (2.7), and (2.10) are easily represented by the dot products of the corresponding potential vectors and the vectors of gray level, gray level co-occurrence, or gray level difference histograms. Let $\mathbf{H}_{\mathrm{pix}} = [H_{\mathrm{pix}}(q|\mathbf{g}) : q \in \mathbf{Q}]$ denote the vector of the gray level histogram (GLH) for a grayscale image $\mathbf{g}$:

$$H_{\mathrm{pix}}(q|\mathbf{g}) = \sum_{i\in\mathbf{R}} \delta(q - g_i).$$

Let $\mathbf{H}_a = [H_a(q, q'|\mathbf{g}) : (q, q' \in \mathbf{Q}^2]$ and $\mathbf{H}_a = [H_a(d|\mathbf{g}) : d \in \mathbf{D}]$ be the vectors of the gray level co-occurrence histogram (GLCH) and gray level difference histogram (GLDH) for a second-order clique family $\mathbf{C}_a$ in an image $\mathbf{g}$, respectively:

$$H_a(q, q'|\mathbf{g}) = \sum_{(i,j)\in\mathbf{C}_a} \delta(q - g_i) \cdot \delta(q' - g_j);$$

$$H_a(d|\mathbf{g}) = \sum_{(i,j)\in\mathbf{C}_a} \delta(d - (g_i - g_j)).$$

Then the partial energies in Eqs. (2.4) and (2.5) can be rewritten as the dot products of the potential and histogram vectors as follows:

$$e(\mathbf{g}^{\mathrm{rf}}|\mathbf{V}_{\mathrm{pix}}) = \mathbf{V}_{\mathrm{pix}} \bullet \mathbf{H}_{\mathrm{pix}}(\mathbf{g}^{\mathrm{rf}}) \equiv \sum_{q\in\mathbf{Q}} V_{\mathrm{pix}}(q) \cdot H_{\mathrm{pix}}(q|\mathbf{g}^{\mathrm{rf}})$$

and

$$\forall a \in \mathbf{A} \ \ e_a(\mathbf{g}^{\mathrm{rf}}|\mathbf{V}_a) = \mathbf{V}_a \bullet \mathbf{H}_a(\mathbf{g}^{\mathrm{rf}}) \equiv \sum_{(q,q')\in\mathbf{Q}^2} V_a(q, q') \cdot H_a(q, q'|\mathbf{g}^{\mathrm{rf}}).$$

These relations result in the following dot-product representation of the total energy:

$$e(\mathbf{g}^{\mathrm{rf}}|\mathbf{V}) = e(\mathbf{g}^{\mathrm{rf}}|\mathbf{V}_{\mathrm{pix}}) + \sum_{a\in\mathbf{A}} e_a^{\mathrm{rf}}(\mathbf{g}|\mathbf{V}) = \mathbf{V} \bullet \mathbf{H}(\mathbf{g}^{\mathrm{rf}}). \tag{2.12}$$

Here, the histogram vector $\mathbf{H}(\mathbf{g}^{\mathrm{rf}})$ for the reference image $\mathbf{g}^{\mathrm{rf}}$ contains the GLH $\mathbf{H}_{\mathrm{pix}}(\mathbf{g}^{\mathrm{rf}})$ and the GLCHs $\mathbf{H}_a(\mathbf{g}^{\mathrm{rf}})$ collected over the clique families $[\mathbf{C}_a : a \in \mathbf{A}]$:

$$\mathbf{H}(\mathbf{g}^{\mathrm{rf}}) = [H_{\mathrm{pix}}(q|\mathbf{g}^{\mathrm{rf}}) : q \in \mathbf{Q}; \ H_a(q, q'|\mathbf{g}^{\mathrm{rf}}) : (q, q') \in \mathbf{Q}^2; \ a \in \mathbf{A}].$$

The GLH and GLCHs are the sufficient statistics for the Gibbs model in Eq. (2.3) as well as the GLH and/or GLDHs are the sufficient statistics for the models in Eqs. (2.7) and (2.10). The total energies for these latter models are also represented by the dot products similar to Eq. (2.12). The Markov/Gibbs model of Eq. (2.7) involves the total energy $e(\mathbf{g}|\mathbf{V}) = \mathbf{V} \bullet \mathbf{H}(\mathbf{g})$ with the vector of the GLDHs for the initial grayscale image $\mathbf{g}$:

$$\mathbf{H}(\mathbf{g}) = [H_a(d|\mathbf{g}) : d \in \mathbf{D}; \ a \in \mathbf{A}].$$

The simplified non-Markov Gibbs model of Eq. (2.10) has the total energy of Eq. (2.12) but with the following vector of the GLH and GLDHs for the normalized reference image $\mathbf{g}^{\mathrm{rf}}$:

$$\mathbf{H}(\mathbf{g}^{\mathrm{rf}}) = [H_{\mathrm{pix}}(q|\mathbf{g}^{\mathrm{rf}}) : q \in \mathbf{Q}; \ H_a(d|\mathbf{g}^{\mathrm{rf}}) : d \in \mathbf{D}; \ a \in \mathbf{A}].$$

Therefore, it is these models that appear implicitly in the well-known approaches to describe the grayscale textures by local features derived from the GLDHs (see, for instance, (Haralick, 1979)).

2.2.5. GLH, GLDH, AND A FEW LESS FAMOUS ABBREVIATIONS

As discussed above, the proposed Gibbs models with pairwise pixel interactions describe each particular grayscale image in terms of the sample histograms of signals and signal pairs collected over the clique families. The gray level, gray level co-occurrence, and gray level difference histograms involved by the Gibbs models of homogeneous grayscale images are in so widespread use that they have deserved the above-mentioned standard abbreviations, GLH, GLCH, and GLDH, respectively. In the subsequent sections, some other histograms will be used to describe the homogeneous region maps and piecewise-homogeneous grayscale images.

Table 2.1 presents all the histograms that describe pixel interactions in homogeneous grayscale images $\mathbf{g}$, region maps $\mathbf{l}$, or piecewise-homogeneous grayscale images $(\mathbf{g}, \mathbf{l})$. The corresponding abbreviations have been listed in Table 1.3.

Generally, the pixelwise interactions in the lattice $\mathbf{R}$ involve the sample frequencies of the gray levels $g_i = q \in \mathbf{Q}$ or/and region labels $l_i = k \in \mathbf{K}$ in the pixels. The pairwise pixel interactions in a second-order clique family $\mathbf{C}_a$ are described by the sample frequencies of co-occurrences of the gray

TABLE 2.1. Histograms for describing multiple pairwise pixel interactions.

GLH	$\mathbf{H}_{\text{pix}}(\mathbf{g}) = [H(q\|\mathbf{g}) : q \in \mathbf{Q}]$
GLCH	$\mathbf{H}_a(\mathbf{g}) = \left[H_a(q, q'\|\mathbf{g}) : q, q' \in \mathbf{Q}^2\right]$
GLDH	$\mathbf{H}_a(\mathbf{g}) = [H_a(d\|\mathbf{g}) : d \in \mathbf{D}]$
RLH	$\mathbf{H}_{\text{pix}}(\mathbf{l}) = [H(k\|\mathbf{l}) : k \in \mathbf{K}]$
RLCH[a]	$\mathbf{H}_a(\mathbf{l}) = \left[H_a(k, k')(\mathbf{l}) : k, k' \in \mathbf{K}^2\right]$
RLCH[b]	$\mathbf{H}_a(\mathbf{l}) = [H_a(k, \alpha)(\mathbf{l}) : k, \alpha \in \mathbf{K} \times \{0, 1\}]$
GL/RLH	$\mathbf{H}_{\text{pix}}(\mathbf{g}, \mathbf{l}) = [H(q, k\|\mathbf{g}, \mathbf{l}) : q, k \in \mathbf{Q} \times \mathbf{K}]$
GLC/RLCH[c]	$\mathbf{H}_a(\mathbf{g}, \mathbf{l}) = \left[H_a(q, k; q', k'\|\mathbf{g}, \mathbf{l}) : q, q', k, k' \in \mathbf{Q}^2 \times \mathbf{K}^2\right]$
GLD/RLCH[d]	$\mathbf{H}_a(\mathbf{g}, \mathbf{l}) = [H_a(d, k, \alpha\|\mathbf{g}, \mathbf{l}) : d, k, \alpha \in \mathbf{D} \times \mathbf{K} \times \{0, 1\}]$

[a] Region Label Co-occurrence Histogram
[b] Region Label Coincidence Histogram
[c] Gray Level and Region Label Co-occurrence Histogram
[d] Gray Level Difference and Region Label Coincidence Histogram

level $(g_i = q, g_j = q')$ or/and region label $(l_i = k, l_j = k')$ pairs in the cliques $(i, j) \in \mathbf{C}_a$.

Simplified particular cases exploit the sample frequencies of the gray level differences $d_{i,j} = g_i - g_j \equiv q - q'$ and indicators of region label coincidences $\alpha = \delta(l_i = k, l_j = k')$ in the cliques. The co-incidence $\alpha \in \{0, 1\}$ of the region labels $l_i = k$ and $l_j = k'$ in the clique means that only a binary relation $\alpha = \delta(k - k')$ between the region labels is taken into account. If the labels coincide $(k = k')$ then it is the intra-region interaction of the gray levels $g_i = q$ and $g_j = q'$, and the indicator $\alpha = 1$. Otherwise it is the inter-region interaction $(k \neq k'$, or $\alpha = 0)$.

For convenience, the correspondences between the components $\mathbf{C}$, $\mathbf{s}$, $\mathbf{u}$, $\mathbf{U}^{|c|}$ of Eqs. (1.13) and (1.14) and components of the histograms in Table 2.1 are summarized in Table 2.2. The sample relative frequency distributions $\mathbf{F}_{...}(\ldots)$ are obtained by normalizing the corresponding histograms $\mathbf{H}_{...}(\ldots)$ of Table 2.1 as follows:

GLH	$\mathbf{F}_{\text{pix}}(\mathbf{g}) = \dfrac{\mathbf{H}_{\text{pix}}(\mathbf{g})}{|\mathbf{R}|}$
GLCH, GLDH	$\mathbf{F}_a(\mathbf{g}) = \dfrac{\mathbf{H}_a(\mathbf{g})}{|\mathbf{C}_a|}$
RLH	$\mathbf{F}_{\text{pix}}(\mathbf{g}) = \dfrac{\mathbf{H}_{\text{pix}}(\mathbf{l})}{|\mathbf{R}|}$
RLCH	$\mathbf{F}_a(\mathbf{l}) = \dfrac{\mathbf{H}_a(\mathbf{l})}{|\mathbf{C}_a|}$
GL/RLH	$\mathbf{F}_{\text{pix}}(\mathbf{g}, \mathbf{l}) = \dfrac{\mathbf{H}_{\text{pix}}(\mathbf{g}, \mathbf{l})}{|\mathbf{R}|}$
GLC/RLCH, GLD/RLCH	$\mathbf{F}_a(\mathbf{g}, \mathbf{l}) = \dfrac{\mathbf{H}_a(\mathbf{g}, \mathbf{l})}{|\mathbf{C}_a|}$

TABLE 2.2. Components of the histograms.

Histogram	**C**	s	u	**U**$^{|c|}$
GLH	$\mathbf{H}_{\text{pix}}(\mathbf{g})$	**R**	g	q	**Q**
GLCH	$\mathbf{H}_a(\mathbf{g})$	$\mathbf{C}_a$	g	q, q'	$\mathbf{Q}^2$
GLDH	$\mathbf{H}_a(\mathbf{g})$	$\mathbf{C}_a$	g	d	**D**
RLH	$\mathbf{H}_{\text{pix}}(\mathbf{l})$	**R**	l	k	**K**
RLCH^a	$\mathbf{H}_a(\mathbf{l})$	$\mathbf{C}_a$	l	k, k'	$\mathbf{K}^2$
RLCH^b	$\mathbf{H}_a(\mathbf{l})$	$\mathbf{C}_a$	l	k, α	$\mathbf{K} \times \{0, 1\}$
GL/RLH	$\mathbf{H}_{\text{pix}}(\mathbf{g}, \mathbf{l})$	**R**	g, l	q, k	$\mathbf{Q} \times \mathbf{K}$
GLC/RLCH^c	$\mathbf{H}_a(\mathbf{g}, \mathbf{l})$	$\mathbf{C}_a$	g, l	q, k, q', k'	$\mathbf{Q}^2 \times \mathbf{K}^2$
GLD/RLCH^d	$\mathbf{H}_a(\mathbf{g}, \mathbf{l})$	$\mathbf{C}_a$	g, l	d, k, α	$\mathbf{D} \times \mathbf{K} \times \{0, 1\}$

^{a}Region Label Co-occurrence Histogram
^{b}Region Label Coincidence Histogram
^{c}Gray Level and Region Label Co-occurrence Histogram
^{d}Gray Level Difference and Region Label Coincidence Histogram

2.3. Prior Markov/Gibbs models of region maps

Gibbs model similar to the model in Eq. (2.7) can also describe the arbitrary region maps. As was indicated in Chapter 1, a region map differs from a grayscale image only in physical meaning of the signals.

Pairwise region label interactions depend, generally, on the label co-occurrences, so that the Markov/Gibbs model of the region maps $\mathbf{l}$ is as follows:

$$\Pr(\mathbf{l}|\mathbf{V}) = \frac{1}{Z_{\mathbf{V}}} \cdot \exp\left(e_{\text{pix}}(\mathbf{l}|\mathbf{V}_{\text{pix}}) + \sum_{a \in \mathbf{A}} e_a(\mathbf{l}|\mathbf{V}_a)\right). \quad (2.13)$$

Here, $e_{\text{pix}}(\mathbf{l}|\mathbf{V}_{\text{pix}})$ is the partial energy of the pixelwise interactions:

$$e_{\text{pix}}(\mathbf{l}|\mathbf{V}_{\text{pix}}) = \sum_{i \in \mathbf{R}} V_{\text{pix}}(l_i) \equiv \mathbf{V}_{\text{pix}} \bullet \mathbf{H}_{\text{pix}}(\mathbf{l}) \quad (2.14)$$

and $e_a(\mathbf{l}|\mathbf{V}_a)$ is the partial energy of the pairwise pixel interactions over a clique family $\mathbf{C}_a$:

$$e_a(\mathbf{l}|\mathbf{V}_a) = \sum_{(i,j) \in \mathbf{C}_a} V_a(l_i, l_j) \equiv \mathbf{V}_a \bullet \mathbf{H}_a(\mathbf{l}). \quad (2.15)$$

Therefore the model in Eq. (2.13) has the following obvious exponential family representation:

$$\Pr(\mathbf{l}|\mathbf{V}) \;=\; \frac{1}{Z_{\mathbf{V}}} \cdot \exp\left(\mathbf{V} \bullet \mathbf{H}(\mathbf{l}))\right). \tag{2.16}$$

with the total Gibbs energy $e(\mathbf{l}|\mathbf{V}) = \mathbf{V} \bullet \mathbf{H}(\mathbf{l})$ where

$$\mathbf{H}(\mathbf{l}) = [\mathbf{H}_{\mathrm{pix}}(\mathbf{l}); \; \mathbf{H}_a(\mathbf{l}) : \; a \in \mathbf{A}]$$

denotes the vector of the RLH and RLCHs collected for the clique families $[\mathbf{C}_a : a \in \mathbf{A}]$ in the map $\mathbf{l}$. This representation shows that the RLH and RLCHs are the sufficient statistic for the region map model of Eq. (2.13).

For simplicity, let us assume that the pixelwise label interactions are the same for all the regions so that the corresponding potentials can be set to zero, $V_{\mathrm{pix}}(k) = 0$, $k \in \mathbf{K}$. Then the model is as follows:

$$\Pr(\mathbf{l}|\mathbf{V}) = \frac{1}{Z_{\mathbf{V}}} \exp\left(\sum_{a \in \mathbf{A}} e_a(\mathbf{l}|\mathbf{V}_a)\right) \tag{2.17}$$

and the RLCHs $\mathbf{H}(\mathbf{l}) = [\mathbf{H}_a(\mathbf{l}) : a \in \mathbf{A}]$ form the sufficient statistics.

The models in Eqs. (2.13) and (2.17) exploit the potential centering:

$$\sum_{k \in \mathbf{K}} V_{\mathrm{pix}}(k) = 0; \quad \forall a \in \mathbf{A} \quad \sum_{(k,k') \in \mathbf{K}^2} V_a(k,k') = 0. \tag{2.18}$$

The model of Eq. (2.17) can be further simplified by assuming that the pairwise interactions in the cliques $(i,j) \in \mathbf{C}_a$ depend only on the region label coincidences. Then, for each region $l_i = k \in \mathbf{K}$, all the "foreign" regions $l_j = k' \neq k$ are equivalent, and we discriminate only between the intra-region ($l_i = l_j$, or $\delta(l_i - l_j) = 1$) and inter-region ($l_i \neq l_j$, or $\delta(l_i - l_j) = 0$) pixel interactions. Under this additional assumption, the partial energy of Eq. (2.15) is as follows:

$$e_a(\mathbf{l}|\mathbf{V}_a) = \sum_{(i,j) \in \mathbf{C}_a} V_a(l_i, \delta(l_i, l_j)) \tag{2.19}$$

where $\mathbf{V}_a = [V_a(k, \alpha) : (k, \alpha) \in \mathbf{K} \times \{0, 1\}]$ is the vector of the centered potential values for the region labels $k = l_i$ in the "guiding" pixel i of a clique $(i,j) \in \mathbf{C}_a$ anf region label coincidences $\alpha = \delta(l_i - l_j)$ of both region labels $k = l_i$ and $k' = l_j$ in the clique. In this case, the region label coincidence histograms (RLCH) are the sufficient statistics, and the potential centering is as follows:

$$\forall a \in \mathbf{A} \quad \sum_{k \in \mathbf{K}} \sum_{\alpha \in \{0,1\}} V_a(k, \alpha) = 0. \tag{2.20}$$

Let us assume that the labels from the different "foreign" regions are mutually independent. Then the potentials for the inter-region interactions in the simplified model of Eq. (2.17) can be set to zero resulting in the centering conditions:

$$\forall a \in \mathbf{A} \ \forall k \in \mathbf{K} \ V_a(k,0) = 0; \quad \sum_{k \in \mathbf{K}} V_{a,1}(k) = 0. \tag{2.21}$$

The introduced Gibbs models of region maps can be used as the prior probability models for simulating or segmenting piecewise-homogeneous grayscale textures. An alternative way of constructing the prior Gibbs models is proposed in (Levitan et al., 1995). A particular GPD describing piecewise-homogeneous tomographic images is built by only choosing the nearest-neighbor pairwise interactions of gray levels and nearest-neighbor 9-fold interaction of region labels. The region label interaction within 3×3 cliques specifies the borders between the homogeneous regions by amplifying the strength of the "ideal" edges or corners compared to the noisy ones. Several similar region map models are also considered in (Gimel'farb and Zalesny, 1993a). The potentials of gray level interactions, chosen on heuristic grounds, depend on the gray level differences as follows[2]: $V_a(d) = \lambda_a \exp(\lambda \cdot d^2)$. Here, the coefficient λ controls the expected smoothness of gray levels within a homogeneous region, and coefficients λ_a specify the interaction strength for the clique families (the model exploits only two families combining the horizontal/vertical and diagonal cliques, respectively).

As is shown in the next section, our Gibbs models with multiple pairwise pixel interaction that describe spatially homogeneous grayscale images $\mathbf{g}$ and region maps l are easily extended to piecewise-homogeneous images $(\mathbf{g}, \mathbf{l})$. This permits us to circumvent the heuristic choice of the interaction structure and potentials to model these images.

2.4. Piecewise-homogeneous textures

A piecewise-homogeneous image texture is represented by the superimposed pair of a grayscale image $\mathbf{g}$ and region map l. The map l specifies $|\mathbf{K}|$ disjoint homogeneous regions $[\mathbf{R}_k(\mathbf{l}) : k \in \mathbf{K}]$ in the lattice:

$$\mathbf{R}_k(\mathbf{l}) = [i : i \in \mathbf{R}; \ l_i = k]; \quad \bigcup_{k \in \mathbf{K}} \mathbf{R}_k(\mathbf{l}) = \mathbf{R}.$$

Each pixel $i \in \mathbf{R}$ of a piecewise-homogeneous image supports the signal pair $(q = g_i, k = l_i)$ of a gray level $q = g_i$ for the grayscale image $\mathbf{g}$ and a

---

[2]This choice suggests that each marginal probability distribution of gray level differences over a clique family is implicitly assumed to be a particular normal distribution (see Chapters 3 and 4 for more detail).

region label $k = l_i$ for the region map l. Therefore, generally the pixelwise interactions are specified by the signal pairs ($q = g_i$, $k = l_i$), and each pairwise interaction in a clique (i, j) of a family $\mathbf{C}_a$ deals with the signal quadruples ($q = g_i$, $q' = g_j$, $k = l_i$, $k' = l_j$).

We introduce here the following interrelated Gibbs models of piecewise-homogeneous textures:

1. joint model for simulating both grayscale images and their region maps,
2. conditional model for simulating grayscale images that have a given region map, and
3. conditional model for simulating region maps corresponding to a given grayscale image (that is, for segmenting this grayscale image).

2.4.1. JOINT MODEL OF GRAYSCALE IMAGES AND REGION MAPS

The joint GPDs describing all possible piecewise-homogeneous textures $\mathbf{g}$ and region maps l on a lattice $\mathbf{R}$ are obtained by generalizing the GPDs for homogeneous textures. Let each pair $(\mathbf{g}, \mathbf{l})$ be considered as a sample of the GRF with multiple pairwise interactions. Then the parent population of such pairs is formed by Cartesian product $\mathcal{G} \times \mathcal{L}$.

Generally, such a model involves a great many signal combinations in the clique families, namely,

$$|\mathbf{Q}| \cdot |\mathbf{K}| + |\mathbf{A}| \cdot |\mathbf{Q}|^2 \cdot |\mathbf{K}|^2 \equiv (Q+1) \cdot (K+1) + |\mathbf{A}| \cdot (Q+1)^2 \cdot (K+1)^2$$

combinations of the label and gray level co-occurrences (the first and second terms correspond to the pixelwise interactions and $|\mathbf{A}|$ pairwise interactions, respectively). The number of possible signal combinations in the cliques is equal to the number of potential values to be learnt. Therefore we simplify this model by additional assumptions that the pairwise pixel interactions depend only (i) on the coincidences of the region labels and (ii) on the gray level differences. In other words, only two types $\alpha \in \{0, 1\}$ of the region label interactions in a clique $(i, j) \in \mathbf{C}_a$ are taken into account for each region $l(i) = k$, namely, the intra-region interaction $(l_i = l_j = k$, or $\alpha = \delta(l_i - l_j) = 1)$ and the inter-region interaction $(l_i = k \neq l_j$, or $\alpha = \delta(l_i - l_j) = 0)$, and in both cases the potential value $V_a(g_i, l_i, g_j, l_j)$ depends only on the gray level difference $d = g_i - g_j$:

$$V_a(g_i, l_i, g_j, l_j) \equiv V_a(g_i - g_j, l_i, \delta(l_i - l_j)).$$

Then in all, only

$$|\mathbf{Q}| \cdot |\mathbf{K}| + 2 \cdot |\mathbf{A}| \cdot (2 \cdot |\mathbf{Q}| - 1) \cdot |\mathbf{K}| \equiv (Q+1) \cdot (K+1) + 2 \cdot |\mathbf{A}| \cdot (2 \cdot Q + 1) \cdot (K+1)$$

TABLE 2.3. Number of signal combinations for a simplified joint Gibbs model of piecewise-homogeneous textures.

$|\mathbf{Q}|$		16			256	
$|\mathbf{K}|$	2	8	16	2	8	16
	1	156	625	1250	2556	10224	20448
$|\mathbf{A}|$ 8	1024	4096	8192	16864	67456	134912
32	3980	15920	31840	69950	279800	559600

potential values have to be learnt per clique family. Table 2.3 presents these latter numbers for some typical values of the parameters $|\mathbf{Q}| = Q+1$, $|\mathbf{K}| = K + 1$, and $|\mathbf{A}|$.

For brevity, we omit below the superscript "rf" for the reference samples of the grayscale images. Let $V_{\text{pix}}(q, k)$ and $V_a(q, q', k, k') \equiv V_a(d = q - q', k, \alpha = \delta(k - k'))$ denote the centered potential values for a signal pair (q, k) in a pixel and for a quadruple (q, q', k, k') in a second-order clique, respectively. Let $\mathbf{V} = [\mathbf{V}_{\text{pix}}, \mathbf{V}_a : a \in \mathbf{A}]$ be the vector of the centered potentials where $\mathbf{V}_{\text{pix}} = [V_{\text{pix}}(q, k) : (q, k) \in \mathbf{Q} \times \mathbf{K}]$ denotes the potential vector for the pixelwise interactions and $\mathbf{V}_a = [V_a(d, k, \alpha) : (d, k, \alpha) \in \mathbf{D} \times \mathbf{K} \times \{0, 1\}]$ is the potential vector for the pairwise pixel interactions in a clique family $\mathbf{C}_a$.

The simplified joint GPD describing the piecewise-homogeneous textures is as follows:

$$\Pr(\mathbf{g}, \mathbf{l}|\mathbf{V}) = \frac{1}{Z_{\mathbf{V}}} \exp\left(e(\mathbf{g}, \mathbf{l}|\mathbf{V}_{\text{pix}}) + \sum_{a \in \mathbf{A}} e_a(\mathbf{g}, \mathbf{l}|\mathbf{V})\right) \qquad (2.22)$$

where $e(\mathbf{g}, \mathbf{l}|\mathbf{V}_{\text{pix}})$ and $e_a(\mathbf{g}, \mathbf{l}|\mathbf{V}_a)$ are the partial energies for the pixelwise and pairwise pixel interactions, respectively:

$$e(\mathbf{g}, \mathbf{l}|\mathbf{V}_{\text{pix}}) = \sum_{i \in \mathbf{R}} V_{\text{pix}}(g_i, l_i);$$

$$e_a(\mathbf{g}, \mathbf{l}|\mathbf{V}_a) = \sum_{(i,j) \in \mathbf{C}_a} V_a(g_i - g_j, l_i, \delta(l_i - l_j)).$$

The potential centering in the joint model of Eq. (2.22) is quite similar to the centering in Eq. (2.6):

$$\sum_{q \in \mathbf{Q}} \sum_{k \in \mathbf{K}} V(q,k) = 0;$$

$$\sum_{d \in \mathbf{D}} \sum_{k \in \mathbf{K}} \sum_{\alpha=0}^{1} V_a(d,k,\alpha) = 0 \quad \forall a \in \mathbf{A}.$$

$$(2.23)$$

This model is represented by the exponential family distribution:

$$\Pr(\mathbf{g}, \mathbf{l} | \mathbf{V}) = \frac{1}{Z_\mathbf{V}} \cdot \exp\left(\mathbf{V} \bullet \mathbf{H}(\mathbf{g}, \mathbf{l})\right) \qquad (2.24)$$

where $\mathbf{H}(\mathbf{g}, \mathbf{l}) = [\mathbf{H}_{\mathrm{pix}}(\mathbf{g}, \mathbf{l}), \mathbf{H}_a(\mathbf{g}, \mathbf{l}) : a \in \mathbf{A}]$ is the vector of the joint GL/RLH and GLD/RLCHs. These histograms are sufficient statistics for the model.

For simplicity, we assume that the characteristic interaction structure $[\mathbf{C}_a : a \in \mathbf{A}]$ is the same in all the regions $k \in \mathbf{K}$. But, it is not too difficult to extend the model in such a way that each region k has its own characteristic structure, that is, a distinct subset $\mathbf{A}_k$ of the clique families with non-zero centered potentials.

2.4.2. CONDITIONAL MODELS OF IMAGES AND REGION MAPS

The joint Gibbs model in Eq. (2.22) can be reduced to the conditional models of grayscale images or region maps by fixing either the region map $\mathbf{l} = \mathbf{l}^\circ$ or the grayscale image $\mathbf{g} = \mathbf{g}^\circ$. The first model assumes that the grayscale images possess a particular region map and can be used for simulating such images. The second model describes the region maps that correspond to a particular grayscale image and can be used for segmenting the grayscale images. For brevity, only the exponential family representations of the conditional models are presented below.

The conditional model of grayscale images, given a particular region map, $\mathbf{l}^\circ$, is as follows:

$$\Pr(\mathbf{g} | \mathbf{V}, \mathbf{l}^\circ) = \frac{1}{Z_{\mathbf{V}, \mathbf{l}^\circ}} \exp\left(\mathbf{V} \bullet \mathbf{H}(\mathbf{g}, \mathbf{l}^\circ)\right). \qquad (2.25)$$

The conditional model of region maps, given a particular grayscale image, $\mathbf{g}^\circ$, has the symmetric form:

$$\Pr(\mathbf{l} | \mathbf{V}, \mathbf{g}^\circ) = \frac{1}{Z_{\mathbf{V}, \mathbf{g}^\circ}} \exp\left(\mathbf{V} \bullet \mathbf{H}(\mathbf{g}^\circ, \mathbf{l})\right). \qquad (2.26)$$

Here, $\mathbf{H}(\mathbf{g}, \mathbf{l}^\circ)$ and $\mathbf{H}(\mathbf{g}^\circ, \mathbf{l})$ are the vectors of the corresponding conditional GL/RLH and GLD/RLCHs that are the sufficient statistics for the models. Both the models of Eqs. (2.25) and (2.26) differ from the joint model (2.24) only in the parent populations ($\mathcal{G}$ and $\mathcal{L}$, respectively), partition functions, and potential centering.

Conditional model of images, given a region map. For the model (2.25), the homogeneous regions, given by the region map $\mathbf{l}^\circ$, are fixed for all the samples $\mathbf{g} \in \mathcal{G}$. Therefore, each clique family $\mathbf{C}_a$ is partitioned onto $|\mathbf{K}| = K + 1$ fixed subfamilies $\mathbf{C}_{a,k^\circ} = [(i,j) : (i,j) \in \mathbf{C}_a; \; l_i^\circ = k^\circ]$. Each subfamily $\mathbf{C}_{a,k^\circ}$ contains the cliques of the family $\mathbf{C}_a$ that are lying in a single region $k^\circ \in \mathbf{K}$ (more specifically, the guiding pixel i of the cliques $(i,j)(\in \mathbf{C}_{a,k^\circ}$ is in this region). Because the subfamilies have the fixed sizes $|\mathbf{C}_{a,k^\circ}|$, we can apply the same considerations that justified the potential centering for each clique family in the Gibbs models of homogeneous images or region maps. Now these considerations result in the individual centering of the potentials for each clique subfamily as follows:

$$\sum_{d \in \mathbf{D}} \sum_{\alpha=0}^{1} V_a(d, k^\circ, \alpha) = 0; \quad \forall a \in \mathbf{A}; \; k^\circ \in \mathbf{K}. \tag{2.27}$$

Generally, each region $k^\circ \in \mathbf{K}$ may have its own characteristic interaction structure $\mathbf{A}_{k^\circ}$ so that the characteristic "star-like" neighborhood for each pixel i is specified by the region label $l_i^\circ = k$:

$$\mathbf{N}_{i,k} = [j : (i,j) \text{ or } (j,i) \in \mathbf{C}_a; \; a \in \mathbf{A}_k].$$

In this case the total interaction energy in the model of Eq. (2.25) can be represented as follows:

$$\mathbf{V} \bullet \mathbf{H}(\mathbf{g}|\mathbf{l}^\circ) \equiv e(\mathbf{g}|\mathbf{V}, \mathbf{l}^\circ) = \sum_{i \in \mathbf{R}} e_i(\mathbf{g}|\mathbf{V}, \mathbf{l}^\circ)$$

where $e_i(\mathbf{g}|\mathbf{V}, \mathbf{l}^\circ)$ denotes the partial interaction energy per pixel i:

$$e_i(\mathbf{g}|\mathbf{V}, \mathbf{l}^\circ) = V_{\text{pix}}(g_i, l_i^\circ) + \sum_{j:(i,j) \in \mathbf{N}_{i,l_i^\circ}} V_a(g_i - g_j, l_i^\circ, \delta(l_i^\circ - l_j^\circ)). \tag{2.28}$$

Let us assume, for simplicity, that the gray levels from the different regions are mutually independent. This permits us to further simplify the model in Eq. (2.25) by setting to zero the potentials for the inter-region pairwise pixel interactions:

$$V_a(d, k^\circ, \alpha) = 0; \quad \forall d \in \mathbf{D}; \; k^\circ \in \mathbf{K}; \; a \in \mathbf{A}.$$

In this case the interaction structure and potentials can be independently learnt in each homogeneous region of the training sample $(\mathbf{g}^\circ, \mathbf{l}^\circ)$, and the piecewise-homogeneous images with a given region map $l°$ can be modelled by adapting first the Gibbs model (2.10) to each type of the homogeneous textures in the training sample. Then the learnt interaction structures and potentials can be used in the conditional model in Eq. (2.25) with the partial energies of Eq. (2.28).

Conditional model of region maps, given a grayscale image. The model of Eq. (2.26) is quite symmetric to the model in Eq. (2.25) in that each clique family $\mathbf{C}_a$ is partitioned onto $|\mathbf{D}| = 2Q + 1$ fixed subfamilies. In this case each subfamily $\mathbf{C}_{a,d°}$ contains the cliques (i, j) that have the constant gray level difference $d° = g_i° - g_j°$ in a given grayscale image $\mathbf{g}°$:

$$\mathbf{C}_{a,d°} = [(i, j) : (i, j) \in \mathbf{C}_a; \; g_i - g_j = d°]; \;\; d° \in \mathbf{D}.$$

Here, the potential centering is also performed for each individual clique subfamily as follows:

$$\sum_{k \in \mathbf{K}} \sum_{\alpha=0}^{1} V_a(d°, k, \alpha) = 0; \;\; \forall a \in \mathbf{A}; \;\; d° \in \mathbf{D}. \tag{2.29}$$

If parameters of the conditional model of Eq. (2.26) are learnt from a given training sample $(\mathbf{g}°, l°)$ then the interaction structure and potential values reflect both the intra-region and inter-region interactions between the gray levels and region labels in this sample. Usually, the mutual arrangement of the training régions represents a very special case of the inter-region interactions relative to other images to be segmented. To avoid it, let us assume, for simplicity, that the different region labels are mutually independent. This assumption restricts our model to only the intra-region interaction between the gray levels and region labels and permits to set to zero all the potentials of the inter-region interactions:

$$V_a(d°, k, 0) = 0; \;\; \forall a \in \mathbf{A}; \; d° \in \mathbf{D}; \; k \in \mathbf{K}.$$

In this case the intra-region potentials possess the following centering:

$$\sum_{k \in \mathbf{K}} V_a(d°, k, 1) = 0; \;\; \forall a \in \mathbf{A}; \; d° \in \mathbf{D}. \tag{2.30}$$

In Chapter 7 we will use the simplified model with the potential centering of Eq. (2.30) to initially segment a given image starting from an arbitrary sample of an IRF, or a "salt-and-pepper" region map. This initial segmentation takes no account of a "typical" mutual arrangement of the different regions in a given training region map and exploits only the typical relations between the gray levels and region labels within each homogeneous region in a training grayscale image. Then the centering of Eq. (2.29) is used for the final segmentation that starts from a region map obtained by the initial segmentation.

Generally, the characteristic interaction structure in the conditional model of Eq. (2.26) may depend on the signal differences $d^\circ = g_i^\circ - g_j^\circ$ in the cliques under consideration. The partial energy of pairwise interactions is computed over a union $\mathbf{A} = \bigcup_{d \in \mathbf{D}} \mathbf{A}_d$ of all the substructures $\mathbf{A}_d$, and each clique $(i, j) \in \mathbf{A}_{d^\circ}$ is taken into account for computing the energy if the gray level difference $g_i^\circ - g_j^\circ = d^\circ$ corresponds to this substructure. In this case the total interaction energy

$$e(\mathbf{l}|\mathbf{V}, \mathbf{g}^\circ) \equiv \mathbf{V} \bullet \mathbf{H}(\mathbf{l}|\mathbf{g}^\circ$$

in the model of Eq. (2.26) takes the following form:

$$
\begin{aligned}
e(\mathbf{l}|\mathbf{V}, \mathbf{g}^\circ) = & \sum_{i \in \mathbf{R}} V_{\text{pix}}(g_i^\circ, l_i) \\
& + \sum_{d \in \mathbf{D}} \sum_{a \in \mathbf{A}_d} \sum_{(i,j) \in \mathbf{C}_a} V_a(d, l_i, \delta(l_i - l_j)) \cdot \delta(d - (g_i^\circ - g_j^\circ)).
\end{aligned}
\tag{2.31}
$$

2.5. Basic features of the models

Gibbs models with multiple pairwise pixel interaction introduced in this chapter have many common features. Each particular type of grayscale images and/or region maps is completely described in terms of a characteristic subset of the signal histograms listed in Table 2.1. These histograms form the sufficient statistics of the GPDs. As will be shown in the subsequent chapters, the centered model potentials resemble closely the scaled and centered signal histograms for a given training sample, and we expect that these histograms vary in the images of a desired type to within a tolerable error with respect to the training ones.

In the context of our models the similarity between the images is treated as the similarity between their signal histograms. It follows that all characteristic signals and signal pairs should be present in a given training sample to ensure that the training histograms reflect basic features of the desired images. What this means is the training samples cannot be arbitrary: the sample size and contents should provide the high-confidence signal histograms. We will consider this in more detail in Chapter 4.

Translational self-similarity of a texture is also treated as similarity between the signal histograms collected within different translation-invariant sublattices. Therefore the stochastic spatial homogeneity of a texture presumed by these models is insensitive to small local signal changes and depends only on most frequent signal combinations.

TABLE 2.4. Numbers of potential values in Gibbs models with multiple pairwise pixel interaction.

Model	Dependence of $|\mathbf{Q}|$, $|\mathbf{K}|$, $|\mathbf{A}|$	Example[a]
(2.3)	$|\mathbf{Q}| + |\mathbf{A}| \cdot |\mathbf{Q}|^2$	8212
(2.7)	$|\mathbf{A}| \cdot (2 \cdot |\mathbf{Q}| - 1)$	992
(2.10)	$|\mathbf{Q}| + |\mathbf{A}| \cdot (2 \cdot |\mathbf{Q}| - 1)$	1008
(2.13)	$\mathbf{K} + |\mathbf{A}| \cdot |\mathbf{K}|^2$	516
(2.17)	$2 \cdot |\mathbf{A}| \cdot |\mathbf{K}|$	256
(2.22)	$|\mathbf{K}| \cdot (|\mathbf{Q}| + 2 \cdot |\mathbf{A}| \cdot (2 \cdot |\mathbf{Q}| - 1))$	7960

[a]This column presents the numbers of potential values in the case of 16 gray levels ($|\mathbf{Q}| = 16$), four regions ($|\mathbf{K}| = 4$), and 32 clique families ($|\mathbf{A}| = 32$).

Compared to traditional Gibbs image models, our models offer more flexible interaction structures and potentials. As we will see later, the model parameters are directly related to signal histograms, and these relations give a more penetrating insight into their physical meaning.

The drawback to these models is that the overall number of scalar parameter values is rather big. Table 2.4 shows the numbers of potential values for our models in relation to the cardinalities of the sets of signal values $|\mathbf{Q}|$, $|\mathbf{K}|$ and clique families $|\mathbf{A}|$. Notice that the conditional models of piecewise-homogeneous textures in Eqs. (2.25) and (2.26) have the same number of parameter values as the joint model of Eq. (2.22). Generally, there could be hundreds or even much more parameter values to be learnt (see, for instance, Table 2.3). But, as will be indicated in Chapter 4, this number can be considerably decreased by approximating the signal histograms by one or another appropriate single- or multimodal function and using such an approximation as a potential.

The most attractive feature of our models is an inherent symmetry that permits us to describe in a similar manner both homogeneous and piecewise-homogeneous grayscale textures. The following chapters will show that this feature permits us to use the same or very similar computational techniques for learning, or estimating the model parameters from a given training sample and for simulating and segmenting various images.

In particular, Chapters 3 and 4 show that the MLE of the potentials for every set of clique families is easily obtained from a given training sample by first analytic and then stochastic approximation. The characteristic interaction structure is recovered using the first approximation of the potentials, and image samples, under a given model, are generated by a computationally feasible CSA technique. Moreover, validity of a model can be visually and quantitatively checked by comparing simulated samples to a training one as well as spatial homogeneity or piecewise homogeneity of a training sample can be quantitatively verified by matching sample relative frequency distributions of signal combinations collected over different patches within this sample.

CHAPTER 3

Supervised MLE-Based Parameter Learning

This chapter shows that the Gibbs models with multiple pairwise pixel interaction proposed in Chapter 2 have almost the same schemes of supervised parameter learning. The learning scheme was proposed first for homogeneous textures in (Gimel'farb, 1996a) and then generalized to piecewise-homogeneous ones in (Gimel'farb, 1996b, 1996c). It recovers both the interaction structure and potentials from a given training sample (grayscale image or/and region map) by starting from an analytic first approximation of the MLE of the potentials. The approximation, computed from signal histograms that are the sufficient statistics of a particular Gibbs model, enables to compare relative strengths of a great many possible pairwise interactions and recover most characteristic clique families for representing a given texture type. Then, the desired MLE of the potentials for the chosen characteristic clique families $[\mathbf{C}_a : a \in \mathbf{A}]$ is obtained by stochastic approximation similar to introduced by Younes (1988).

The learning scheme presumes that the GPDs under consideration belong to the exponential family of probability distributions. As indicated in Chapter 1, in such a case both the potential and sample histogram vectors have to be affinely independent (Barndorff-Nielsen, 1978). In the next sections we prove first the affine independence for the models of homogeneous grayscale textures and region maps in Eqs. (2.10) and Eq. (2.17), respectively. The proofs for other models can be obtained in a similar manner. Then we describe both the analytic and stochastic approximation stages of the learning scheme. It is shown that the refinement of the potentials by stochastic approximation can be considered as an alternative image simulation technique, called *Controllable Simulated Annealing* (CSA). We have already discussed that the image modelling scenario based on CSA overcomes efficiently main drawbacks of the traditional image simulation by stochastic relaxation. In this chapter, the CSA-based scenario will be explained in more detail.

3.1. Affine independence of sample histograms

Affine independence of the centered potential vectors $\mathbf{V}$ in our image models is obvious because they are only restricted by the centering conditions.

Therefore, only the independence of the centered histogram vectors $\mathbf{H}(\mathbf{s})$ over the parent population $\mathcal{S}$ has to be proven in each particular case.

3.1.1. GIBBS MODEL OF HOMOGENEOUS TEXTURES

The affine independence of the histogram vectors for the Gibbs image model in Eq. (2.10.) is established by the following

Lemma 3.1 *Let the lattice* $\mathbf{R}$ *contain, at least,* $2 \cdot Q$ *cliques of each family* $\mathbf{C}_a$; $a \in \mathbf{A}$, *and allow to arrange them into separate pairs sharing each the same pixel. Then the centered histogram vectors* $\mathbf{H}(\mathbf{g}) = [\mathbf{H}_{\text{pix}}(\mathbf{g}), \mathbf{H}_a(\mathbf{g})$: $a \in \mathbf{A}]$ *for all the samples* $\mathbf{g} \in \mathcal{G}$ *are affinely independent in the subspace* $\mathcal{R}^G$ *of the centered vectors. Here,* $G = (|\mathbf{Q}| - 1) + |\mathbf{A}|(2|\mathbf{Q}| - 2)$, *or* $G = (2|\mathbf{A}| + 1)Q$.

Each histogram $\mathbf{H}(\mathbf{g})$ has $|\mathbf{Q}| + |\mathbf{A}|(2|\mathbf{Q}| - 1)$, or $Q + 1 + |\mathbf{A}|(2Q + 1)$ scalar components, but only G components of them are independent because of the following obvious relations:

$$\sum_{q \in \mathbf{Q}} H_{\text{pix}}(q|\mathbf{g}) = |\mathbf{R}|;$$

$$\sum_{d \in \mathbf{D}} H_a(d|\mathbf{g}) = |\mathbf{C}_a|; \quad a \in \mathbf{A}.$$

For proving Lemma 3.1, let us form in the subspace $\mathcal{R}^G$ an orthogonal basis

$$[\mathbf{r}_1, \ldots, \mathbf{r}_Q, \mathbf{r}_{1,1}, \ldots, \mathbf{r}_{1,2Q}, \ldots, \mathbf{r}_{|\mathbf{A}|,1}, \ldots, \mathbf{r}_{|\mathbf{A}|,2Q}]$$

with G vectors $\mathbf{r}$ stratified into $1 + |\mathbf{A}|$ groups. The first group, corresponding to the GLH-part $\mathbf{H}_{\text{pix}}(\mathbf{g})$ of the histogram vector, has Q vectors $[\mathbf{r}_1, \ldots, \mathbf{r}_Q]$. The other $\mathbf{A}$ groups correspond to the GLDH-parts $\mathbf{H}_a(\mathbf{g})$ for the different clique families $\mathbf{C}_a$ and each have $2Q$ vectors $[\mathbf{r}_{a,1}, \ldots, \mathbf{r}_{a,2Q}]$.

Each basis vector is formed by concatenating one subvector of length $Q + 1$ that represents the GLH-part of the histogram vectors and $|\mathbf{A}|$ subvectors of length $2Q + 1$ that represent the GLDH-parts. In so doing, all the subvectors are zero-valued, except for the subvector that represents the same part of the histograms as the basis vector. In other words, each basis vector of the first group representing the GLH-part is as follows:

$$\mathbf{r}_t = [\rho_t, \mathbf{0}, ..., \mathbf{0}]; \quad t = 1, \ldots, Q$$

where ρ_t is the corresponding subvector of length $Q + 1$. Each basis vector $\mathbf{r}_{a,t}$ of the group a representing the GLDH-part contains all the zero-valued subvectors, except for the corresponding subvector $\rho_{\mathbf{a},\mathbf{t}}$ of length $2Q + 1$ in the position a:

$$\mathbf{r}_{a,t} = [\mathbf{0}, ..., \mathbf{0}, \rho_{a,t}, \mathbf{0}, ..., \mathbf{0}]; \quad t = 1, \ldots, 2Q.$$

Below, we consider only the GLDH-parts of the histogram and basis vectors because the proof for the GLH-part is quite similar. Table 3.1 shows a set of $2 \cdot Q$ possible nontrivial subvectors $\{\mathbf{b}_{a,t}, \mathbf{c}_{a,t} : t = 1, \ldots, Q\}$ that form the orthogonal (sub)basis in $\mathcal{R}^G$.

TABLE 3.1. Basis subvectors to prove Lemma 3.1

Diffe-rence d	$-Q$	$1\text{-}Q$	$2\text{-}Q$	...	-1	0	1	...	$Q\text{-}2$	$Q\text{-}1$	Q
$\mathbf{b}_{a,1}$	-1	0	0	...	0	0	0	...	0	0	1
$\mathbf{b}_{a,2}$	0	-1	0	...	0	0	0	...	0	1	0
$\mathbf{b}_{a,3}$	0	0	-1	...	0	0	0	...	1	0	0
...	...	...	...	...	...	...	...	...	...	...	...
$\mathbf{b}_{a,Q}$	0	0	0	...	-1	0	1	...	0	0	0
$\mathbf{c}_{a,1}$	-1	1	0	...	0	0	0	...	0	1	-1
$\mathbf{c}_{a,2}$	-1	-1	2	...	0	0	0	...	2	-1	-1
...	...	...	...	...	...	...	...	...	...	...	...
$\mathbf{c}_{a,Q-1}$	-1	-1	-1	...	$Q\text{-}1$	0	$Q\text{-}1$	...	-1	-1	-1
$\mathbf{c}_{a,Q}$	-1	-1	-1	...	-1	$2Q$	-1	...	-1	-1	-1

To prove the desired affine independence of the histogram vectors, it is sufficient to show that it holds for the difference vectors. It is easily shown that all the difference histogram vectors $\mathbf{\Delta}(\mathbf{g}, \mathbf{g}') = \mathbf{H}(\mathbf{g}) - \mathbf{H}(\mathbf{g}')$ for the pairs of the reference image samples lie in the vector subspace $\mathcal{R}^G$, too. Now we will show that all the basis subvectors $\{\mathbf{b}_{a,t}, \mathbf{c}_{a,t} : t = 1, \ldots, Q\}$ from Table 3.1 appear in the difference vectors for each clique family $\mathbf{C}_a$.

The following image samples permit us to form the desired difference subvectors.

(i) Let the reference image $\mathbf{g}$ have two contiguous regions with the constant signals q° and $q^\circ + q$ in the pixels, respectively, with the exception of the two pixels with the maximum (Q) and minimum (0) values within the region with the signal q°. Let the sample $\mathbf{g}'$ have just the same two regions but with the signals q° and $q^\circ - q$, respectively. The value q° has to be chosen so that all three values $q^\circ - q$, q°, and $q^\circ + q$ are in the set $\mathbf{Q}$. Then, for $q = 1, \ldots, Q$ the difference subvectors of these sample pairs are, to within a certain scaling factor, the same as the basis subvectors $\{\mathbf{b}_{a,q} : q = 1, \ldots, Q\}$.

(ii) Let the reference sample $\mathbf{g}$ contain all zero–valued signals, except for q pixels with the same signal $Q - q$ and one pixel with the signal Q. The

q pixels are arranged in such a way that each pixel belongs to its "own" pair of the cliques from the family $\mathbf{C}_a$ having the signal configurations $(0, Q - q)$ and $(Q - q, 0)$, respectively. The sample $\mathbf{g}'$ has the same form but the signals in the above q pixels possess the successive values $Q, \ldots, Q - q + 1$. For $q = 1, \ldots, Q$ the difference subvectors are the same as the basis subvectors $\{\mathbf{c}_{a,q} : q = 1, \ldots, Q\}$.

All the basis subvectors from Table 3.1 take part in the non–zero difference vectors $\mathbf{\Delta}(\ldots)$ of each second–order clique family. Therefore the GLDH-parts of the centered histogram vectors are affinely independent in the vector subspace $\mathcal{R}^G$. The independence of the GLH-part is proved in a similar way. The proof can be also adapted to the conditional models of Eq. (2.25) and (2.28) of the grayscale images under a given region map.

3.1.2. GIBBS MODEL OF REGION MAPS

The same proof exists for the region map model with the partial energies of Eq. (2.19).

Lemma 3.2 *Let the lattice $\mathbf{R}$ contains, at least, $3 \cdot K$ cliques of each family and allow to arrange them into separate pairs each sharing the same pixels. Then the centered RLCH vectors $\mathbf{H}(\mathbf{l})$; $\mathbf{l} \in \mathcal{L}$, are affinely independent in the subspace $\mathcal{R}^L$ of the centered vectors where $L = (2K + 1) \cdot |\mathbf{A}|$.*

In this case the orthogonal basis in the subspace $\mathcal{R}^L$ contains $L = (2K + 1) \cdot |\mathbf{A}|$ centered basis vectors. For proving Lemma 3.2, the basis vectors are stratified into $|\mathbf{A}|$ groups (one group per clique family) having each $2K + 1$ vectors. Each basis vector concatenates $|\mathbf{A}|$ subvectors of length $2 \cdot (K + 1)$. The basis vector $\mathbf{r}_{a,t}$ of a group a contains all the zero-valued subvectors, except for the subvector $\rho_{a,t}$ that corresponds to the clique family $\mathbf{C}_a$:

$$\mathbf{r}_{a,t} = [\mathbf{0}, ..., \mathbf{0}, \rho_{a,t}, \mathbf{0}, ..., \mathbf{0}]; \quad t = 1, \ldots, 2K + 1.$$

Table 3.2 presents $2K + 1$ possible subvectors $\{\mathbf{b}_{a,t} : t = 0, \ldots, K; \mathbf{c}_{a,t} : t = 1, \ldots, K\}$ that specify the desired orthogonal (sub)basis.

Once again, let us show that all the basis subvectors from Table 3.2 appear in the difference histogram vectors for each clique family $\mathbf{C}_a$. For brevity, the positions i and j of a clique (i, j) be called the guiding and secondary position, respectively. The region maps resulting in the desired difference vectors are as follows.

(*i*) Let all the cliques $(i, j) \in \mathbf{C}_a$ in the map $\mathbf{l}$ contain only the labels k, that is, $l(i) = l(j) = k$.
Let the map $\mathbf{l}'$ differ by only one clique (i', j') so that $l(i) = k$; $l(j) \neq k$ and the pixel j' is on the forward lattice border (that is, there is

TABLE 3.2. Basis subvectors to prove Lemma 3.2

Region	0	1	2	...	K-1	K	0	1	2	...	K-1	K
α	1	1	1	...	1	1	0	0	0	...	0	0
$\mathbf{b}_{a,0}$	1	0	0	...	0	0	-1	0	0	...	0	0
$\mathbf{b}_{a,1}$	0	1	0	...	0	0	0	-1	0	...	0	0
...	...	...	...	...	...	...	...	...	...	...	...	...
$\mathbf{b}_{a,K-1}$	0	0	0	...	1	0	0	0	0	...	-1	0
$\mathbf{b}_{a,K}$	0	0	0	...	0	1	0	0	0	...	0	-1
$\mathbf{c}_{a,1}$	-1	1	0	...	0	0	-1	1	0	...	0	0
$\mathbf{c}_{a,2}$	-1	-1	2	...	0	0	-1	-1	2	...	0	0
...	...	...	...	...	...	...	...	...	...	...	...	...
$\mathbf{c}_{a,K-1}$	-1	-1	-1	...	K-1	0	-1	-1	-1	...	K-1	0
$\mathbf{c}_{a,K}$	-1	-1	-1	...	-1	K	-1	-1	-1	...	-1	K

no clique with the guiding position j'). Then, for $k = 0, \ldots, K$ the difference histogram subvectors for these map pairs are equal to the basis subvectors $\{\mathbf{b}_{a,k} : k = 0, \ldots, K\}$.

(ii) Let the cliques $(i, j) \in \mathbf{C}_a$ in the map l contain only the labels k, that is, $l(i) = l(j) = k$, except for k separate pairs of the concatenated cliques $(i_\beta, j_\beta); (j_\beta, m_\beta); \beta = 0, \ldots, k - 1$. Let the pixel i_β be on the backward lattice border (that is, there is no clique with the secondary position i_β). Let $l(i_\beta) = l(j_\beta) = \beta$ and $l(m_\beta) = k$.

Let the cliques in the map l' contain only the labels k, except for k cliques $(i_\beta, j_\beta) : \beta = 0, \ldots, k - 1$, so that the pixel j_β is on the forward lattice border (that is, there is no clique with the guiding position j_β). Let $l(i_\beta) = k$ and $l(j_\beta) = \beta$. Then for $k = 1, \ldots, K$ the difference histogram subvectors for these map pairs, are equal to the basis subvectors $\{\mathbf{c}_{a,k} : k = 1, \ldots, K\}$.

Because all the basis subvectors from Table 3.2 for any second-order clique family $\mathbf{C}_a$ take part in the non-zero difference histogram vectors $\Delta(\ldots)$, the centered RLCH vectors are affinely independent in the vector subspace $\mathcal{R}^L$. This proof can be also adapted to the conditional Gibbs model in Eq. (2.26) of the region maps under a given image.

3.2. MLE of Gibbs potentials

The above Lemmas 3.1 and 3.2 show that the GPDs for the Gibbs models with multiple pairwise pixel interactions meet the requirements introduced in (Barndorff-Nielsen, 1978; Jacobsen, 1989) for a strictly log-concavity, or strong unimodality of a GPD with respect to the potentials $\mathbf{V}$. General forms of these requirements are considered in Chapter 1. Here, we discuss them in more detail with respect to the non-Markov Gibbs model of homogeneous grayscale textures in Eq. (2.10) and joint and conditional models of piecewise-homogeneous textures of Eqs. (2.22), (2.25, and (2.26).

The log-likelihood functions of the potential vector $\mathbf{V}$ for a given training sample $\mathbf{g}°$ or $(\mathbf{g}°, \mathbf{l}°)$ for these models, respectively, are as follows:

$$L(\mathbf{V}|\mathbf{g}°) = \frac{1}{|\mathbf{R}|} \ln\left(\Pr(\mathbf{g}°|\mathbf{V})\right) \tag{3.1}$$

and

$$L(\mathbf{V}|\mathbf{g}°, \mathbf{l}°) = \frac{1}{|\mathbf{R}|} \ln\left(\Pr(\mathbf{g}°, \mathbf{l}°|\mathbf{V})\right). \tag{3.2}$$

of the potential vector $\mathbf{V}$ The likelihood function of Eq. (3.1) is strictly concave and has the unique finite maximum if the training sample is not singular with respect to any gray level $q \in \mathbf{Q}$ or gray level difference $d \in \mathbf{D}$ (see Theorem 1.3 in Chapter 1). What this means is that the finite MLE of the potentials exists if and only if all the signals and differences are present in the sample so that the following conditions hold for the sample relative frequency distributions $\mathbf{F}_{\mathrm{pix}}(\mathbf{g}°)$ and $\mathbf{F}_a(\mathbf{g}°)$ of gray levels and gray level differences, respectively:

$$\begin{array}{ll} \forall q \in \mathbf{Q} & 0 < F_{\mathrm{pix}}(q|\mathbf{g}°) < 1; \\ \forall a \in \mathbf{A};\ d \in \mathbf{D} & 0 < F_a(d|\mathbf{g}°) < 1. \end{array} \tag{3.3}$$

The sample relative frequency distributions for a given training sample are expected to closely approximate the corresponding marginal probability distributions of gray levels and gray level differences in the texture to be modelled.

The sample relative frequency distributions $\mathbf{F}_{\mathrm{pix}}(\mathbf{g}°, \mathbf{l}°)$ and $\mathbf{F}_a(\mathbf{g}°, \mathbf{l}°)$ of gray levels, region labels, gray level differences, and region label coincidences have the similar conditions that ensure the unique finite maximum of the log-likelihood function in Eq. (3.2):

$$\begin{array}{ll} \forall q \in \mathbf{Q};\ k \in \mathbf{K} & 0 < F_{\mathrm{pix}}(q, k|\mathbf{g}°, \mathbf{l}°) < 1; \\ \forall a \in \mathbf{A};\ d \in \mathbf{D};\ k \in \mathbf{K};\ \alpha \in \{0, 1\} & 0 < F_a(d, k, \alpha|\mathbf{g}°, \mathbf{l}°) < 1. \end{array} \tag{3.4}$$

The conditions in Eqs. (3.3) and (3.4) are not too restrictive in practice. For each training sample, we can easily avoid singularities by substituting the well-known Bayesian estimates of the marginal probabilities for the collected sample relative frequencies. In particular, the Bayesian estimates of the marginal probability of gray levels and gray level differences are obtained from the GLH and GLDHs, respectively, as follows (Lloyd, 1984): $F_{\text{pix}}(q|\mathbf{g}^\circ) = \frac{H_{\text{pix}}(q|\mathbf{g}^\circ)+|\mathbf{Q}|}{|\mathbf{R}|+|\mathbf{Q}|}$ and $F_a(d|\mathbf{g}^\circ) = \frac{H_a(d|\mathbf{g}^\circ)+|\mathbf{D}|}{|\mathbf{C}_a|+|\mathbf{D}|}$. These estimates never reach the limit bounds 0 or 1, and such a substitution can be treated as an implicit transformation of a given singular training sample into a very similar but non-singular one.

Specific features of the log-likelihood functions of Eq. (3.1) and (3.2) allow for learning both the interaction structure and potentials from a given training sample. The learning technique is outlined below for the likelihood function in Eq. (3.1) but it is easily adapted for learning the parameters of the models in Eqs. (2.25), (2.28), and (2.26) using the log-likelihood function of Eq. (3.2).

Maximum Likelihood. The log-likelihood function in Eq. (3.1) has the maximum in the point $\mathbf{V}^* \in \mathcal{R}^G$ where the gradient of this function is equal to zero, the potentials being centered according to Eq. (2.9). Let $M_{\text{pix}}(q|\mathbf{V})$ denote the marginal probability of a gray level q in the pixels for the Gibbs model of Eq. (2.10):

$$M_{\text{pix}}(q|\mathbf{V}) \equiv \mathcal{E}\{F_{\text{pix}}(q|\mathbf{g})|\mathbf{V}\} = \sum_{\mathbf{g}\in\mathcal{G}} F_{\text{pix}}(q|\mathbf{g}) \cdot \Pr(\mathbf{g}|\mathbf{V}).$$

Let $M_a(d|\mathbf{V})$ be the marginal probability of a gray level difference d in the cliques of family $\mathbf{C}_a$ for the Gibbs model of Eq. (2.10):

$$M_a(d|\mathbf{V}) \equiv \mathcal{E}\{F_a(d|\mathbf{g})|\mathbf{V}\} = \sum_{\mathbf{g}\in\mathcal{G}} F_a(d|\mathbf{g}) \cdot \Pr(\mathbf{g}|\mathbf{V}).$$

The marginal probabilities are the mathematical expectations of the corresponding relative sample frequencies under the GPD of Eq. (2.10) with the potential vector $\mathbf{V}$.

In this notation the gradient components of the log-likelihood function in Eq. (3.1) are as follows :

$$\frac{\partial L(\mathbf{V}|\mathbf{g}^\circ)}{\partial V_{\text{pix}}(q)} = F_{\text{pix}}(q|\mathbf{g}^\circ) - M_{\text{pix}}(q|\mathbf{V});$$
$$\frac{\partial L(\mathbf{V}|\mathbf{g}^\circ)}{\partial V_a(d)} = \rho_a \cdot (F_a(d|\mathbf{g}^\circ) - M_a(d|\mathbf{V}).)$$

Thus the gradient lies in the subspace $\mathcal{R}^G$ of the centered potential vectors, too. The factor $\rho_a = |\mathbf{C}_a|/|\mathbf{R}|$ gives the relative size of the clique family $\mathbf{C}_a$ with respect to the lattice size.

The following system of stochastic equations:

$$\begin{array}{ll}
\forall q \in \mathbf{Q} & F_{\text{pix}}(q|\mathbf{g}^\circ) = M_{\text{pix}}(q|\mathbf{V}^*); \\
\forall a \in \mathbf{A}; \ d \in \mathbf{D} & F_a(d|\mathbf{g}^\circ) = M_a(d|\mathbf{V}^*)
\end{array} \tag{3.5}$$

holds at the unique maximum point of the likelihood function. The samples $\mathbf{g}$ under a given GPD of Eq. (2.10) can be generated by the pixelwise stochastic relaxation outlined in Chapter 1. It follows that the desired MLE of the potentials $\mathbf{V}^*$ can be found, as is shown, for example, in (Younes, 1988), by solving the system in Eq. (3.5) with stochastic approximation.

In our case, as was originally shown in (Gimel'farb, 1996a), the first approximation of the potentials can be found analytically, and the stochastic approximation is only used to refine the MLE starting from the first analytic approximation of the potentials.

3.3. Analytic first approximation of potentials

The analytic first approximation of the MLE $\mathbf{V}^*$ is derived by expanding the log-likelihood function into a truncated Taylor's series about the zero point $\mathbf{V} = \mathbf{0}$. Under the potential centering, this point corresponds to the singular case of the GRF, namely, to the IRF with equiprobable and independent signals over the lattice. For the GPD in Eq. (2.10) the first approximation is obtained as follows.

Let us expand the log-likelihood function in Eq. (3.1) into a truncated Taylor's series about the null potential vector $\mathbf{V} = \mathbf{0}$:

$$\begin{array}{ll}
L(\mathbf{V}|\mathbf{g}^\circ) & \approx \ L(\mathbf{0}|\mathbf{g}^\circ)+ \\
& \mathbf{V} \bullet \left.\dfrac{\partial L(\mathbf{V}|\mathbf{g}^\circ)}{\partial \mathbf{V}}\right|_{\mathbf{V}=0} + \tfrac{1}{2} \cdot \mathbf{V}^T \left.\dfrac{\partial^2 L(\mathbf{V}|\mathbf{g}^\circ)}{\partial \mathbf{V}^2}\right|_{\mathbf{V}=0} \mathbf{V}
\end{array} \tag{3.6}$$

where $\left.\dfrac{\partial L(\mathbf{V}|\mathbf{g}^\circ)}{\partial \mathbf{V}}\right|_{\mathbf{V}=0}$ and $\left.\dfrac{\partial^2 L(\mathbf{V}|\mathbf{g}^\circ)}{\partial \mathbf{V}^2}\right|_{\mathbf{V}=0}$ denote the gradient vector and Hessian matrix, respectively, in the point $\mathbf{V} = \mathbf{0}$ (see also Eqs. (1.17) and (1.18) and related discussion in Chapter 1 concerning the log-likelihood maximization). Let the potential vector be directed along the gradient: $\mathbf{V} = \lambda \cdot \left.\dfrac{\partial L(\mathbf{V}|\mathbf{g}^\circ)}{\partial \mathbf{V}}\right|_{\mathbf{V}=0}$. Very weak interdependencies of the pairwise gray level differences (and co-occurrences) in the IRF permit to approximate the Hessian of Eq. (1.18) by a diagonal matrix

$$\mathbf{diag}\{[-\phi_{\text{pix}}(q) : \ q \in \mathbf{Q}; \ -\rho_a \phi_a(d) : \ d \in \mathbf{D}; \ a \in \mathbf{A}]\}$$

with the components which are proportional to the variances of the sample relative frequencies taken with the negative sign:

$$\begin{array}{ll}
\forall q \in \mathbf{Q} & \phi_{\text{irf}}(q) = M_{\text{irf}}(q)(1 - M_{\text{irf}}(q)); \\
\forall a \in \mathbf{A} \ d \in \mathbf{D} & \phi_{\text{dif}}(d) = M_{\text{dif}}(d)(1 - M_{\text{dif}}(d)).
\end{array} \tag{3.7}$$

Here, $M_{\mathrm{irf}}(q)$ and $M_{\mathrm{dif}}(d)$ are the marginal probabilities of the gray levels and gray level differences for the IRF, respectively. They have the following analytic forms:

$$
\forall q \in \mathbf{Q} \qquad M_{\mathrm{irf}}(q) = \frac{1}{|\mathbf{Q}|} ;
$$
$$
\forall a \in \mathbf{A} \ d \in \mathbf{D} \quad M_{\mathrm{dif}}(d) = \frac{|\mathbf{Q} - |d||}{|\mathbf{Q}|^2} .
$$

(3.8)

For brevity, let us denote $\Delta_0(q) = F_{\mathrm{pix}}(q|\mathbf{g}^\circ) - M_{\mathrm{irf}}(q)$ and $\Delta_{a,0}(d) = F_a(d|\mathbf{g}^\circ) - M_{\mathrm{dif}}(d)$ the differences between the sample relative frequencies for the training sample and marginal probabilities for the IRF that specify the gradient components of Eq. (3.5). Then it is easily shown that the truncated Taylor's expansion of Eq. (3.6) for the potential vector

$$
\mathbf{V}_\lambda = \lambda \cdot \left.\frac{\partial L(\mathbf{V}|\mathbf{g}^\circ)}{\partial \mathbf{V}}\right|_{\mathbf{V}=0} \equiv \lambda [\Delta_0(q) : q \in \mathbf{Q}; \ \rho_a \Delta_{a,0}(d) : a \in \mathbf{A}; \ d \in \mathbf{D}]
$$

is as follows:

$$
L(\mathbf{V}_\lambda|\mathbf{g}^\circ) \approx L(\mathbf{0}|\mathbf{g}^\circ) + \lambda \left(\sum_{q \in \mathbf{Q}} \Delta_0^2(q) + \sum_{a \in \mathbf{A}} \rho_a^2 \sum_{d \in \mathbf{D}} \Delta_{a,0}^2(d) \right) -
$$
$$
\frac{\lambda^2}{2} \left(\sum_{q \in \mathbf{Q}} \phi_{\mathrm{irf}}(q) \Delta_0^2(q) + \sum_{a \in \mathbf{A}} \rho_a^3 \sum_{d \in \mathbf{D}} \phi_{\mathrm{dif}}(d) \Delta_{a,0}^2(d) \right) .
$$

(3.9)

Such expansion depends only on the scaling factor λ that defines a particular step along the gradient from the zero point $\mathbf{V}_0 = \mathbf{0}$. Therefore, the following first approximation of the MLE

$$
\mathbf{V}_0^* = \lambda_0 [\Delta_0(q) : q \in \mathbf{Q}; \ \rho_a \Delta_{a,0}(d) : a \in \mathbf{A}; \ d \in \mathbf{D}] \qquad (3.10)
$$

is obtained by maximizing the univariate function in Eq. (3.9) with respect to the factor λ. It is easily shown that the scaling factor λ_0 that maximizes the function of Eq. (3.9) has the form:

$$
\lambda_0 = \frac{\displaystyle\sum_{q \in \mathbf{Q}} \Delta_0^2(q) + \sum_{a \in \mathbf{A}} \rho_a^2 \sum_{d \in \mathbf{D}} \Delta_{a,0}^2(d)}{\displaystyle\sum_{q \in \mathbf{Q}} \phi_{\mathrm{irf}}(q) \Delta_0^2(q) + \sum_{a \in \mathbf{A}} \rho_a^3 \sum_{d \in \mathbf{D}} \phi_{\mathrm{dif}}(d) \Delta_{a,0}^2(d)}
$$

(3.11)

It is worth noting that the larger the lattice, the closer the factors ρ_a to unity. Therefore, the approximate potential estimate in Eqs. (3.10) and (3.11) is almost independent of the lattice size.

The GPD in Eq. (2.10) with $|\mathbf{A}|$ clique families has $G = |\mathbf{Q}| + |\mathbf{A}| \cdot (2 \cdot |\mathbf{Q}| - 1)$ independent potential values to be estimated from $|\mathbf{R}|$ signals for a training sample (see Table 2.4). Therefore to ensure the asymptotic consistency of the MLEs, it is necessary that $G \ll |\mathbf{R}|$.

3.4. Most characteristic interaction structure

3.4.1. MODEL-BASED INTERACTION MAP

The analytic initial potential estimates in Eqs.(3.10) and (3.11) show that the relative interaction strength of each clique family $\mathbf{C}_a$ in the models of Eqs. (2.7) and (2.10) depends directly on the deviation $[\Delta_{a,0}(d) = F_a(d|\mathbf{g}^\circ) - M_{\text{dif}}(d); d \in \mathbf{D}]$ of the marginal sample relative frequencies of gray level differences for a given training sample from the corresponding marginal probabilities for the IRF. Therefore, any distance between the two distributions, for instance, the chi-square distance, can represent a relative interaction strength (Gimel'farb, 1996a). The lesser the distance (that is, the closer the potential estimates to the zeroth point), the weaker the interaction. The clique families with a sufficiently weak interaction strength can be excluded from the models or, what is the same, the potential values for them can be set to zero.

Instead of the distances, the relative Gibbs energies for the training sample, showing contributions of each clique family to the total exponent of the GPD in Eqs. (2.7) or (2.10) can be used directly for comparing the clique families by their interaction strengths. In this case, the higher the relative energy of a clique family, the stronger the interaction. Let

$$\epsilon_{a,0}(\mathbf{g}^\circ) = \rho_a \sum_{d \in \mathbf{D}} (F_a(d|\mathbf{g}^\circ) - M_{\text{dif}}(d)) F_a(d|\mathbf{g}^\circ) \qquad (3.12)$$

denote the relative energy per pixel in the sample $\mathbf{g}^\circ$ for the clique family $\mathbf{C}_a$, given the initial potential estimates of Eq. (3.10). Let $\mathbf{W}$ be a search set of neighbors of a pixel covering a given large range of possible inter-pixel shifts in the clique families:

$$\mathbf{W} = [(\mu, \nu) : |\mu| \le \mu_{\max}; |\nu| \le \nu_{\max}].$$

A rich variety of the clique families, defined by the search set $\mathbf{W}$, are compared by their relative energies $\epsilon_{a=(\mu,\nu),0}(\mathbf{g}^\circ)$ computed for a given training sample $\mathbf{g}^\circ$.

In so doing, the relative energy values over the search set $\mathbf{W}$ are represented in planar cartesian co-ordinates (μ, ν) as a following 2D "energy function graph":

$$\mathbf{E}_0(\mathbf{g}^\circ) = [\epsilon_{(\mu,\nu),0}(\mathbf{g}^\circ) : (\mu, \nu) \in \mathbf{W}]. \qquad (3.13)$$

This representation constitutes a *model-based interaction map* which shows relative contributions of each clique family to the total energy and can be displayed, for a visual analysis, in a grayscale or color form. It should be noted that the very similar interaction maps are obtained when the relative energies $\epsilon_{(\mu,\nu),0}(\mathbf{g}^\circ)$ are replaced by particular distances $\mathrm{Dist}_{(\mu,\nu),0}(\mathbf{g}^\circ)$ between the same distributions.

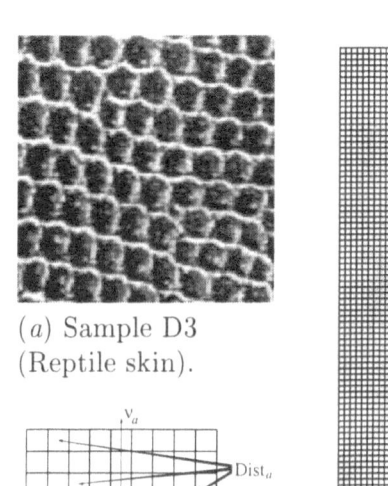

(*a*) Sample D3
(Reptile skin).

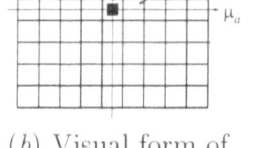

(*b*) Visual form of
the interaction map.

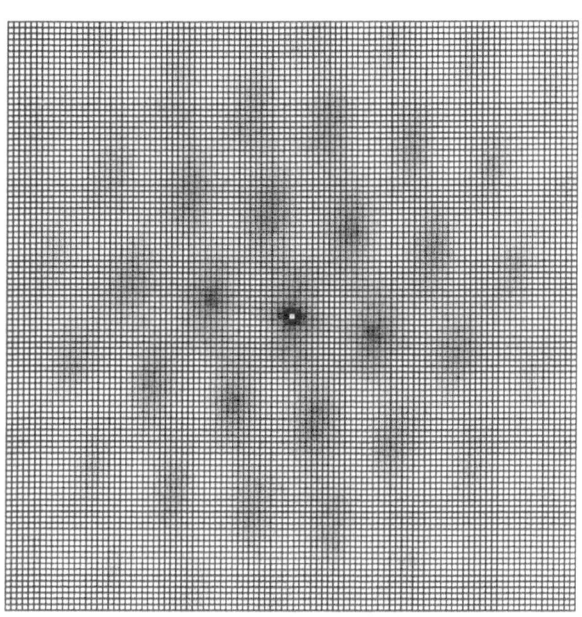

(*c*) Interaction map on an enlarged scale
for the sample (*a*): *5100* clique families.

Figure 3.1. Model-based interaction map to recover most characteristic structure of the pairwise pixel interactions.

An example of the interaction map of Eq, (3.13) for the natural image texture D3 "Reptile skin" from (Brodatz, 1966) is displayed in Figure 3.1. Figures 3.1,*a* and 3.1,*c* show a digitized fragment 128×128 of this texture and the corresponding map of pairwise pixel interactions, respectively. A visual form of the interaction map is explained in Figure 3.1,*b*. This interaction map illustrates the interaction strengths in terms of the chi-square distances between the GLDHs collected for a given training sample and the marginal probability distribution of gray level differences for the IRF,

$$\mathrm{Dist}_a = \sum_{d \in \mathbf{D}} \frac{(F_a(d) - M_{\mathrm{dif}}(d))^2}{M_{\mathrm{dif}}(d)}, \tag{3.14}$$

for all the clique families within a given window $\mathbf{W}$, that is, all inter-pixel shifts (μ, ν), within the window (see Figure 3.1,(b)). Each clique family is represented in the interaction map of Figure 3.1,(c) by two small square boxes with relative positions (μ, ν) and $(-\mu, -\nu)$ with respect to the origin $(0, 0)$. The relative interaction strength is gray-coded in each box: the darker the box, the higher the strength of the interaction and the more essential the corresponding clique family. The center of the map indicates the origin $(0, 0)$; all other boxes correspond to the clique families in the coordinates of the relative horizontal and vertical inter-pixel shift (μ, ν).

Here, as well as in all our experiments described in the subsequent chapters, the grayscale images are quantized to have a reduced signal set $\mathbf{Q}$ that contains only 16 gray levels ($q_{\max} = 15$). To visualize these images, the gray levels are equally spaced within the gray range $[0, 255]$.

The interaction map exhibits the relative contributions of the different clique families to an imaginary Gibbs image model that contains all the clique families. This allows for approximating a given texture type by a reduced model with zero-valued potentials for the families with too weal interaction strengths. Generally, this suggests that all the models have the same interaction structure corresponding to the largest possible search set $\mathbf{W}$ and differ by only the potentials: the non-zero values for the characteristic clique families and zero values for all the other families. It is this feature that simplifies comparisons of different interaction structures and strengths for a query-by-image texture retrieval in Chapter 6.

3.4.2. INTERACTION MAPS AND GIBBS POTENTIALS

Figures 3.2 and 3.3 show the interaction maps that are constructed for the Gibbs model of Eq. (2.10) and describe the four 128×128 fragments of the textures D29 (Beach sand) and D101 (Cane) from (Brodatz, 1966) and "Fabrics 2" and "Fabrics 8" from (Pickard et al., 1995). These interaction maps display 3280 clique families in the square search window $\mathbf{W}$ with $\mu_{\max} = \nu_{\max} = 40$. For simplicity, the above boxes are not displayed but the coordinate axes (μ_a, ν_a) are included to point out the origin $(0, 0)$ of the search window $\mathbf{W}$.

Figure 3.4 shows the first approximations $\mathbf{V}_{a,0} = [V_a(d) : d \in \mathbf{D}]$ of the potentials for the two clique families $\mathbf{C}_0$ and $\mathbf{C}_{80}$ that are at top rank in the interaction map for the texture D29 in Figure 3.2 by the interaction strength of Eq. (3.14). These families have the inter-pixel shifts $\mathrm{const}_a = [1, 0]$ and $\mathrm{const}_a = [0, 1]$, respectively. The potentials depend only on gray level differences d having 31 integer values $\mathbf{D} = \{-15, \ldots, 0, \ldots, 15\}$ in our particular case of 16 gray levels per image.

Although one might expect very different potential functions, it is of

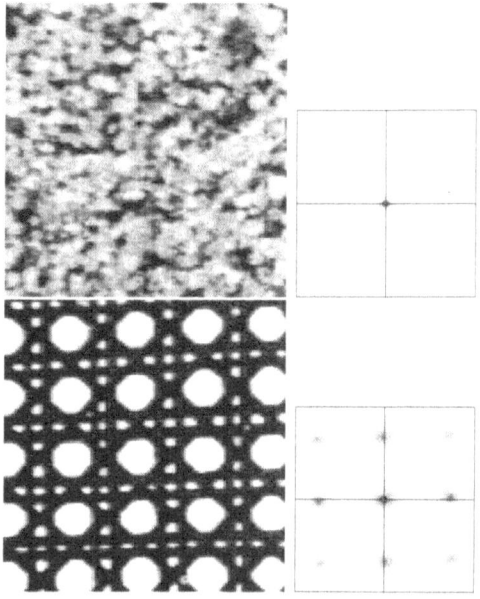

Figure 3.2. Training samples D29 and D101 and their interaction maps.

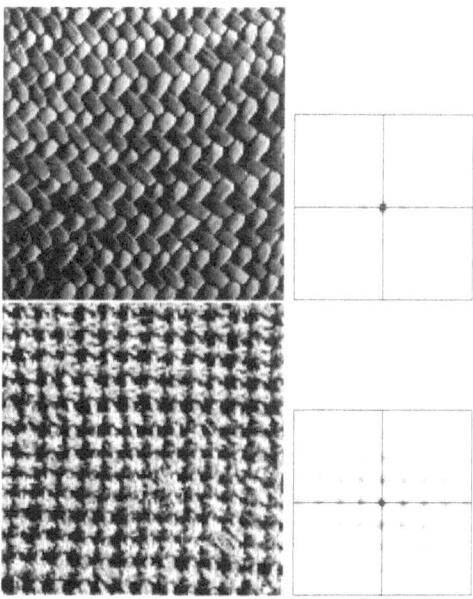

Figure 3.3. Training samples "Fabrics 2" and "Fabrics 8" and their interaction maps.

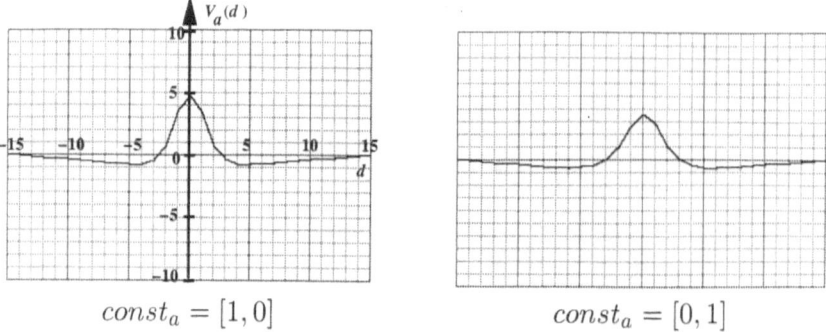

$$const_a = [1, 0] \qquad\qquad const_a = [0, 1]$$

Figure 3.4. First approximation of the potentials for the clique families $\mathbf{C}_0$ and $\mathbf{C}_{80}$ having the top-rank interaction strengths in the interaction map of the texture D29. The ranges of the x-coordinate axis of gray level differences and the y-coordinate axis of potential values are $[-15, 15]$ and $[-10, 10]$, respectively.

interest that most of our experiments in modelling various textures with the Gibbs model of Eq. (2.10) have resulted in very similar potential estimates. For almost all the clique families $a \in \mathbf{A}$, the GLDHs $\mathbf{H}_a(\mathbf{g}^\circ)$ and hence the sample relative frequency distributions $\mathbf{F}_a(\mathbf{g}^\circ)$ agree well with Gaussian-like curves, centered either on zero difference $d = 0$ or at times on the neighboring differences $1 \leq |d| \leq 2$. As a result, the initial potential estimates, obtained by subtracting the triangular marginal distribution $\mathbf{M}_{\text{dif}} = [M_{\text{dif}}(d) : d \in \mathbf{D}]$ in Eq. (3.10), were shaped like a symmetric or slightly asymmetric "Mexican hat" and differ only by their absolute values[1]. Thus, as our experiments suggest, it can be used also for approximation of Gibbs potentials. We will see later that the potential refinement by stochastic approximation in many cases preserves the form of a potential function. Figures 3.5 – 3.9 present several examples of symmetric and asymmetric potentials $\mathbf{V}_{a,0}$ for the top-rank clique families in the interaction maps describing the textures in Figures 3.2 – 3.3.

3.4.3. RECOVERING THE INTERACTION STRUCTURE

Most characteristic clique families comprising a desired interaction structure of the models (2.7) and (2.10) can be found by directly thresholding the interaction map:

$$\mathbf{A} = \{a : a \in \mathbf{W}; \epsilon_{a,0}(\mathbf{g}^\circ) \geq \theta\} \qquad (3.15)$$

[1]This Laplacian–of–Gaussian curve of type $f(x) = (1 - \gamma x^2) \exp\left(0.5\gamma x^2\right)$ is well-known but in the context of visual image filtering (Marr, 1982) and is closely approximated by a difference of two Gaussian curves.

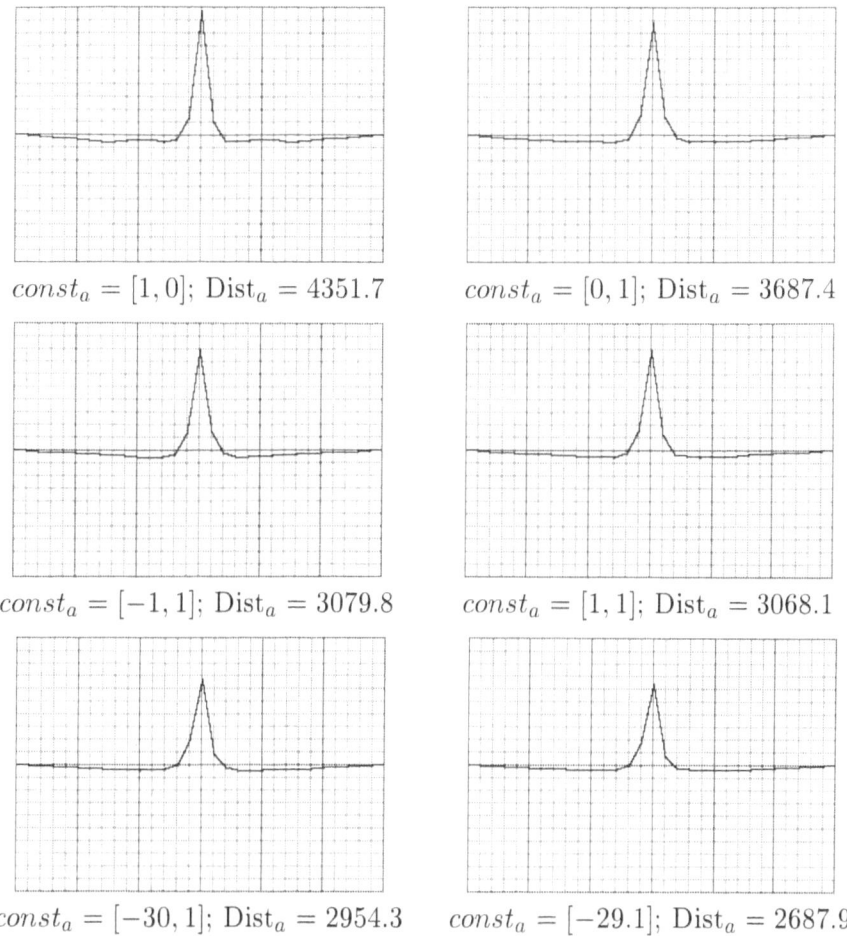

$const_a = [1, 0]$; $Dist_a = 4351.7$ $const_a = [0, 1]$; $Dist_a = 3687.4$

$const_a = [-1, 1]$; $Dist_a = 3079.8$ $const_a = [1, 1]$; $Dist_a = 3068.1$

$const_a = [-30, 1]$; $Dist_a = 2954.3$ $const_a = [-29.1]$; $Dist_a = 2687.9$

Figure 3.5.   First approximation of the potentials for the 6 top-rank clique families in the interaction map of the texture D101 (the same coordinate ranges as in Figure 3.4).

where θ denotes a given threshold of the characteristic relative energies. In the simplest case, the threshold can be chosen as a function of the mean relative energy $\overline{E}$ and standard deviation σ_E in the interaction map:

$$\theta = \overline{E} + c \cdot \sigma_E \qquad (3.16)$$

where, for instance, $c = 3 \dots 4$.

To justify this heuristic for finding the interaction structure, let us consider the resulting characteristic clique families found for the texture in Figure 3.1 by thresholding the interaction map of Figure 3.1,c with the different thresholds. The families chosen by the thresholds of Eq. (3.16) with the factor $c = 3$, 3.5, and 4 are shown in Figure 3.10,a - c, respectively.

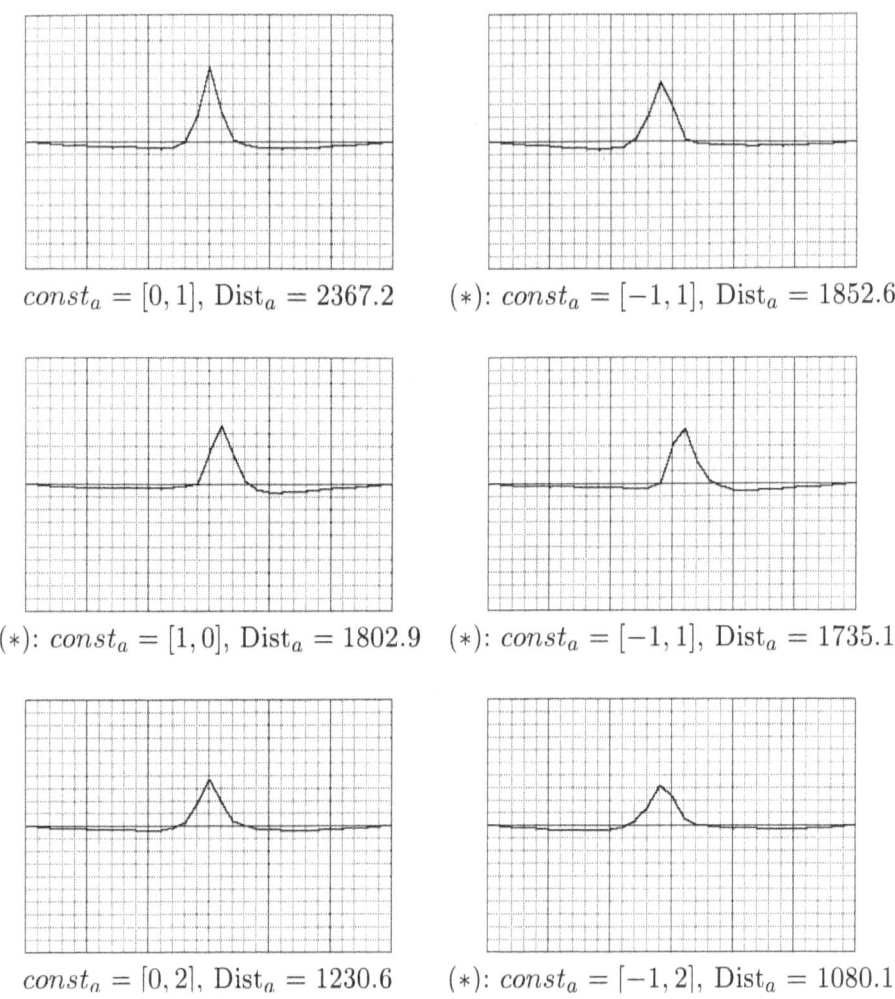

$const_a = [0, 1]$, $Dist_a = 2367.2$ $\quad$ (∗): $const_a = [-1, 1]$, $Dist_a = 1852.6$

(∗): $const_a = [1, 0]$, $Dist_a = 1802.9$ $\quad$ (∗): $const_a = [-1, 1]$, $Dist_a = 1735.1$

$const_a = [0, 2]$, $Dist_a = 1230.6$ $\quad$ (∗): $const_a = [-1, 2]$, $Dist_a = 1080.1$

Figure 3.6. First approximation of the potentials for the 6 top-rank clique families in the interaction map of the texture "Fabrics 2". The coordinate ranges are the same as in Figure 3.4 and the asymmetric potential functions are labelled by (∗).

Here, the gray boxes demonstrate the chosen characteristic clique families in the search window **W**, and white boxes correspond to all other, non-characteristic families with the interaction strength below the threshold.

The like choices for the textures in Figures 3.2 – 3.3 are shown, respectively, in Figures 3.11 – 3.14. For simplicity, here and below the boxes are replaced by the coordinate axes (μ_a, ν_a) in Figures 3.12 – 3.14.

We use the simple thresholding of Eq. (3.16) for choosing the interaction structures in texture simulation, retrieval, and segmentation experiments

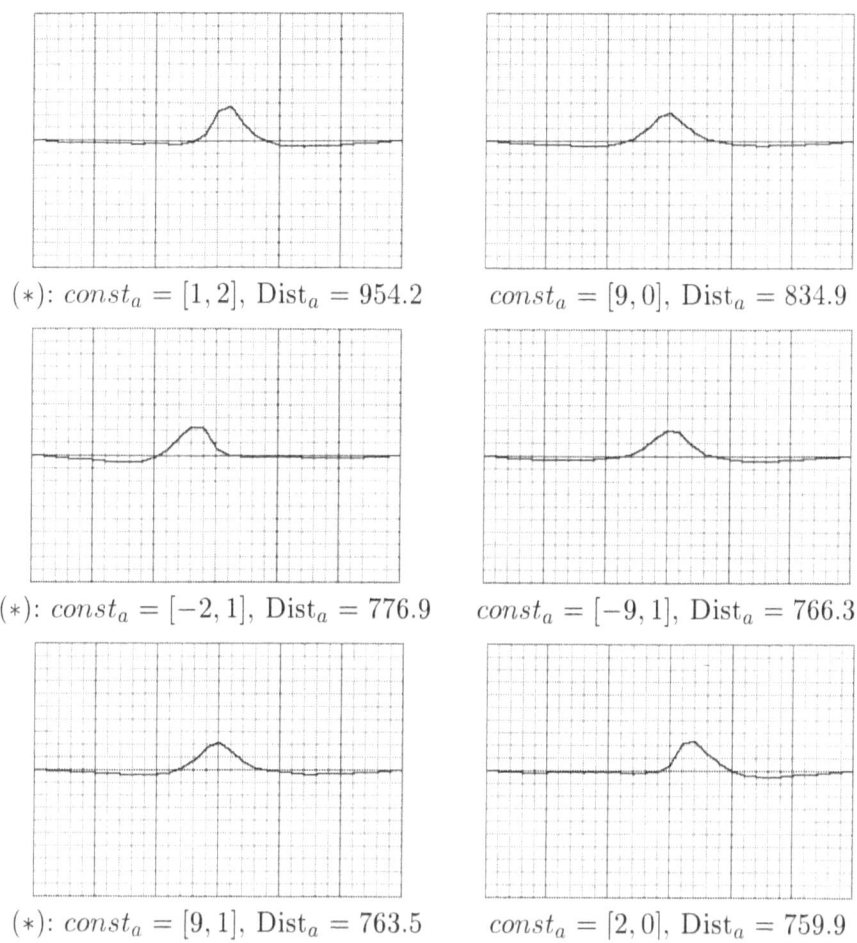

(∗): $const_a = [1, 2]$, $Dist_a = 954.2$ $const_a = [9, 0]$, $Dist_a = 834.9$

(∗): $const_a = [-2, 1]$, $Dist_a = 776.9$ $const_a = [-9, 1]$, $Dist_a = 766.3$

(∗): $const_a = [9, 1]$, $Dist_a = 763.5$ $const_a = [2, 0]$, $Dist_a = 759.9$

Figure 3.7.   First approximation of the potentials for the next 6 top-rank clique families in the interaction map of the texture "Fabrics 2" (the same notation as in Figure 3.6).

described in Chapters 5 – 7. But further theoretical investigation is needed for optimizing such a search.

Let $\mathbf{E}(\mathbf{A}, \mathbf{V}) = \left[\epsilon_{[\mu, \nu]}(\mathbf{A}, \mathbf{V}) : [\mu, \nu] \in \mathbf{W} \right]$ denote an "ideal" model-based interaction map for a particular Gibbs model of Eq. (2.10) having a fixed subset $\mathbf{A} \subset \mathbf{W}$ of characteristic clique families in the search window $\mathbf{W}$ and a corresponding potential vector $\mathbf{V}$. Here, $\epsilon_{[\mu, \nu]}(\mathbf{A}, \mathbf{V})$ denotes the relative energy of pairwise pixel interactions over the clique family with the inter-pixel shift $[\mu, \nu]$:

$$\epsilon_{(\mu, \nu)}(\mathbf{A}, \mathbf{V}) = \rho_{(\mu, \nu)} \sum_{d \in \mathbf{D}} \left(M_{(\mu, \nu)}(d | \mathbf{V}) - M_{\text{dif}}(d) \right) M_{(\mu, \nu)}(d | \mathbf{V}). \quad (3.17)$$

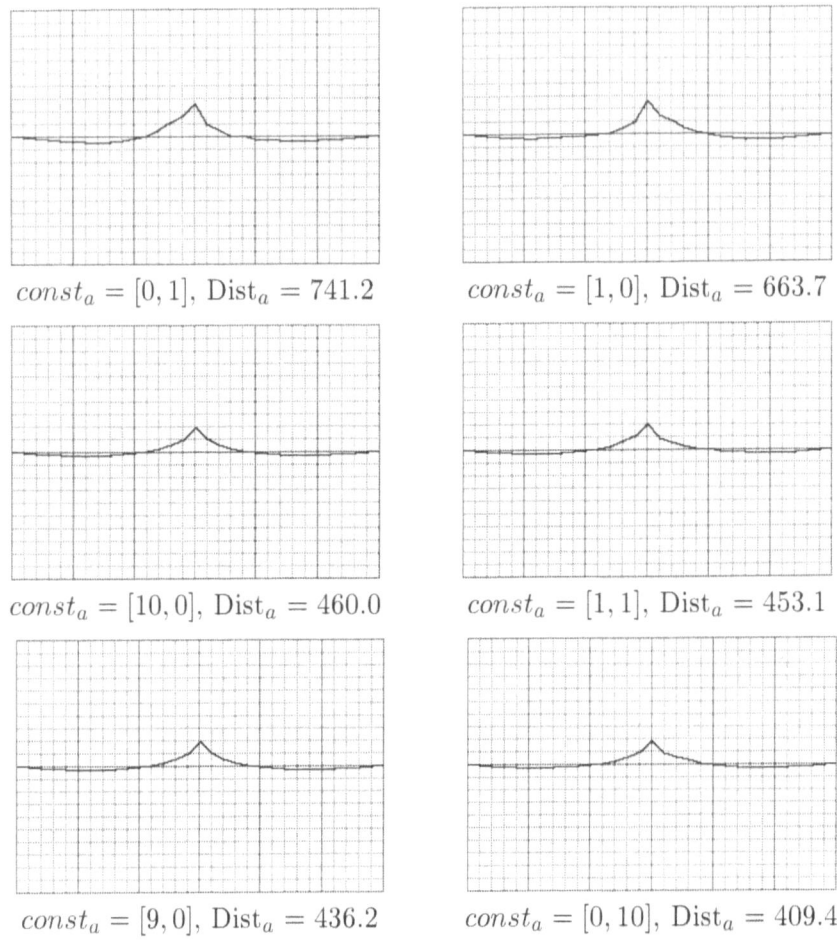

$$const_a = [0,1], \text{Dist}_a = 741.2 \qquad const_a = [1,0], \text{Dist}_a = 663.7$$

$$const_a = [10,0], \text{Dist}_a = 460.0 \qquad const_a = [1,1], \text{Dist}_a = 453.1$$

$$const_a = [9,0], \text{Dist}_a = 436.2 \qquad const_a = [0,10], \text{Dist}_a = 409.4$$

Figure 3.8. First approximation of the potentials for the 6 top-rank clique families in the interaction map of the texture "Fabrics 8" (the same notation as in Figure 3.6).

The energy is specified by the marginal probabilities

$$\mathbf{M}(\mathbf{V}) = [M_{(\mu,\nu)}(d|\mathbf{V}) : \ d \in \mathbf{D}; \ [\mu,\nu] \in \mathbf{W}]$$

of gray level differences in the clique families:

$$M_{(\mu,\nu)}(d|\mathbf{V}) = \sum_{\mathbf{g} \in \mathcal{G}} F_{[\mu,\nu]}(d|\mathbf{g}) \Pr(\mathbf{g}|\mathbf{V})$$

where the GPD $\Pr(\mathbf{g}|\mathbf{V})$ represents the Gibbs model of Eq. (2.10).

Generally, what we need to recover is a characteristic subset $\mathbf{A}$ of the clique families so that the ideal interaction map $\mathbf{E}(\mathbf{A}, \mathbf{V})$ approximates

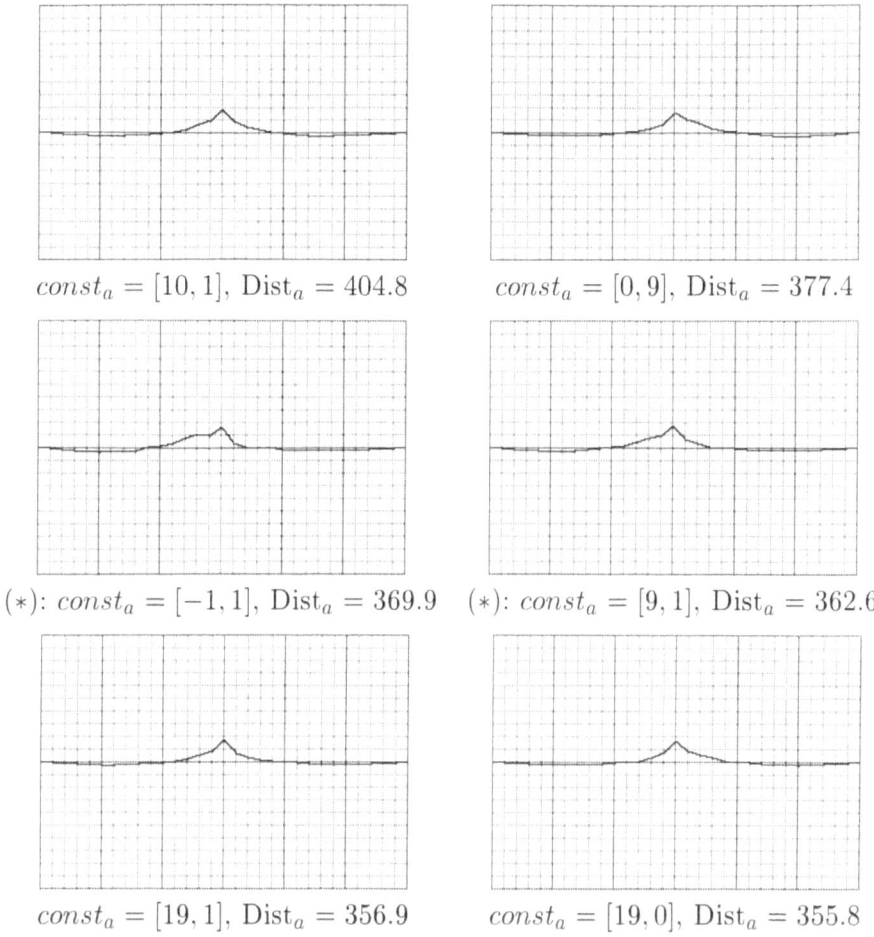

$const_a = [10, 1]$, $Dist_a = 404.8$ $const_a = [0, 9]$, $Dist_a = 377.4$

$(*)$: $const_a = [-1, 1]$, $Dist_a = 369.9$ $(*)$: $const_a = [9, 1]$, $Dist_a = 362.6$

$const_a = [19, 1]$, $Dist_a = 356.9$ $const_a = [19, 0]$, $Dist_a = 355.8$

Figure 3.9. First approximation of the potentials for the next 6 top-rank clique families in the interaction map of the texture "Fabrics 8" (the same notation as in Figure 3.6).

closely the interaction map of Eq. (3.13) for a given training sample. Unfortunately, it is still unclear how to analytically compute an expected change in a particular set of the marginal signal probabilities after adding one or a few clique families to a current Gibbs model.

But because the Gibbs models permit to simulate the image samples, the analytic approach can be replaced by an experimental one. The Gibbs models resemble usually the δ-function in that they isolate very small subsets of the images with sufficiently high relative probabilities with respect to a given training sample. Therefore, the sample relative frequency distributions for these latter images could be used to represent the marginal probability distributions $\mathbf{M}(\mathbf{V})$ in the ideal interaction map.

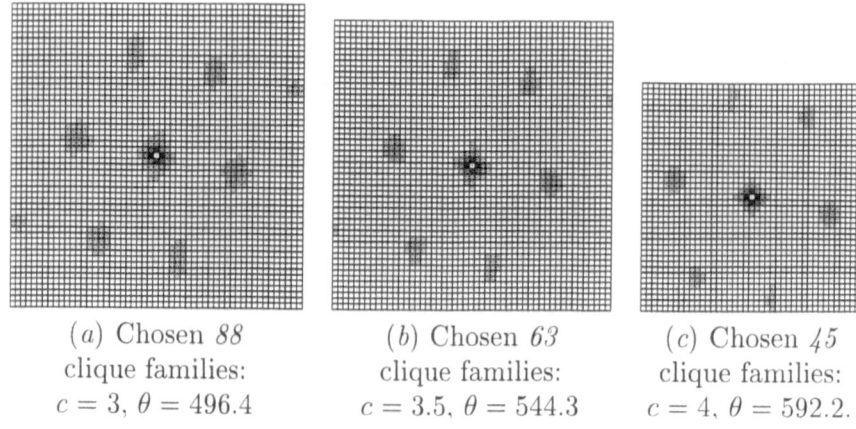

(a) Chosen *88*	(b) Chosen *63*	(c) Chosen *45*
clique families:	clique families:	clique families:
$c = 3, \theta = 496.4$	$c = 3.5, \theta = 544.3$	$c = 4, \theta = 592.2.$

Figure 3.10.   Thresholding of the interaction map in Figure 3.1,c.

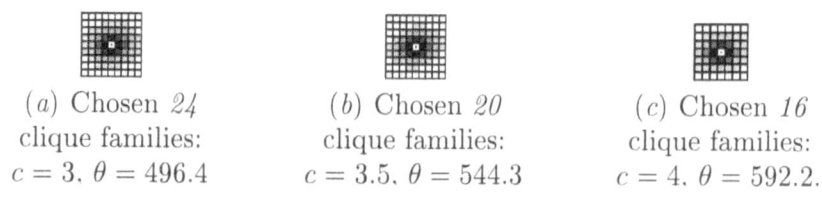

(a) Chosen *24*	(b) Chosen *20*	(c) Chosen *16*
clique families:	clique families:	clique families:
$c = 3, \theta = 496.4$	$c = 3.5, \theta = 544.3$	$c = 4, \theta = 592.2.$

Figure 3.11.   Thresholding of the interaction map for the texture D29 in Figure 3.2.

What this means is that the sample relative frequency distributions $\mathbf{F}(\mathbf{g}) = [\mathbf{F}_{\mu,\nu}(\mathbf{g}) : (\mu,\nu) \in \mathbf{W}]$ of gray level differences for an image simulated under a chosen Gibbs model should approximate the similar distributions for the training sample. Therefore, the minimum subset of the characteristic clique families could be chosen by a following heuristic iterative procedure based on a sequential approximation of the "training" distributions $\mathbf{F}(\mathbf{g}^\circ)$ by the simulated ones (Zalesny, 1996).

(i) At each iteration $t = 0, 1, \ldots$, a sample $\mathbf{g}_t$ is simulated under a current Gibbs model $\Pr(\mathbf{g}|\mathbf{V}_t)$ with the t characteristic families, and the distributions $\mathbf{F}(\mathbf{g}_t)$ are compared to the training ones.

(ii) The clique family with the inter-pixel shift μ_t, ν_t which provide the maximum relative energy of Eq. (3.17) is included into a current subset of the characteristic clique families.

The interaction maps and structures for different artificial and natural image textures will be discussed in more detail in Chapters 5 – 7.

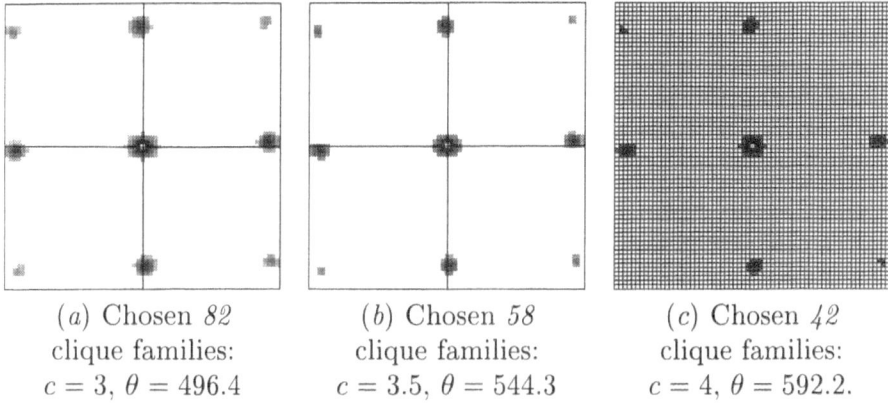

(a) Chosen 82
clique families:
$c = 3, \theta = 496.4$

(b) Chosen 58
clique families:
$c = 3.5, \theta = 544.3$

(c) Chosen 42
clique families:
$c = 4, \theta = 592.2$.

Figure 3.12.   Thresholding of the interaction map for the texture D101 in Figure 3.2.

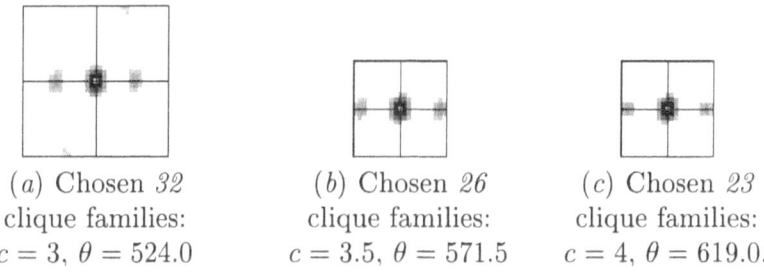

(a) Chosen 32
clique families:
$c = 3, \theta = 524.0$

(b) Chosen 26
clique families:
$c = 3.5, \theta = 571.5$

(c) Chosen 23
clique families:
$c = 4, \theta = 619.0$.

Figure 3.13.   Thresholding of the interaction map for the texture "Fabrics 2" in Figure 3.3.

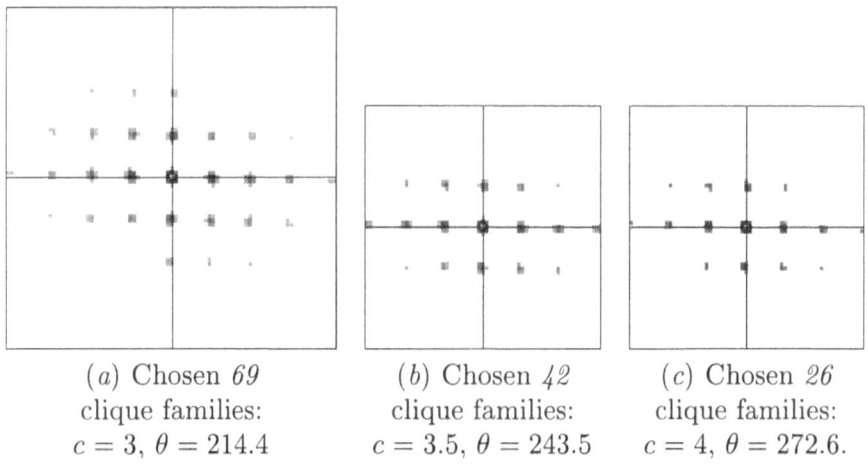

(a) Chosen 69
clique families:
$c = 3, \theta = 214.4$

(b) Chosen 42
clique families:
$c = 3.5, \theta = 243.5$

(c) Chosen 26
clique families:
$c = 4, \theta = 272.6$.

Figure 3.14.   Thresholding of the interaction map for the texture "Fabrics 8" in Figure 3.3.

More general model of pixel interactions. Potentials in the general model of grayscale textures in Eq. (2.3) depend on gray level co-occurrences and have much more diverse forms in comparison with potentials for the simplified model of Eq. (2.10). Figure 3.15 shows a fragment 128×128 of the texture D101 on a reduced scale and the corresponding interaction map constructed for the Gibbs model in Eq. (2.3). Potentials for the two top-rank clique families, $\mathbf{C}_{37}$ and $\mathbf{C}_0$, with the inter-pixel shift $[0, 14]$ and $[1, 0]$, respectively, are displayed in Figures 3.16 and 3.17. For simplicity, the potential values $V_a(g_i, g_j)$; $g_i, g_j \in \mathbf{Q} = \{0, \ldots, 15\}$, are represented by gray-coded squares in such a way that the minimum value is white and the maximum one is black. Two right plots to the right of Figures 3.16 and 3.17 are the cross-sections of the potential function for $g_i = 2$ and $g_i = 15$. Here, the potential function is bi-modal because both white and black co-occurrences are most frequent in such a texture.

Figures 3.18 – 3.26 demonstrate the interaction maps constructed for the model in Eq. (2.3) and the unimodal or multimodal first approximations of the potentials for the top-rank clique families describing the textures D3 in Figure 3.1, D29 in Figure 3.2, and "Fabrics 8" in Figure 3.3, respectively.

It should be mentioned that the interaction maps and hence the recovered characteristic interaction structures are quite similar for both the models of Eqs. (2.3) and (2.10). For example, Figures 3.27 and 3.28 show the structures obtained by thresholding the interaction maps for the textures D101, D3, D29, and "Fabrics 8" in Figures 3.15, 3.18, 3.21, and 3.24, respectively. Here, the threshold of Eq. (3.16) with $c = 3$ is used.

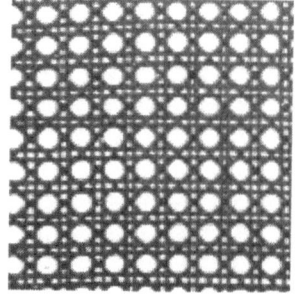
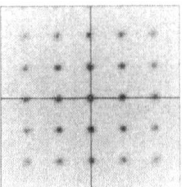

Figure 3.15. Training sample D101 and its interaction map for the Gibbs model of Eq. (2.3).

3.5. Stochastic approximation to refine potentials

After finding most characteristic interaction structure $\mathbf{A} \subset \mathbf{W}$, the initial potential estimates in Eq. (3.10) for the chosen clique families of Eq. (3.15)

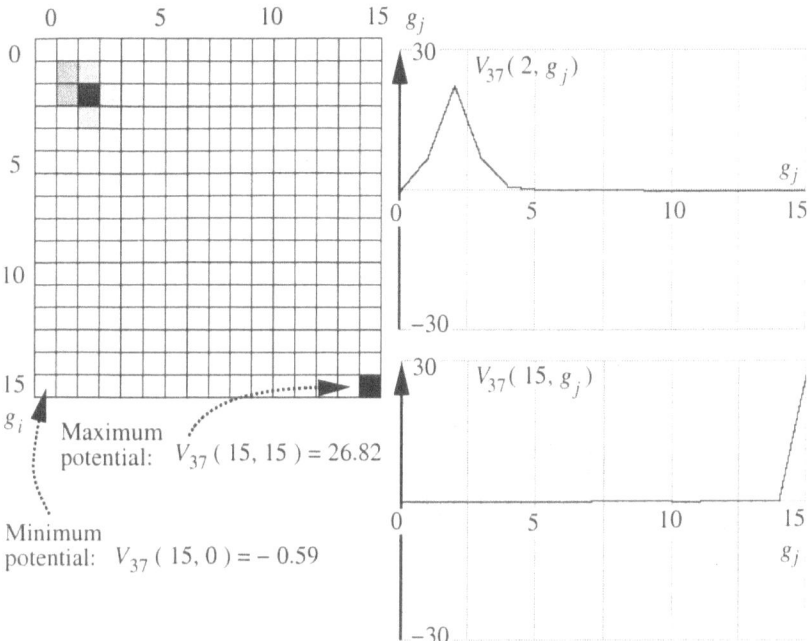

Figure 3.16. First approximation of the potential for the top-rank clique family $\mathbf{C}_{37}$ with the inter-pixel shift $[0, 14]$ in the interaction map of Figure 3.15.

can be, in principle, refined by stochastic approximation. As was discussed in Chapter 1, such a refinement, proposed by Younes (1988), is based on a stochastic relaxation generation of an inhomogeneous Markov chain of images of a gradually changing GPD, during which the initial potential estimates converge almost surely to the desired MLEs.

At each (macro)step t of stochastic relaxation, a current image $\mathbf{g}^t$ is generated from a previous one, $\mathbf{g}^{t-1}$, under a current GPD $\Pr(\mathbf{g}|\mathbf{V}_{t-1})$. Potential estimates $\mathbf{V}_{[t]}$ are updated using the signal histograms for a current generated image, for instance, the GLH and GLDHs in the case of the Gibbs model of Eq. (2.10). The potentials are changed in line with the differences between the marginal sample relative frequencies for the training and simulated samples:

$$\forall q \in \mathbf{Q}$$
$$V_{\text{pix},t+1}(q) \;=\; V_{\text{pix},t}(q) + \lambda_{t+1} \left(F_{\text{pix}}(q|\mathbf{g}^\circ) - F_{\text{pix}}(q|\mathbf{g}_t)\right)$$

$$\forall a \in \mathbf{A}; \qquad d \in \mathbf{D}$$
$$V_{a,t+1}(d) \;=\; V_{a,t}(d) + \lambda_{t+1}\rho_a \left(F_a(d|\mathbf{g}^\circ) - F_a(d|\mathbf{g}_t)\right)$$

$$(3.18)$$

where t is a macrostep ($t = 0$ for the initial estimates of Eq. (3.10)), and

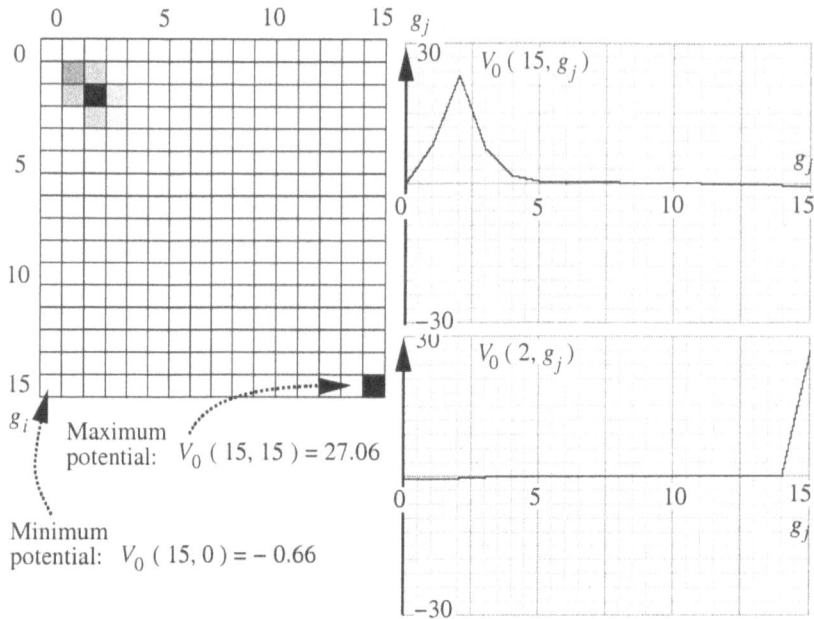

Figure 3.17. First approximation of the potential for the second top-rank clique family $\mathbf{C}_0$ with the inter-pixel shift $[1,0]$ in the interaction map of Figure 3.15.

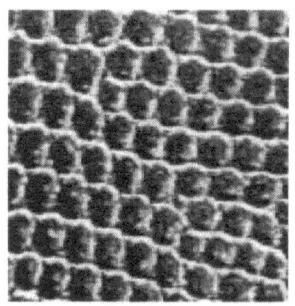
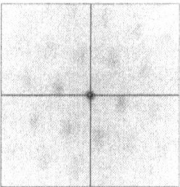

Figure 3.18. Training sample D3 and its interaction map for the Gibbs model of Eq. (2.3).

$\mathbf{g}_t$ is a sample generated under the current GPD $\Pr(\mathbf{g}|\mathbf{V}_t)$ by stochastic relaxation. The starting sample, $\mathbf{g}_0$, is generated using the first analytic approximation of the potentials in Eq. (3.10) and a sample of the IRF.

The scaling factor λ_t determines a contracted step along a current approximation of the gradient of Eq. (3.5) in the point $\mathbf{V}_t$. We use the scaling

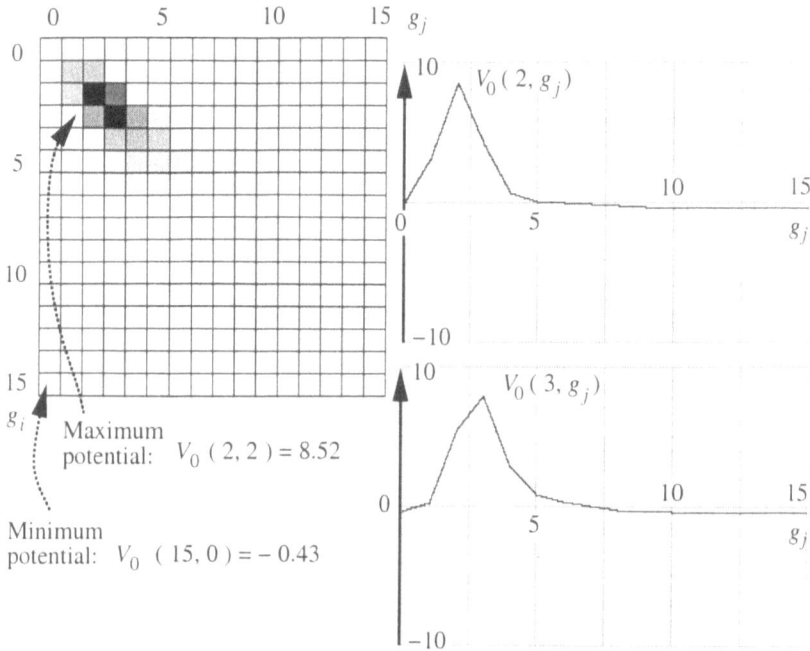

Figure 3.19. First approximation of the potential for the top-rank clique family C_0 with the inter-pixel shift $[1, 0]$ in the interaction map of Figure 3.18.

factor

$$\lambda_t = \lambda_0 \frac{c_0 + 1}{c_1 + c_2 t}$$

that decreases from the starting value λ_0 in (3.11) as was proposed in (Younes, 1988). But along with Younes, we have also found in our experiments that the theoretically justified control parameters c_0, c_1, c_2 discussed in Chapter 1 result in too slow convergence of the updating process of Eq. (3.18) to the desired MLE of the potentials and should be replaced by some empirically found values. Such a heuristic choice $c_0 = 0$, $c_1 = 1$, $c_2 = 0.001$ is used in most of the experiments in Chapters 5 and 7.

Figure 3.29 shows the potential estimates for the texture D29 in Figure 3.2 obtained by 300 and 2000 macrosteps of stochastic approximation, and Figures 3.30 – 3.33 display the similar refined estimates for the texture "Fabrics 8". The initial estimates are presented in Figures 3.4 and 3.8 – 3.9, respectively. In this experiment, as well as in many other experiments, the potential refinement results in notable changes of the potential values but mostly preserves the overall shape of this function except when the po-

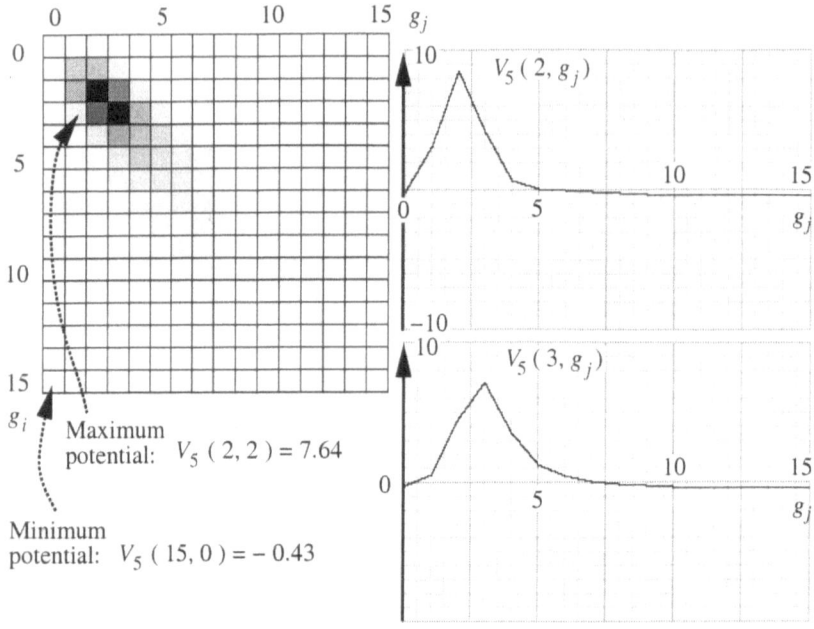

Figure 3.20. First approximation of the potential for the second top-rank clique family C_5 with the inter-pixel shift $[0, 1]$ in the interaction map of Figure 3.18.

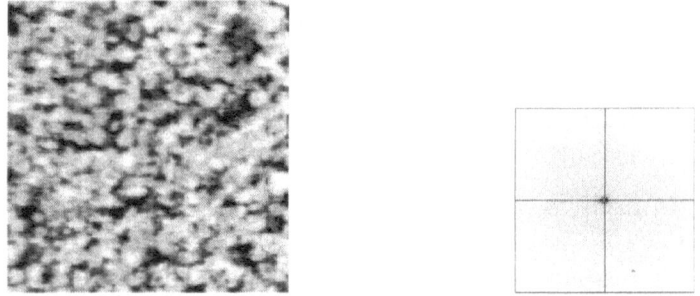

Figure 3.21. Training sample D29 and its interaction map for the Gibbs model of Eq. (2.3).

tential simply tends to zero because the corresponding clique family turns to be noncharacteristic for the texture under consideration.

The described learning scheme is easily extended to Gibbs models of region maps and joint and conditional models of the piecewise-homogeneous grayscale images introduced in Chapter 2. As will be shown in Chapter 7, the learnt interaction structure and initial analytic estimates of the poten-

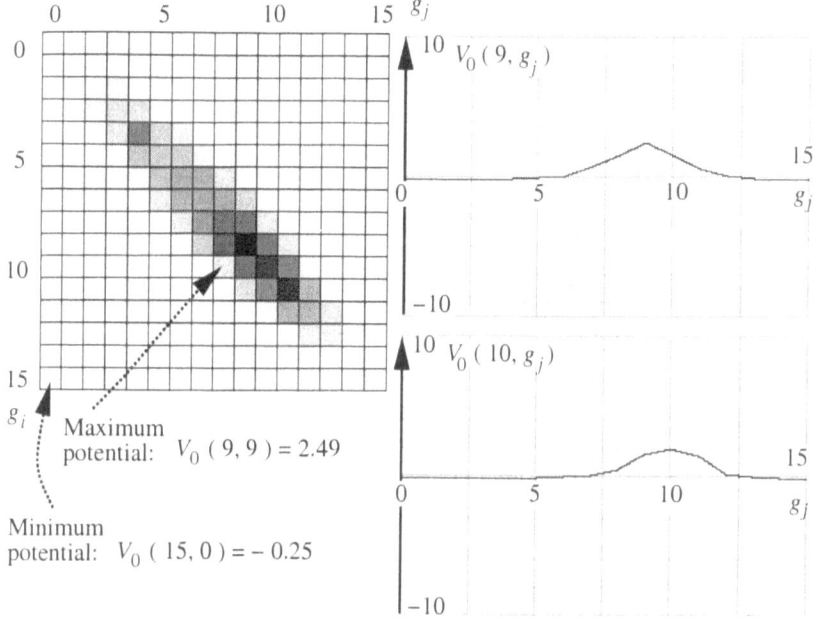

Figure 3.22. First approximation of the potential for the top-rank clique family C_0 with the inter-pixel shift $[1, 0]$ in the interaction map of Figure 3.21.

tials allow to both simulate and segment the piecewise-homogeneous image textures.

In theory, this stochastic approximation refinement of potentials needs a sizable number of the (macro)steps for ensuring the convergence to the desired maximum of the likelihood function, that is, to the solution of the system of Eq. (3.5). In practice, we are much more interested in generating the images with desired textural features than in getting precise model parameters. Therefore, the modelling scenario can be simplified as indicated in the next section, by simply changing the aim of the potential refinement.

First of all, it is worthy to note that the GPDs at hand are closely similar to the δ–function in that the images with the significantly non-zero probabilities form a very small subset concentrated around the maximum probable image(s), and we expect that a given training sample should be in such a subset. Therefore the stochastic approximation of Eq. (3.18) that brings close together the relative frequency distributions of the selected signal combinations in the generated and training samples can also be regarded as an adaptive image generation technique. As was indicated earlier, we call such a modelling technique *Controllable Simulated Annealing* (CSA).

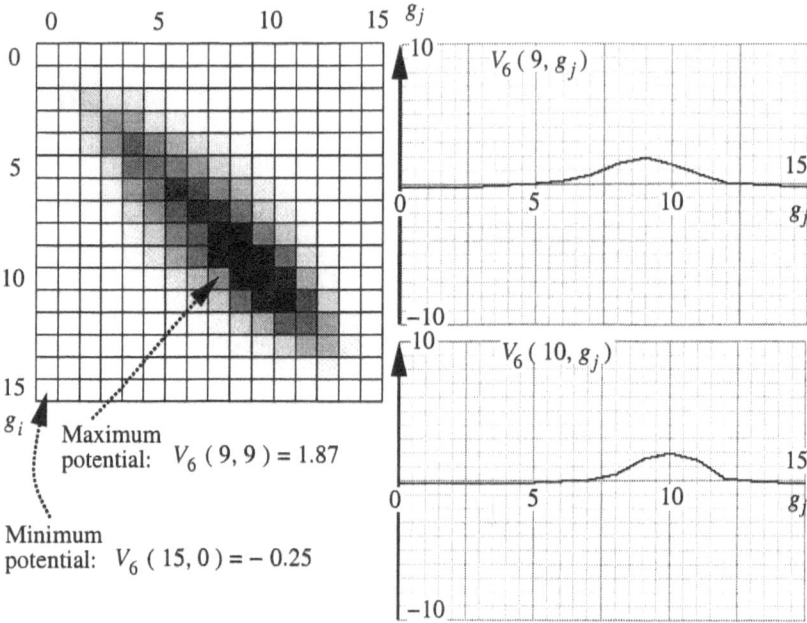

Figure 3.23. First approximation of the potential for the second top-rank clique family C_6 with the inter-pixel shift $[0, 1]$ in the interaction map of Figure 3.21.

Figure 3.24. Training sample "Fabrics 8" and its interaction map for the Gibbs model of Eq. (2.3).

3.5.1. ALTERNATIVE SCENARIO OF TEXTURE MODELLING

Usually, a traditional scenario of signal modelling by Gibbs random fields is directed at obtaining the samples that follow the desired probability distribution. It usually contains the following stages.

1. Choose a particular class of models, or Gibbs probability distributions, to describe signals under consideration.

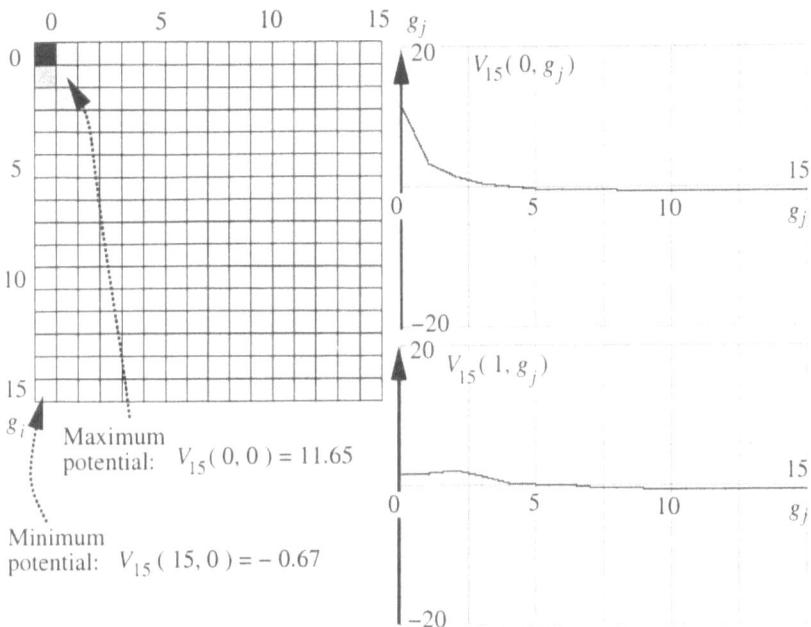

Figure 3.25.   First approximation of the potential for the top-rank clique family $\mathbf{C}_{15}$ with the inter-pixel shift $[0, 1]$ in the interaction map of Figure 3.24.

2. Estimate the quantitative model parameters from a given training sample to select a particular model.

3. Generate samples under the estimated model using a well known pixel-wise stochastic relaxation technique such as the Metropolis or Gibbs sampler. As was indicated in Chapter 1, a Markov chain of samples generated by stochastic relaxation has in an equilibrium, that is, in the limit, the desired Gibbs distribution.

In a favorable case, the parameters are obtained analytically. But mostly they have to be iteratively adapted by stochastic approximation. The adaptation is based on matching the samples, successively generated by stochastic relaxation, to a given training sample and on updating the model parameters at each step according to the current result of matching.

This scenario traditionally extends to image modelling although in this case it is impracticable due to the following drawbacks.

Inaccessible equilibrium states.   The third and the adaptive second stage lack formal criteria to test whether the Markov chain has reached an equilibrium so that the samples are generated in line with a distribution that closely approximates the desired one. The known statistical estimates cited

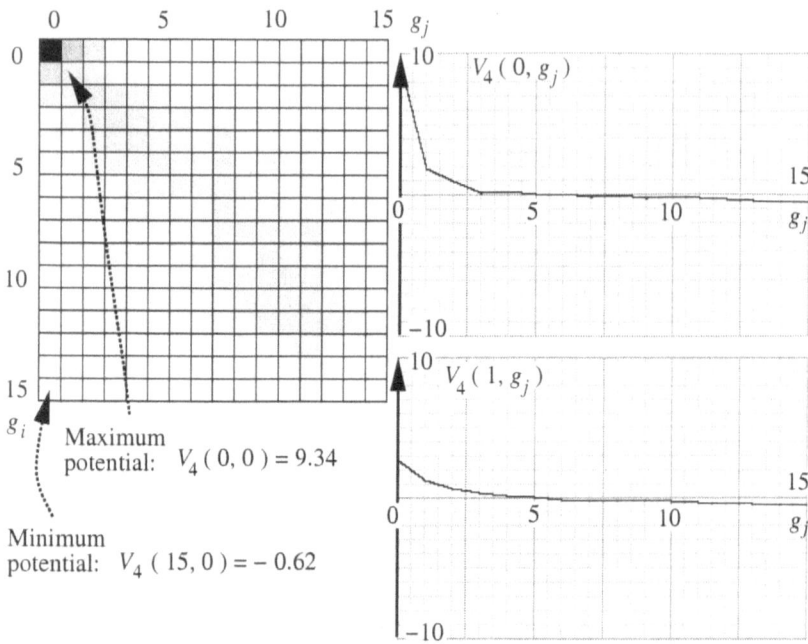

Figure 3.26. First approximation of the potential for the second top-rank clique family C_4 with the inter-pixel shift $[10, 0]$ in the interaction map of Figure 3.24.

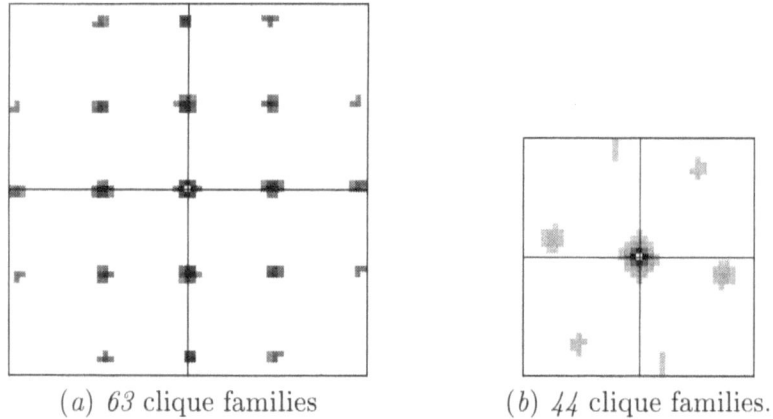

(*a*) *63* clique families   (*b*) *44* clique families.

Figure 3.27. Thresholding of the interaction maps for the textures D101 (*a*) and D3 (*b*) described by the model of Eq. (2.3).

in Chapter 1 suggest that to reach an equilibrium the length of a chain should be at least quadratic in the size of the parent population of all possible samples.

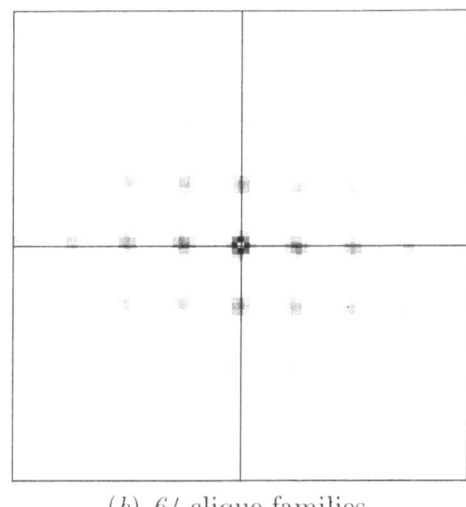

(a) *16* clique families   (b) *64* clique families.

Figure 3.28. Thresholding of the interaction maps for the textures D29 (a) and "Fabrics 8" (b) described by the model of Eq. (2.3).

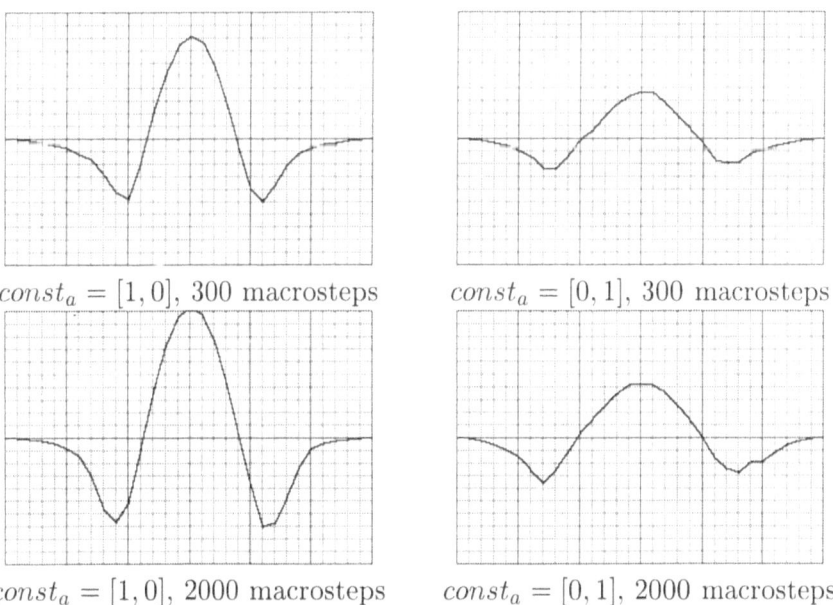

$const_a = [1, 0]$, 300 macrosteps $const_a = [0, 1]$, 300 macrosteps

$const_a = [1, 0]$, 2000 macrosteps $const_a = [0, 1]$, 2000 macrosteps

Figure 3.29. Potentials refined by stochastic approximation starting from the initial estimates in Figure 3.4. The ranges of the x-coordinate axis of gray level differences and the y-coordinate axis of potential values are $[-15, 15]$ and $[-10, 10]$, respectively.

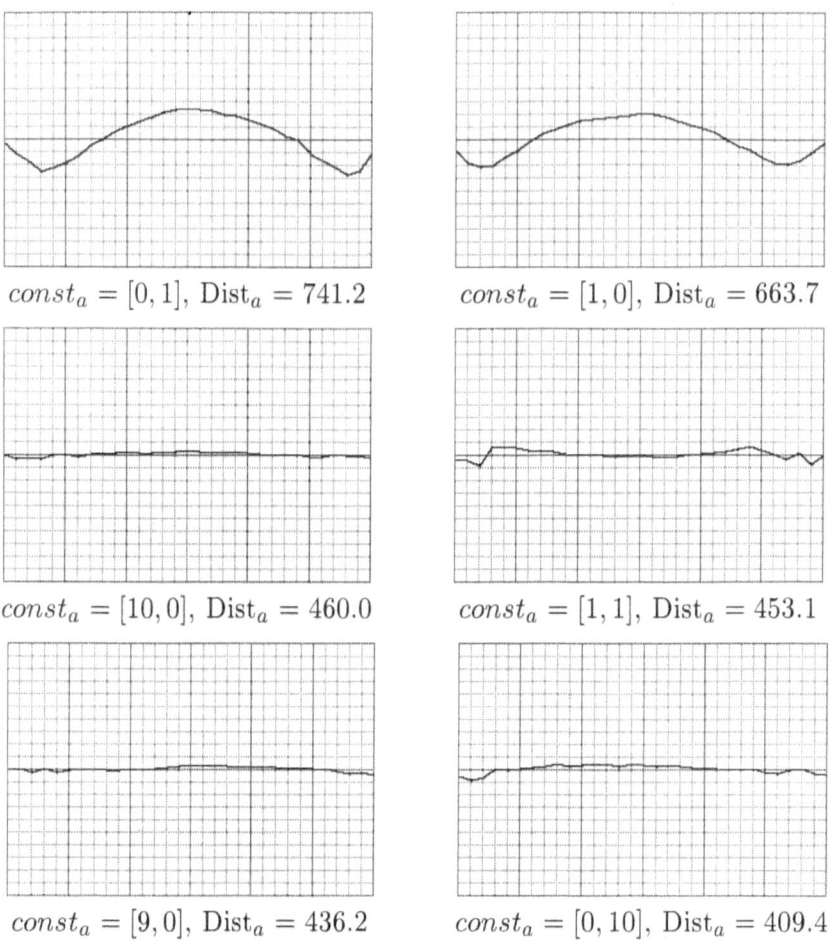

Figure 3.30. Refined potentials for the 6 top-rank clique families in the interaction map of the texture "Fabrics 8" after 300 macrosteps of stochastic relaxation (the same notation as in Figure 3.29).

But the parent population of images is of combinatorial size, say, 10^{100} – $10^{10,000}$ digital images even for small 10×10 – 100×100 lattices with only 10 gray levels per pixel. Therefore, in practice it is impossible to expect that a Markov chain of generated images will tend to an equilibrium.

Usually, image generation is terminated by using one or another heuristic stopping rule based on matching certain integral features of a generated sample to the same features of the training one. Such a heuristic guarantees that all the generated samples possess the similar features but, of course, cannot ensure that the sample frequency distribution does correspond to a given probability model or that the parameter estimates for the model are

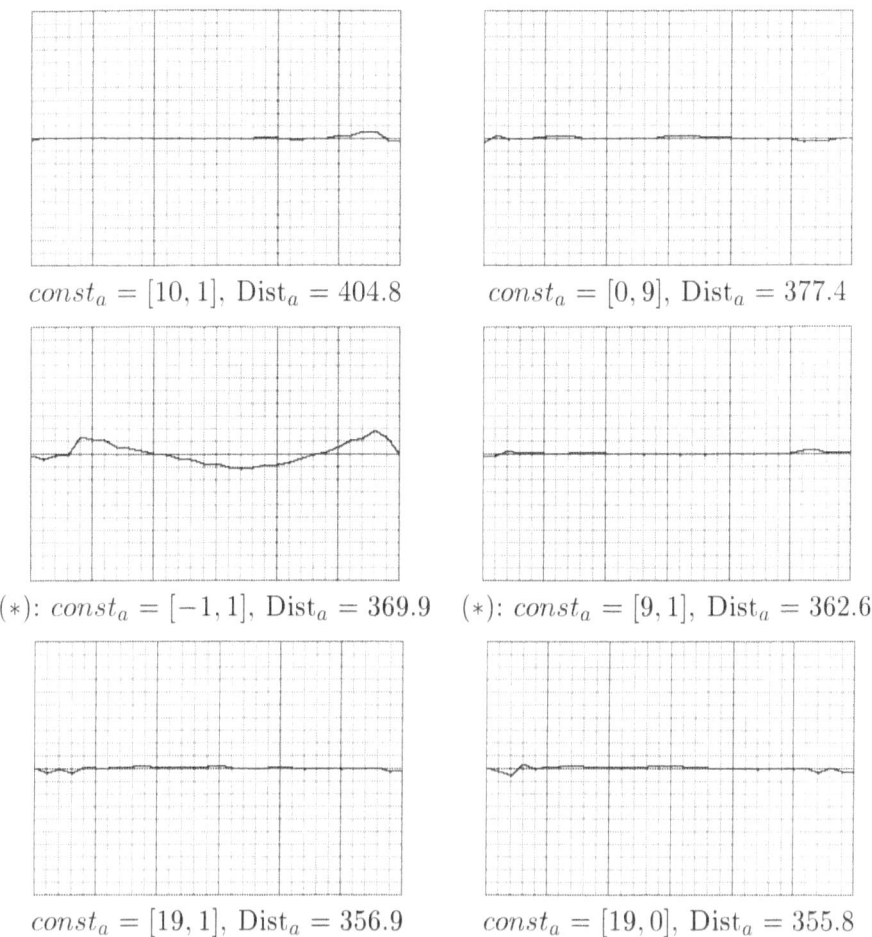

$const_a = [10, 1]$, $Dist_a = 404.8$ $const_a = [0, 9]$, $Dist_a = 377.4$

$(*)$: $const_a = [-1, 1]$, $Dist_a = 369.9$ $(*)$: $const_a = [9, 1]$, $Dist_a = 362.6$

$const_a = [19, 1]$, $Dist_a = 356.9$ $const_a = [19, 0]$, $Dist_a = 355.8$

Figure 3.31.   Refined potentials for the next 6 top-rank clique families in the interaction map of the texture "Fabrics 8" after 300 steps of stochastic approximation (the same notation as in Figure 3.29).

really in a close vicinity of their true values.

Specific modelling goals.   On the other hand, true probability distributions or model parameters do not constitute our main goal in image modelling. Parameter estimation ranks next to generation of images having a given training sample as a "typical" representative. A typical experiment on image simulation results in a single sample or a small number of samples which are expected to closely resemble a given training sample. Perhaps, nobody would give much thought to a probability distribution of all the images if the desired resemblance be really achieved...

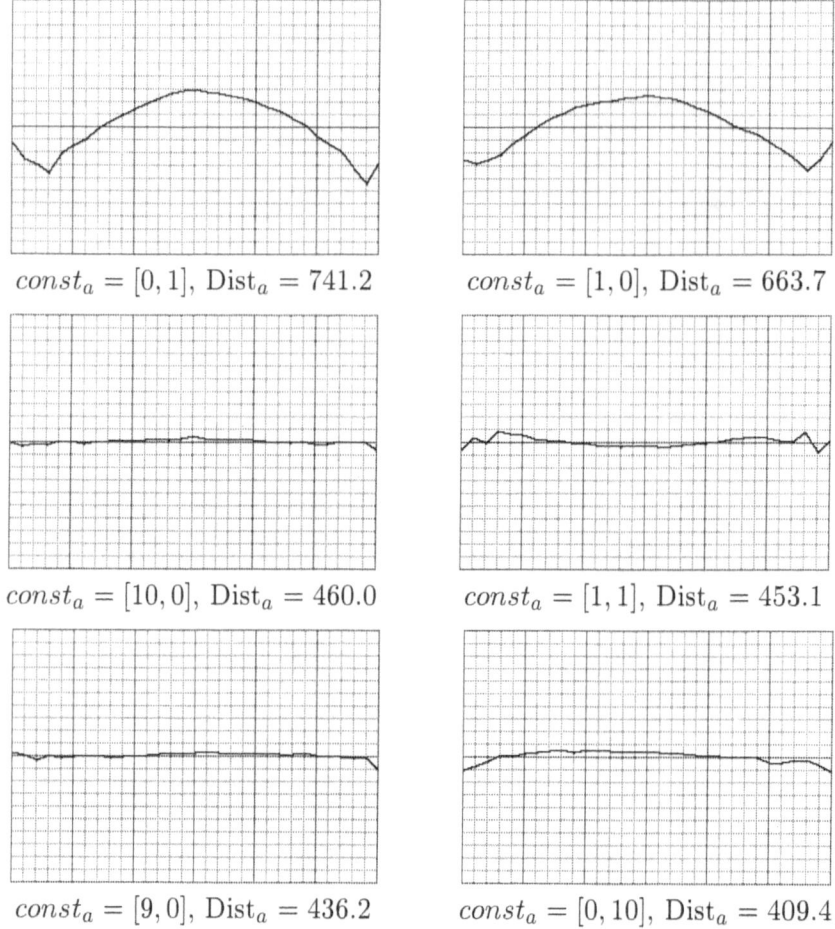

$const_a = [0, 1]$, $Dist_a = 741.2$ $const_a = [1, 0]$, $Dist_a = 663.7$

$const_a = [10, 0]$, $Dist_a = 460.0$ $const_a = [1, 1]$, $Dist_a = 453.1$

$const_a = [9, 0]$, $Dist_a = 436.2$ $const_a = [0, 10]$, $Dist_a = 409.4$

Figure 3.32. Refined potentials for the 6 top-rank clique families in the interaction map of the texture "Fabrics 8" after 2000 macrosteps of stochastic relaxation (the same notation as in Figure 3.29).

Alternative scenario. Thus, the traditional scenario does much more than it is required for image modelling. In practice, the modelling scenario has to only provide both the training and generated samples with sufficiently high relative probabilities under the model at hand.

Our alternative scenario, suggested by the Gibbs models proposed in Chapter 2, combines the parameter estimation and image generation into a single stochastic approximation process. It is called CSA because it resembles in many respects the conventional simulated annealing[2] but overcomes

---

[2]Simulated annealing is described in detail in (Kirkpatrick et al., 1983; Geman and Geman, 1984; Kirkpatrick and Swendsen, 1985; van Laarhoven, 1988).

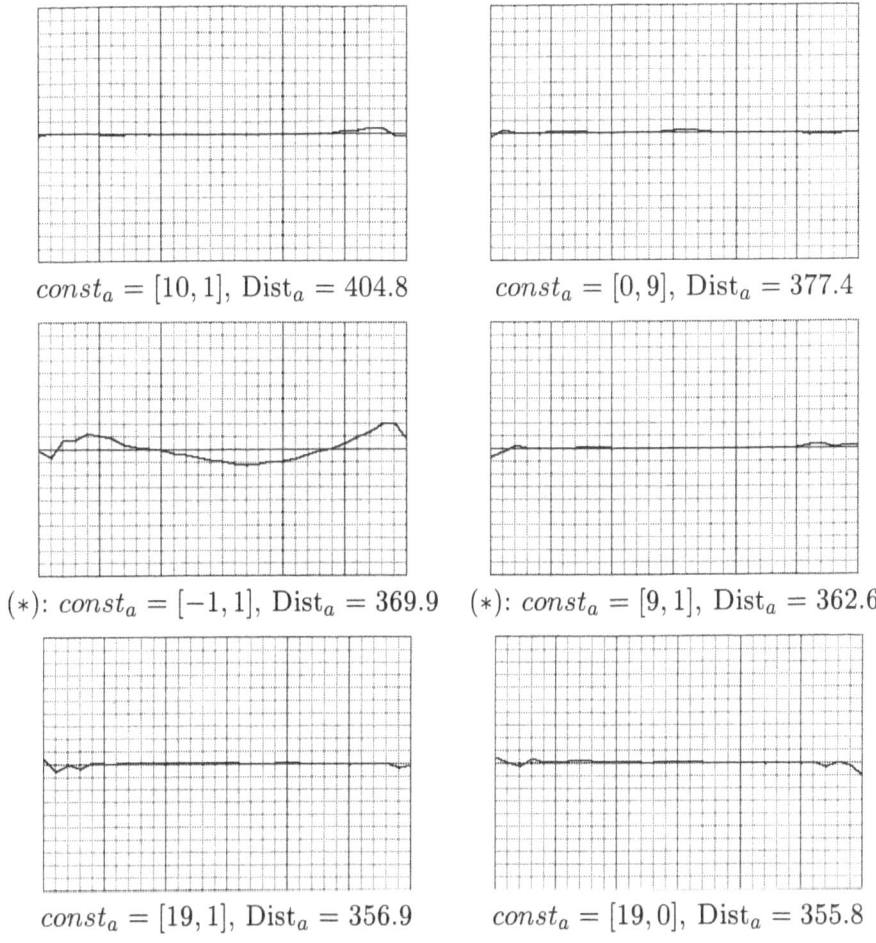

$const_a = [10, 1]$, $Dist_a = 404.8$ $const_a = [0, 9]$, $Dist_a = 377.4$

$(*)$: $const_a = [-1, 1]$, $Dist_a = 369.9$ $(*)$: $const_a = [9, 1]$, $Dist_a = 362.6$

$const_a = [19, 1]$, $Dist_a = 356.9$ $const_a = [19, 0]$, $Dist_a = 355.8$

Figure 3.33. Refined potentials for the next 6 top-rank clique families in the interaction map of the texture "Fabrics 8" after 2000 steps of stochastic approximation (the same notation as in Figure 3.29).

the drawbacks of this latter with respect to image simulation. The CSA-based modelling scenario is as follows.

1. Choose a class of models with multiple pairwise pixel interactions.
2. Compute analytically the first approximations of potentials and select most characteristic interaction structure using the computed model-based interaction map.
3. Generate samples using the CSA with a stopping rule based on a total distance between the signal histograms (or the Gibbs energies which depend on these histograms) for a current generated sample and the training one.

In this case, the goal of modelling is restricted to only generating samples having relative probabilities "around" the probability of the training sample because the CSA tends to minimize, on average, the total distance between the histograms. But as was indicated above, these are precisely the samples which are of the main practical interest in image modelling.

3.5.2. CONTROLLABLE SIMULATED ANNEALING

The CSA differs from the stochastic approximation refinement of the initial potential values only in its final aim to get the images themselves instead of the model parameters. At each macrostep of the CSA, the samples are generated by a pixelwise stochastic relaxation, for instance, by the Metropolis or Gibbs sampler (Metropolis et al., 1953; Geman and Geman, 1984), and the potentials are changing just as in the stochastic approximation process of Eq. 3.18. But, in this case we exploit only a by-product of this process, namely, a close proximity between the sample relative frequencies of gray level differences in the training sample and the samples generated with the refined potentials after a relatively small number of macrosteps.

Under the GPD $\Pr(\mathbf{g}|\mathbf{V}^*)$ with the MLE $\mathbf{V}^*$ of the potentials, the mathematical expectation of the total energy over the parent population is genuinely equal to the total energy for the training sample (see Eq. (3.5) relating the training and expected marginals for the MLE):

$$\mathcal{E}\{\mathbf{V}^* \bullet \mathbf{H}(\mathbf{g})\} = \mathbf{V}^* \bullet \mathcal{E}\{\mathbf{H}(\mathbf{g})\} = \mathbf{V}^* \bullet \mathbf{H}(\mathbf{g}^\circ).$$

The variance of the total energy is usually rather low, and if the total energies of simulated images are close to the total energy of the training sample then these images have high probabilities in relation to the GPD under consideration an can be considered as the final goal of image simulation. Such a proximity yields, by and large, a fairly good visual similarity between the generated and training homogeneous image textures.

The images simulated by the CSA can approach more closely the training sample, as regarding the signal histograms, than the images generated by using the traditional scenario with fixed learnt potentials and usual stochastic relaxation. Also, the CSA gives much more freedom in choosing the control parameters of the stochastic approximation because we need only the proximity between the sample relative frequency distributions in the training and the final generated sample and no true convergence to the MLE of the potentials itself.

Compared to the conventional simulated annealing used for image modelling and processing in (Geman and Geman, 1984; Geman et al., 1993; Gelfand and Mitter, 1993), the CSA differs in that the system of Eq. (3.5) presents an explicit unimodal measure of the proximity between the current generated sample and the goal, or training one in terms of their selected signal histograms. Also, the process of approaching the maximum proximity is equivalent to the one of maximizing the unimodal likelihood function and this results in a good convergence of the CSA. When using the CSA, we avoid the fairly long potential refinement stage and can simulate the model samples just after choosing the characteristic interaction structure.

3.5.3. TEXTURE SIMULATION AND SEGMENTATION

Gibbs models under consideration allow to embed simulation and segmentation of the image textures into a unified framework based on analytic and stochastic approximation of the signal histograms that are most characteristic for the image samples under a given model.

Usually, the problems of image simulation and segmentation are considered within the Bayesian framework. In most cases it implements either the simple MAP-decision with the maximum a posteriori probability of a desired sample:

$$\mathbf{s}^* = \left\{ \arg \max_{\mathbf{s} \in \mathcal{S}} \Pr(\mathbf{s}|\Theta) \right\} \qquad (3.19)$$

or the compound MPM-decision with the maximum posterior marginal probabilities of the sample components introduced first in (Abend, 1966):

$$\forall i \in \mathbf{R} \ \ \mathbf{s}^*(i) = \left\{ \arg \max_{u \in \mathbf{U}} \Pr_i(\mathbf{s}(i) = u|\Theta) \right\}. \qquad (3.20)$$

Here, Θ represents fixed parameters of a GPD, $\mathbf{s}$ is a sample (an image, region map, or image–region map pair) from the parent population $\mathcal{S}$, and u denotes a signal in the pixel (a gray level, region label, or their pair) taken from a finite set of the signal values $\mathbf{U}$.

The MAP-decision exploits the conventional simulated annealing techniques for searching the most probable sample. The potentials are changing as in Eq. (3.18) but by such a schedule $\lambda_{t+1} \propto \frac{1}{\log(1+t)}$ that permits to stay in a vicinity of the desired maximum during while the GPD is shrinking from the IRF with equiprobable samples up to the δ-function in the limit, when the number of macrosteps t approaches the infinity (Geman and Geman, 1984; Younes, 1988; Chellappa and Jain, 1993).

The MPM-decision uses the conventional stochastic relaxation with the fixed potentials to get a set of the samples under a given GPD and estimate from them the marginal signal probabilities $[\Pr_i(u|\Theta) : u \in \mathbf{U}]$. But, in this case the generated homogeneous Markov chain of the images has to reach an equilibrium state and one can ensure this only in the limit, when the number of the obtained samples is about $|\mathcal{S}|^2$ (van Laarhoven, 1988).

Both the cases need, in principle, a great many relaxation macrosteps whereas in practice we can afford only a moderately small number of them. Theoretically justified simulated annealing schedules of changing the potentials are derived as to ensure the convergence to the maximum point even in the worst cases rarely met in practice. Usually they are replaced by some empirically found schedules, but then it is hard to say whether the final decision is really the MAP-one. The MPM-case needs a relatively small number of steps to get good frequency estimates of the marginals if the Markov chain of the generated images is really in equilibrium. But, in practice it is hard to verify whether the chain has reached this state.

Instead of implementing the Bayesian decisions we exploit in our alternative modelling scenario a "rejecting" feature of the GPDs, namely, that in most cases the introduced GPDs are fairly close to the δ-function with respect to the total energies of the images. Therefore, the following conjecture seems to be valid:

Conjecture 3.1 *The images giving the MAP- or the MPM-decision have the partial Gibbs energies and, therefore, possess the sample relative frequency distributions of the signals and signal pairs which differ little from the corresponding energies and distributions for the training sample(s).*

Under this conjecture, the CSA allows to approximate the desired Bayesian decisions by simply generating the images with the desired signal histograms, say, with the GLH, or RLH, or joint GL/RLH and with a selected subset of the GLDHs, or RLCHs, or joint GLD/RLCHs, which are close to the given training histograms.

It is worth noting that some heuristics used in practice to implement the traditional scenarios do actually bring them close to our alternative CSA-based scenario.

CHAPTER 4

Supervised Conditional MLE-Based Learning

The learning scheme introduced in Chapter 3 involves a sizable number of the unknown potential values for the Gibbs models to be refined by stochastic approximation (see Table 2.4). This number can be considerably reduced by exploiting, instead of the unconditional MLE of Eq. (3.3), the conditional MLE of potentials described in this chapter. In particular, only $|\mathbf{A}| + 1$ unknown parameters have to be refined by stochastic approximation for the models of homogeneous textures in Eqs. (2.3) and (2.10). The conditional MLE gives, to within a scaling factor, an analytical form of a Gibbs potential to be learnt. Such a form permits, in particular, to choose the most appropriate parametric functions approximating the potentials for a given natural texture[1].

4.1. The least upper bound condition

We will mostly consider below the Gibbs model of Eq. (2.10) describing the homogeneous grayscale images. But the similar conditional MLE of potentials exists for each Gibbs model introduced in Chapter 2.

Each potential $\mathbf{V} = [\mathbf{V}_{\text{pix}}, \mathbf{V}_a : a \in \mathbf{A}]$ yields a particular ranking of the parent population $\mathcal{G}$ by the total energy, $e(\mathbf{g}|\mathbf{V}) = \mathbf{V} \bullet \mathbf{H}(\mathbf{g})$. Let $\text{rank}(\mathbf{g}|\mathbf{V}) \in \{1, \ldots, |\mathcal{G}|\}$ denote a rank of the image $\mathbf{g}$ produced by the potential $\mathbf{V}$:

$$\text{rank}(\mathbf{g}'|\mathbf{V}) > \text{rank}(\mathbf{g}''|\mathbf{V}) \ \text{ if } \ e(\mathbf{g}'|\mathbf{V}) > e(\mathbf{g}''|\mathbf{V}).$$

The conditional MLE of potentials is obtained under an additional constraint that the rank of a given training sample, $\mathbf{g}°$), has to reach the least upper bound of the possible ranks $\text{rank}(\mathbf{g}°|\mathbf{V})$. Because we expect that the learnt potentials should bring the total energy of the training sample as close as possible to the maximum energy, such a constraint seems to be quite natural.

The following lemma holds for the potentials:

---

[1] Recall that one such function, the Laplacian-of-Gaussian, has already been mentioned in Chapter 3 with respect to the potentials shown in Figures 3.4 – 3.9.

Lemma 4.1 *Let the samples* $\mathbf{g} \in \mathbf{G}$ *be ranked in the ascending order of the total energies* $e(\mathbf{g}|\mathbf{V})$ *for each potential vector* $\mathbf{V}$*:*

$$\operatorname{rank}(\mathbf{g}_1|\mathbf{V}) \geq \operatorname{rank}(\mathbf{g}_2|\mathbf{V}) \geq \ldots \geq \operatorname{rank}(\mathbf{g}_{|\mathcal{G}|}|\mathbf{V}).$$

Then all the potential vectors that rank a given training sample $\mathbf{g}^\circ$ *to the least upper bound in the total energy possess the following explicit, except for scaling factors* $\mathbf{\Lambda}$*, form:*

$$\mathbf{V}^\circ(\mathbf{\Lambda}) = [\lambda \cdot \mathbf{F}_{\mathrm{cn,pix}}(\mathbf{g}^\circ); \ \lambda_a \cdot \mathbf{F}_{\mathrm{cn},a}(\mathbf{g}^\circ) : \ a \in \mathbf{A}]. \tag{4.1}$$

Here, $\mathbf{\Lambda} = [\lambda, \lambda_a : a \in \mathbf{A}; \ \lambda, \lambda_a > 0]$ *denotes a vector of the arbitrary positive scaling factors, and the subscript "cn" indicates the centering of the corresponding sample relative frequencies:*

$$\begin{aligned} F_{\mathrm{cn,pix}}(q|\mathbf{g}^\circ) &= F(q|\mathbf{g}^\circ) - \frac{1}{|\mathbf{Q}|}; \\ F_{\mathrm{cn},a}(d|\mathbf{g}^\circ) &= F_a(d|\mathbf{g}^\circ) - \frac{1}{|\mathbf{D}|}. \end{aligned} \tag{4.2}$$

To prove Lemma 4.1, let us notice that, for the first-order clique family $\mathbf{R}$ and every second-order family $\mathbf{C}_a$, $a \in \mathbf{A}$, the ranking of the samples $\mathbf{g}$ in the partial energy, $e_{\mathrm{pix}}(\mathbf{g}|\mathbf{V}_{\mathrm{pix}}) = \mathbf{V}_{\mathrm{pix}} \bullet \mathbf{H}_{\mathrm{pix}}(\mathbf{g})$ and $e_{\mathrm{pix}}(\mathbf{g}|\mathbf{V}_{\mathrm{pix}}) = \mathbf{V}_a \bullet \mathbf{H}_a(\mathbf{g})$, respectively, is invariant to the potential (and energy) transformation that replaces the corresponding potential subvector, $\mathbf{V}_{\mathrm{pix}}$ or $\mathbf{V}_a$, by the unit subvector, $\tilde{\mathbf{v}}_{\mathrm{pix}} = \frac{\mathbf{V}_{\mathrm{pix}}}{|\mathbf{V}_{\mathrm{pix}}|}$ or $\tilde{\mathbf{v}}_a = \frac{\mathbf{V}_a}{|\mathbf{V}_a|}$.

Let $\mathbf{F}_{\mathrm{cn,pix}}(\mathbf{g}^\circ)$ and $\mathbf{F}_{\mathrm{cn},a}(\mathbf{g}^\circ)$ denote the centered vectors of the sample relative frequencies of gray levels and gray level differences in the clique family $\mathbf{C}_a$, respectively:

$$\begin{aligned} \mathbf{F}_{\mathrm{cn,pix}}(\mathbf{g}^\circ) &= \{F_{\mathrm{cn,pix}}(q|\mathbf{g}^\circ) : q \in \mathbf{Q}\} \\ \mathbf{F}_{\mathrm{cn},a}(\mathbf{g}^\circ) &= \{F_{\mathrm{cn},a}(d|\mathbf{g}^\circ) : d \in \mathbf{D}\}. \end{aligned}$$

It is easy to see that the following unit subvectors:

$$\tilde{\mathbf{v}}_{\mathrm{pix}}^\circ = \frac{\mathbf{F}_{\mathrm{cn,pix}}(\mathbf{g}^\circ)}{|\mathbf{F}_{\mathrm{cn,pix}}(\mathbf{g}^\circ)|} \quad \text{and} \quad \tilde{\mathbf{v}}_a^\circ = \frac{\mathbf{F}_{\mathrm{cn},a}(\mathbf{g}^\circ)}{|\mathbf{F}_{\mathrm{cn},a}(\mathbf{g}^\circ)|}$$

maximize the transformed partial energies $e_{\mathrm{pix}}(\mathbf{g}^\circ|\tilde{\mathbf{v}}_{\mathrm{pix}})$ and $e_a(\mathbf{g}^\circ|\tilde{\mathbf{v}}_a)$, respectively.

Therefore, every arbitrary potential subvector obtained by scaling such a unit subvector, provides for the same top rank of the training sample $\mathbf{g}^\circ$ in the corresponding partial energy among the samples $\mathbf{g} \in \mathbf{G}$ of the parent population as compared to any other potential subvector. Therefore, the potential estimates of Eq. (4.1) ensure that the training sample reaches the least upper bound in ranking by the total energy summed over all the clique families. Although the partial energies $e_{\text{pix}}(\mathbf{g}^\circ|\mathbf{V}_{\text{pix}})$ and $e_a(\mathbf{g}^\circ|\mathbf{V}_a)$ vary for the different potential subvectors $\mathbf{V}_{\text{pix}}$ and $\mathbf{V}_a$, respectively, the training sample has the feasible top rank in each partial energy and consequently in the total energy if the potentials are proportional to the centered marginal relative frequency distributions collected over this sample.

Therefore, the following statement is valid:

Lemma 4.2 *The conditional MLE* $\mathbf{V}^\star \equiv \mathbf{V}^\circ(\mathbf{\Lambda}^\star)$ *of the potentials such that* $\mathbf{\Lambda}^\star = \arg\max_{\mathbf{\Lambda}} \Pr(\mathbf{g}^\circ|\mathbf{V}^\circ(\mathbf{\Lambda}))$ *yields the maximum probability of the training sample* $\mathbf{g}^\circ$ *provided that its rank reaches the least upper bound within the parent population $\mathcal{G}$ ordered in the total energies.*

The obtained conditional MLE has the following components:

$$\begin{aligned} \forall q \in \mathbf{Q} \quad V_{\text{pix}}^\star(q) &= \lambda^\star F_{\text{cn,pix}}(q|\mathbf{g}^\circ); \\ \forall a \in \mathbf{A}; \quad d \in \mathbf{D} \quad V_a^\star(d) &= \lambda_a^\star F_{\text{cn},a}(d|\mathbf{g}^\circ) \end{aligned} \tag{4.3}$$

Therefore, the introduced *least upper bound principle* produces the explicit analytic forms of the Gibbs potentials given in Eq. (4.3) with only the scaling factors $\mathbf{\Lambda}^\star = [\lambda^\star, \lambda_\mathbf{a}^\star : \mathbf{a} \in \mathbf{A}]$ to be computed for each clique family by maximizing the likelihood function.

The desired factors $\mathbf{\Lambda}^\star$ are learnt in a similar way as the potentials themselves in Chapter 3. First, they are approximated analytically to find most characteristic interaction structure. Then, the factors for the chosen clique families are refined by stochastic approximation.

4.2. Potentials in analytic form

Generally, the potential estimate of Eq. (4.3) may differ from the true un-conditional MLE in Eq. (3.5). But for the GPDs introduced in Chapter 2 some plausible considerations exist that suggest that both the estimates are, at least, fairly close if not equivalent. This conjecture needs further theoretical investigations. The supporting considerations are based on a close similarity between the analytic first approximation of the uncondi-tional MLE and the conditional MLE in Eqs. (3.10) and (4.3), respectively, and on the fixed ranking of the samples under both the potential scaling of Eq. (4.1) and, by symmetry, the uniform scaling of the sample histograms.

The IRFs obtained from our Gibbs models when the potentials are zero-valued, $\mathbf{V} = \mathbf{0}$, have the uniform marginal probability distributions of signals (gray leveles or/and region labels) and signal co-occurrences. Therefore, the first analytic approximation of the potentials for the models in Eqs. (2.3) and (2.13) as well as for the joint and conditional GPDs that combine these latter models is given by the scaled centered sample frequency distributions for a training sample similar to those in Eq. (4.1). In particular, the Gibbs models of homogeneous grayscale textures in Eq. (2.3) and of region maps in Eq. (2.13) have the following first approximations of the unconditional MLE of potentials:

$$\forall q \in \mathbf{Q}$$
$$V^*_{\text{pix},0}(q) = \lambda F_{\text{cn,pix}}(q|\mathbf{g}^\circ) \equiv \lambda \left(F_{\text{pix}}(q|\mathbf{g}^\circ - \tfrac{1}{|\mathbf{Q}|} \right);$$

$$\forall a \in \mathbf{A}; \quad (q, q') \in \mathbf{Q}^2$$
$$V^*_a(q, q') = \lambda F_{\text{cn},a}(q, q'|\mathbf{g}^\circ) \equiv \lambda \left(F_a(q, q'|\mathbf{g}^\circ - \tfrac{1}{|\mathbf{Q}|^2} \right)$$

(4.4)

and

$$\forall k \in \mathbf{K}$$
$$V^*_{\text{pix},0}(q) = \lambda F_{\text{cn,pix}}(k|\mathbf{l}^\circ) \equiv \lambda \left(F_{\text{pix}}(k|\mathbf{l}^\circ - \tfrac{1}{|\mathbf{K}|} \right);$$

$$\forall a \in \mathbf{A}; \quad (k, k') \in \mathbf{K}^2$$
$$V^*_a(k, k') = \lambda F_{\text{cn},a}(k, k'|\mathbf{l}^\circ) \equiv \lambda \left(F_a(k, k'|\mathbf{l}^\circ - \tfrac{1}{|\mathbf{K}|^2} \right)$$

(4.5)

respectively, and these estimates differ from the conditional MLEs for the same models only in the uniform scaling. The simplified Gibbs models of Eqs. (2.7), (2.10), (2.22), (2.25), and (2.26) have slightly more differences between the two MLEs because the marginal probability distributions of gray level differences and region label coincidences are not uniform. But the resulting potential vectors that approximate the unconditional MLE in the vicinity of the zero point $\mathbf{V} = \mathbf{0}$ have rather small angular deviations from the relevant vectors of centered sample frequency distributions representing the conditional MLE.

4.2.1. FIRST APPROXIMATION OF SCALING FACTORS

The first approximation of the factors $\mathbf{\Lambda}^\star$ is obtained by the same technique of maximizing a truncated Taylor's series expansion of the log-likelihood function about the zero point $\mathbf{\Lambda} = \mathbf{0}$ as that in Chapter 3. We obtain the following approximation:

$$\lambda_0 = \alpha_0 \cdot e_0(\mathbf{g}^\circ); \quad \forall_{a \in \mathbf{A}} \; \lambda_{a,0} = \alpha_0 \cdot e_{a,0}(\mathbf{g}^\circ)$$
(4.6)

where $e_0(\mathbf{g}^\circ) = \sum_{q \in \mathbf{Q}} F^2_{\text{cn}}(q|\mathbf{g}^\circ)$ is the relative first-order energy and $e_{a,0}(\mathbf{g}^\circ)$ is the second-order one of (3.12). The factor α_0 depends also on these energies:

$$\alpha_0 = \frac{e_0^2(\mathbf{g}^\circ) + \sum\limits_{a \in \mathbf{A}} e_{a,0}^2(\mathbf{g}^\circ)}{e_0^2(\mathbf{g}^\circ) \cdot U_0(\mathbf{g}^\circ) + \sum\limits_{a \in \mathbf{A}} e_{a,0}^2(\mathbf{g}^\circ) \cdot U_{a,0}(\mathbf{g}^\circ)}$$
(4.7)

where

$$U_0(\mathbf{g}^\circ) = \sum_{q \in \mathbf{Q}} \left(F_{\mathrm{cn}}^2(q|\mathbf{g}^\circ) \cdot \sigma_{\mathrm{irf}}(q) \right)$$

$$U_{a,0}(\mathbf{g}^\circ) = \rho_a \sum_{d \in \mathbf{D}} \left(F_{\mathrm{cn},a}^2(d|\mathbf{g}^\circ) \cdot \sigma_{\mathrm{dif}}(d) \right) \tag{4.8}$$

4.2.2. SEARCH FOR THE INTERACTION STRUCTURE

The search exploits, just as in Chapter 3, the model-based interaction map. It is formed in this case by using the weighted relative energies of Eq. (3.12):

$$\mathbf{E}_0(\mathbf{g}^\circ) = [\rho_a \omega_{a,0} e_{a,0}(\mathbf{g}^\circ) : a \in \mathbf{A_W}]$$

where the weight $\omega_{a,0} = \sum_{d \in \mathbf{D}} F_{\mathrm{cn},a}^2(d|\mathbf{g}^\circ)$. Any previous strategy of recovering the characteristic clique families is applicable to this map, too.

4.2.3. STOCHASTIC APPROXIMATION TO REFINE FACTORS

This process exploits the differences of the relative energies to evaluate the proximity between the marginal gray level difference frequencies for each clique family in the training and generated samples. At each macrostep t of the stochastic approximation the current factors Λ_t are updated using the GLDHs for the sample $\mathbf{g}_t$ generated at this step as follows:

$$\forall a \in \mathbf{A} \quad \begin{aligned} \lambda_{\mathrm{pix},t+1} &= \lambda_{\mathrm{pix},t} + \alpha_t \cdot \Delta_t(\mathbf{g}^\circ, \mathbf{g}_t); \\ \lambda_{a,[t+1]} &= \lambda_{a,t} + \alpha_t \cdot \Delta_{a,t}(\mathbf{g}^\circ, \mathbf{g}_t) \end{aligned} \tag{4.9}$$

where α_t is the current scaling factor and $\Delta_t(\mathbf{g}^\circ, \mathbf{g}_t)$ and $\Delta_{a,t}(\mathbf{g}^\circ, \mathbf{g}_t)$ denote the current differences of the relative partial energies:

$$\begin{aligned} \Delta_t(\mathbf{g}^\circ, \mathbf{g}_t) &= \sum_{q \in \mathbf{Q}} (F_{\mathrm{cn}}(q|\mathbf{g}^\circ) - F_{\mathrm{cn}}(q|\mathbf{g}_t)) \cdot F_{\mathrm{cn}}(q|\mathbf{g}^\circ); \\ \Delta_{a,t}(\mathbf{g}^\circ, \mathbf{g}_t) &= \rho_a \sum_{d \in \mathbf{D}} (F_{\mathrm{cn},a}(d|\mathbf{g}^\circ) - F_{\mathrm{cn},a}(d|\mathbf{g}_t)) \cdot F_{\mathrm{cn},a}(d|\mathbf{g}^\circ). \end{aligned} \tag{4.10}$$

4.3. Practical consistency of the MLEs

Although in theory both the unconditional and conditional MLEs are consistent in a statistical sense, the system of equations that specifies the unconditional MLE of Eq. (3.5) and especially the conditional MLE of Eq. (4.1) show the main reasons for the possible "practical inconsistency".

TABLE 4.1. Number of the gray level combinations
in the clique families of different order.

Family's order $\|c_a\|$	Number of signal values $\|\mathbf{Q}\| = Q + 1$			
	4	8	16	256
2	16	64	256	65536
3	64	512	4096	2^{24}
4	256	4096	65536	2^{32}

The training samples are expected to display an ergodicity in that the sample relative frequencies of the signal combinations in the cliques over the lattice are, at least, close to the marginal probabilities of these combinations for the parent population of the images. In other words, the combinations having sufficiently high probabilities to be met in the test samples under the given GPD are expected to be present and have the like frequencies in the training sample.

4.3.1. SIGNAL COMBINATIONS IN THE CLIQUES

Generally, there are $\|\mathbf{Q}\|^{\|c_a\|}$ different gray level combinations $\mathbf{q}_a = (q_j : j \in c_a; q_j \in \mathbf{Q})$ in the clique of the family $\mathbf{C}_a$ and, therefore, one less number of the potential values $V_a(q_j : j \in c_a)$ to be learnt (see Table 4.1).

In most cases the training sample contains few cliques of a single family $\left(\|\mathbf{C}_a\| \ll \|\mathbf{Q}\|^{\|c_a\|} \right)$ so that the absence of certain configurations in the training sample do not ensure that the corresponding marginals are really close to zero. Thus, the higher the family's order, the less the potential values may be directly estimated from the small training sample. In this case, there is a need to interpolate the absent values using the obtained ones and a prior knowledge about the images.

Also, the search for the characteristic interaction structure by exhausting all the possible clique families $\left(\dfrac{\|\mathbf{W}\|}{\|c_a\| - 1} \right)$ in the search window $\mathbf{W}$ can be implemented in practice only for the pairwise pixel interactions. For the higher-order families, the search is to be reduced from other considerations specifying a reasonably small subset of the families to be exhausted.

4.3.2. POTENTIAL APPROXIMATION AND INTERPOLATION

The explicit potential estimates in Eq. (4.1) suggest that the potential estimation is quite similar to a reconstruction of an unknown marginal probability distribution from incomplete sample frequency distributions. Of

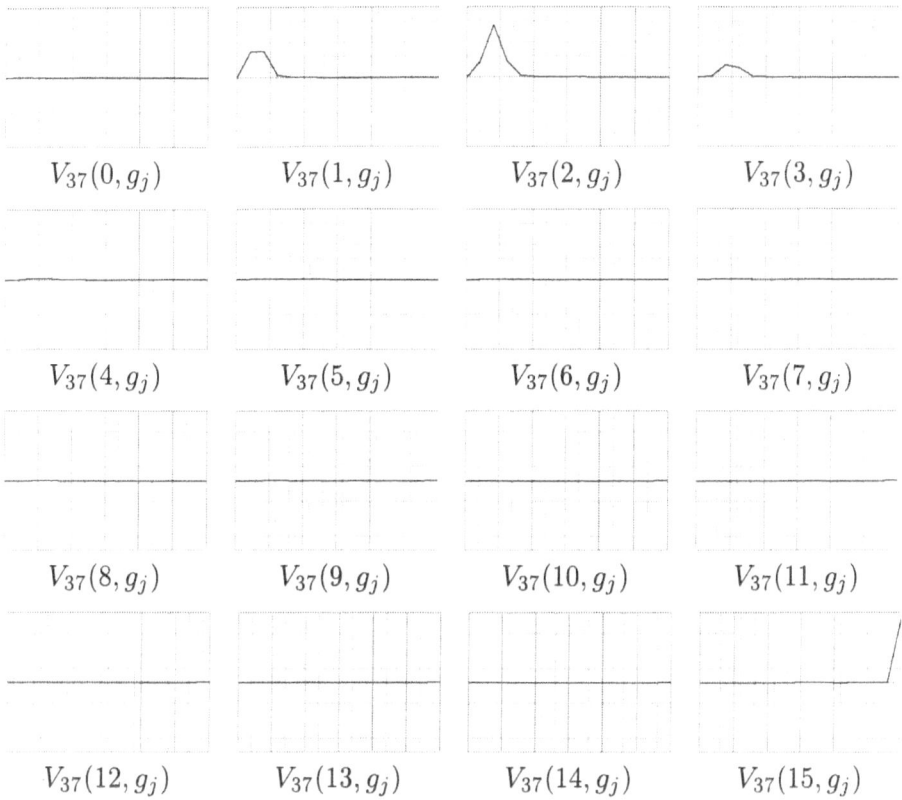

Figure 4.1.   Successive cross-sections for the first approximation of the potential $\mathbf{V}_{37} = [V_{37}(g_i, g_j) : g_i, g_j \in \mathbf{Q}]$ for the top-rank clique family $\mathbf{C}_{37}$ with the inter-pixel shift $[0, 14]$ in the interaction map of the texture D101.

course, it is possible to *a priori* define the potentials as one or another simple function of the signal combinations that depends on a few parameters and to then estimate only these parameters. Well-known auto-binomial and auto-normal Gibbs models considered in brief in Chapter 2 exemplify this way. But, these pre-defined functions reflect little or no specific features of the textures represented by a training sample.

More adequate potential functions are obtained by approximation or interpolation of relative sample frequency distributions $\mathbf{F}_{\text{pix}}(\mathbf{g}^\circ)$ before using them in the conditional MLE of Eq. (4.1).

Approximation of the potentials.   To approximate the potentials, we can smooth a "noisy" sample frequency distribution or histogram (e.g., GLCH or GLDH) collected for a given training sample as to bring it closer to the marginal probability distributions of signal combinations.

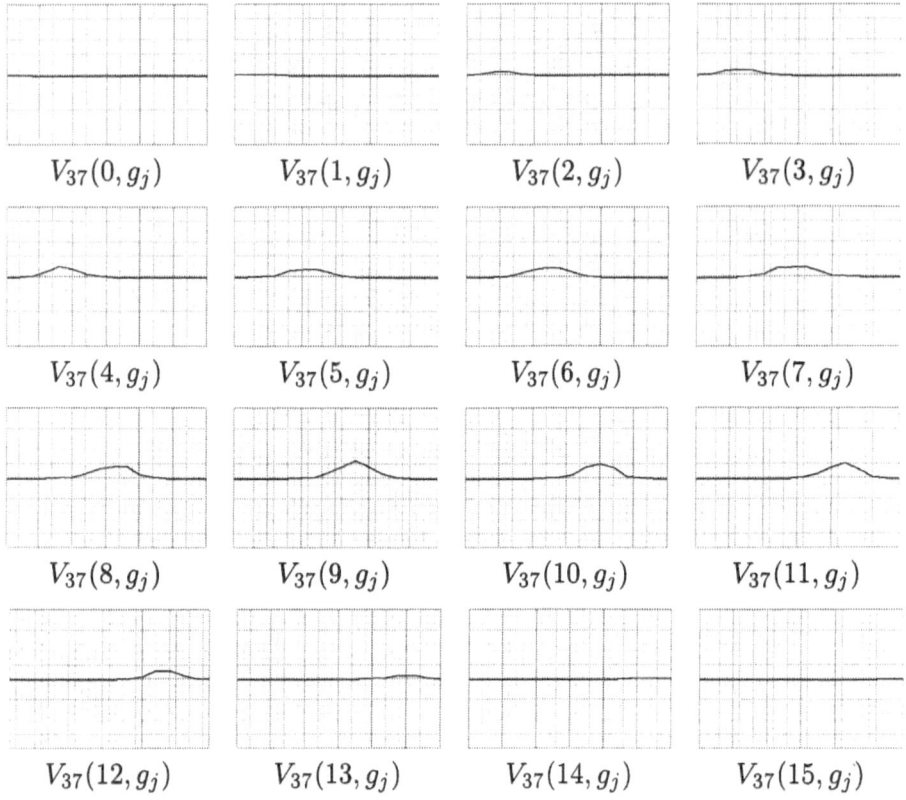

$$V_{37}(0, g_j) \quad V_{37}(1, g_j) \quad V_{37}(2, g_j) \quad V_{37}(3, g_j)$$

$$V_{37}(4, g_j) \quad V_{37}(5, g_j) \quad V_{37}(6, g_j) \quad V_{37}(7, g_j)$$

$$V_{37}(8, g_j) \quad V_{37}(9, g_j) \quad V_{37}(10, g_j) \quad V_{37}(11, g_j)$$

$$V_{37}(12, g_j) \quad V_{37}(13, g_j) \quad V_{37}(14, g_j) \quad V_{37}(15, g_j)$$

Figure 4.2.   Successive cross-sections for the first approximation of the potential $\mathbf{V}_0 = [V_0(g_i, g_j) : g_i, g_j \in \mathbf{Q}]$ for the top-rank clique family $\mathbf{C}_0$ with the inter-pixel shift $[1, 0]$ in the interaction map of the texture D29.

For example, let us look more closely at the potentials for the Gibbs model of Eq. (2.3) in Figures 4.1 and 4.2 (see also Figures 3.15, 3.16, Figures 3.21, and 3.22). We may readily check that the initial GLCHs can be closely approximated either by a single Gaussian or by a mixture of a few (in these cases, $N = 2..3$) Gaussians:

$$V_a(\mathbf{u}) = \sum_{n=1}^{N} \beta_n \exp\left(-\mathbf{u}^{\mathrm{T}} \mathbf{B}_n \mathbf{u}\right)$$

Here, $\mathbf{u} = [q_i, q_j]^{\mathrm{T}}$ denotes a gray level combination as the vector–column, and the approximation parameters are given by the scalar factors β_n and 2×2 matrices $\mathbf{B}_n$.

It should be emphasized that such an approximation can be done by directly fitting the above function to every collected GLCH, and we may not

include the approximation parameters into the Gibbs model itself. Moreover, the approximation permits also to interpolate the potential values for
the signal combinations which are absent in a given training sample.

Interpolation of the potentials. Let $\mathbf{Q}'_a \subset \mathbf{Q}_a = \mathbf{Q}^{|\mathbf{c}_a|}$ be a subset of the
signal combinations $\mathbf{q}_a$ in a clique family $\mathbf{C}_a$ which are present in the training sample. Generally, a "representative" relative frequency distribution
$\mathbf{F}_a(\mathbf{g}^\circ)$ of all the signal combinations in a given texture can be interpolated from the real frequency distribution $\mathbf{F}'_a(\mathbf{g}^\circ)$ collected for the training
sample. The interpolation takes account of the distances between the com-
binations that are present and absent in the training sample as follows
(Bezdek and Dunn, 1975):

$$F_a(\mathbf{q}_a|g^\circ) = \sum_{\mathbf{q}'_a \in \mathbf{Q}'_a} F'_a(\mathbf{q}'_a|g^\circ) \cdot \beta(\mathbf{q}'_a, \mathbf{q}_a). \tag{4.11}$$

Here, the weights $\beta(\mathbf{q}'_a, \mathbf{q}_a)$ depend on the Cartesian distances between the
combinations, $d(\mathbf{q}'_a, \mathbf{q}_a) = \|\mathbf{q}'_a - \mathbf{q}_a\|^2$, and the following constraints hold:

$$\sum_{\mathbf{q}'_a \in \mathbf{Q}'_a} F'_a(\mathbf{q}'_a|g^\circ) = 1; \quad \sum_{\mathbf{q}_a \in \mathbf{Q}_a} F_a(\mathbf{q}^a|g^\circ) = 1. \tag{4.12}$$

Due to Eq. (4.1), the same interpolation can be applied to the potentials for
getting the potential values for each signal combination absent in the training sample. This interpolation takes into account that the combinations at
a short distance should have the close potential values because the overall
visual appearance of the images is almost the same if such a combination
replaces another one.

Experiments in Chapters 5–7 with the Gibbs models of Chapter 2, hav-
ing $|\mathbf{c}_a| = 2$, assume the simplest weights:

$$\beta(\mathbf{q}'_a, \mathbf{q}_a) = \begin{cases} 1 & \text{if } d(\mathbf{q}'_a, \mathbf{q}_a) = 0; \\ 0 & \text{if } d(\mathbf{q}'_a, \mathbf{q}_a) \neq 0, \end{cases} \tag{4.13}$$

giving no interpolation at all. Instead, only the reduced set of gray levels
$Q = \{0, \ldots, 15\}$ and the second-order potentials depending on gray level
differences were used in the experiments.

The interpolation based on a fuzzy clustering scheme in (Bezdek and
Dunn, 1975) allows, in principle, to exploit the full gray range, such as
$[0, \ldots, 255]$. This scheme represents a frequency distribution (and, therefore, the related potential) as the above mixture of Gaussians, or normal
distributions with the following weights:

$$\beta(\mathbf{q}'_a, \mathbf{q}_a) = \begin{cases} 1 & \text{if } d(\mathbf{q}'_a, \mathbf{q}_a) = 0; \\ \dfrac{d^{-1}(\mathbf{q}'_a, \mathbf{q}_a)}{\displaystyle\sum_{\mathbf{t}_a \in \mathbf{Q}_a} d^{-1}(\mathbf{q}'_a, \mathbf{t}_a)} & \text{if } d(\mathbf{q}'_a, \mathbf{q}_a) \neq 0. \end{cases} \tag{4.14}$$

Generally, such interpolation exploits $|\mathbf{Q}'_a|$ signals per interpolated potential value.

To reduce the amount of computations, the interpolation may be restricted to only the combinations at a relatively short distance θ from the combinations in the training sample, the potentials for all other combinations being set to zero. The resulting interpolation is as follows:

$$\beta(\mathbf{q}'_a, \mathbf{q}_a) = \begin{cases} 1 & \text{if } d(\mathbf{q}'_a, \mathbf{q}_a) = 0; \\ \dfrac{d^{-1}(\mathbf{q}'_a, \mathbf{q}_a)}{\displaystyle\sum_{\mathbf{t}_a \in \mathbf{S}_a;\, d(\mathbf{q}'_a, \mathbf{t}_a) \leq \theta} d^{-1}(\mathbf{q}'_a, \mathbf{t}_a)} & \text{if } 0 < d(\mathbf{q}'_a, \mathbf{q}_a) \leq \theta; \\ 0 & \text{otherwise.} \end{cases} \qquad (4.15)$$

Practical consistency of the MLE. Generally, the proposed Gibbs models can be tailored to the expected sizes of the training samples either (i) by reducing the number of possible signal combinations or (ii) by isolating signal combinations with almost zero marginal probabilities from the combinations absent only because of a small training sample. The first way is most appropriate in the case of the unconditional MLE of potentials because it is rather difficult to deduce the proper convergence conditions for stochastic approximation if the above interpolation is embedded into the potential refinement process. Thus this way was accepted for the experiments on texture simulation and segmentation in Chapters 5–7.

But the experimental results gave grounds to expect that in the case of the conditional MLE of potentials the adequate interpolation of the potentials is also feasible. The stochastic approximation refinement of the conditional MLE, as opposed to the unconditional one, does not influence the overall forms of the potential functions and only changes their scaling factors. Therefore only the first analytic approximation of the potentials has to be interpolated.

The interpolation is alleviated because small changes of the gray levels usually have no effect on the visual appearance of the image. Therefore, the closer the distance between the signal combinations, the closer the potential values so that the distances between the combinations may guide the interpolation for those combinations which are absent in the training sample. In such a way, the theoretically consistent potential MLEs may gain the "practical consistency".

CHAPTER 5

Experiments in Simulating Natural Textures

From here on, we start experimental investigations of the proposed Gibbs image models with multiple pairwise pixel interaction and parameter learning schemes. This chapter presents results of simulation of various natural image textures taken from the MIT Media Laboratory "VisTex" image database (Pickard et al., 1995) or obtained by digitizing some examples from the album of Brodatz (1966). First, we consider the self-similarity between the different patches of visually homogeneous textures that can be revealed. It follows that the self-similarity is quantitatively described by the model-based interaction maps and resulting interaction structures and Gibbs potentials. Then multiple simulation results obtained by the CSA are presented to show that homogeneous, or stochastic textures may be separated from textures which are weakly homogeneous and inhomogeneous in terms of our models by comparing the training and simulated samples.

5.1. Comparison of natural and simulated textures

In this section we use different fragments of the natural textures D3 (Reptile Skin) and D14 (Woven Aluminum Wire) from (Brodatz, 1966) to analyze to which extent the non-Markov Gibbs model in Eq. (2.10) reflects the self-similarity within such a texture. Visually, the D14 fragments seem to be quite similar whereas the D3 fragments have notable relative geometric distortions and possess only a limited visual self-similarity.

Figures 5.1 – 5.4 display both the training and simulated fragments 128×128 of these textures, their interaction maps, and interaction structures recovered by simple thresholding of the maps. All the interaction maps represent 3280 clique families in the search window $\mathbf{W}$. The window parameters $\mu_{max} = 40$, $\nu_{max} = 40$ indicate the longest range of pixel interaction to be taken into account for learning the model parameters.

Each interaction map contains two square boxes 2×2 with relative coordinates (μ_a, ν_a) and $(-\mu_a, -\nu_a)$ per clique family $\mathbf{C}_a$ with the interpixel shift $(\mu_a, \nu_a) \in \mathbf{W}$. The origin of the Cartesian coordinates (μ, ν) is marked by a white square. The relative energies $\epsilon_{a,[0]}(\mathbf{g}°)$ of Eq. (3.12) are gray-coded; the darker the box, the stronger the interaction.

Here, the most characteristic clique families are found by the simple

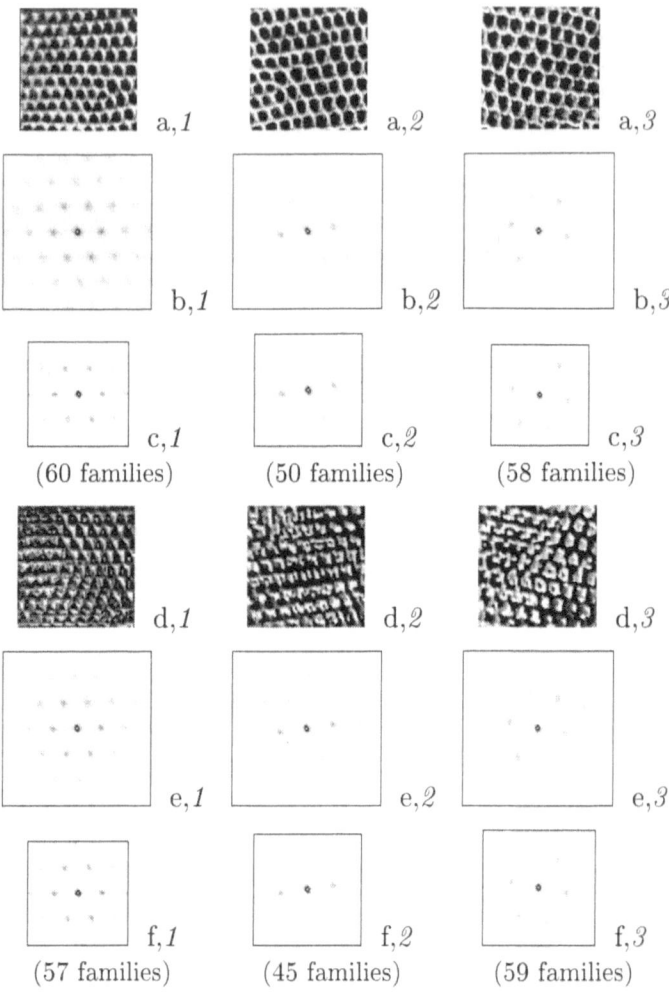

Figure 5.1. Fragments *1 – 3* of the natural image texture D3 (Reptile skin): training (a) and generated (d) samples, their interaction maps (b, e), and recovered structures (c, f) with the number of chosen clique families.

thresholding of the interaction map in Eq. (3.15) with the threshold of Eq. (3.16). All the interaction structures in Figures 5.1 – 5.4 are learnt with the thresholding parameter $c = 3$. It is evident that the different samples produce very similar interaction maps and structures reflecting the basic visually perceived hexagonal or tetragonal patterns of the textures. Also, the relative scaling and rotation of a texture result in the similar geometric transformations of the interaction maps and structures. The structures of the most similar samples are formed by approximately the same number

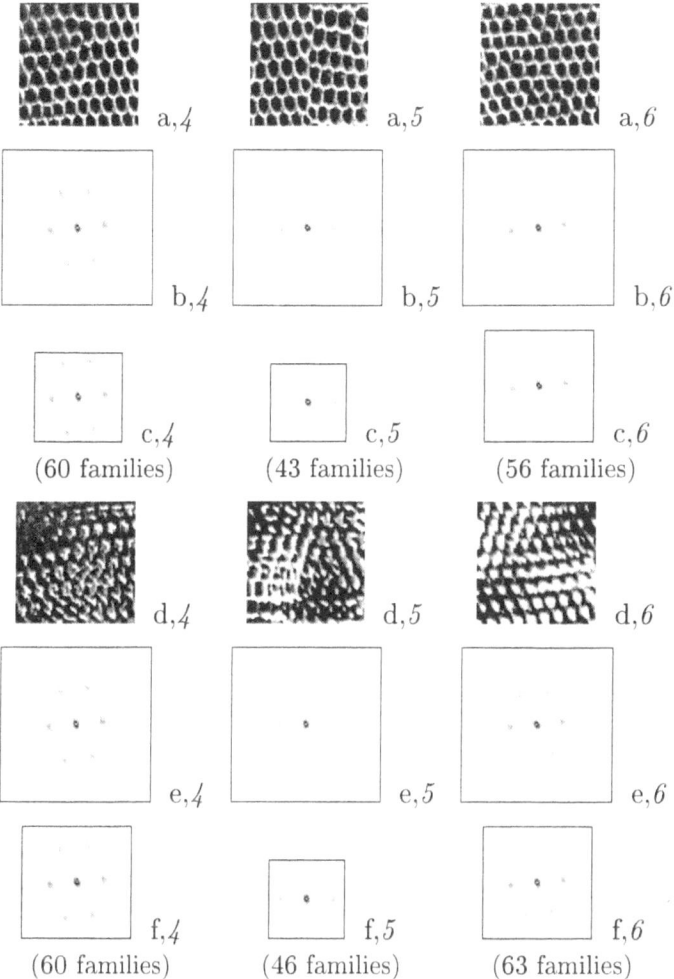

Figure 5.2.   Fragments *4 – 6* of the natural image texture D3; the same notation as in Figure 5.1.

of the clique families; in particular, 56 ... 60 families in Figures 5.1 and 5.2 for the samples D3-*1*, *3*, *4*, and *6* or 71 ... 79 families in Figures 5.3 and 5.3 for all the samples D14-em 1 – *6*. But the local inhomogeneities may notably affect the learnt structures. For instance, due to considerable local changes of the cell sizes within the sample D3-*2* in Figure 5.1 or relative cell positions within the sample D3-*5* in Figure 5.2 some characteristic long-range interactions are missing from the resulting interaction structures.been

All the fragments of the same natural texture are visually self-similar, to a greater or lesser extent. The same holds for the simulated fragments

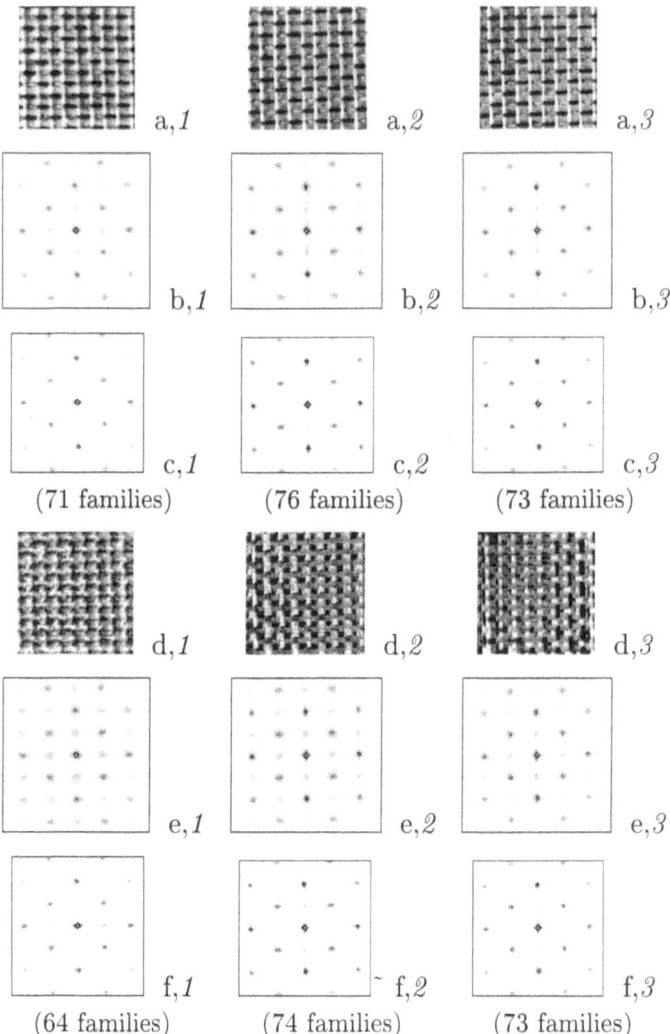

Figure 5.3. Fragments *1 – 3* of the natural image texture D14 (Woven aluminum wire); the same notation as in Figure 5.1.

in Figures 5.1,d – 5.4,d because their size 128×128 is sufficiently large to collect the GLDHs and learn the most characteristic interaction structures that describe the basic repetitive parts of these patterns. Nonetheless, the homogeneous textures in Figures 5.1,d – 5.4,d simulated by the CSA under the model in Eq. (2.10) with the learnt interaction structures and potentials show that only the pairwise pixel interactions do not reflect all the features which are perceived visually in these textures. Nonetheless, the

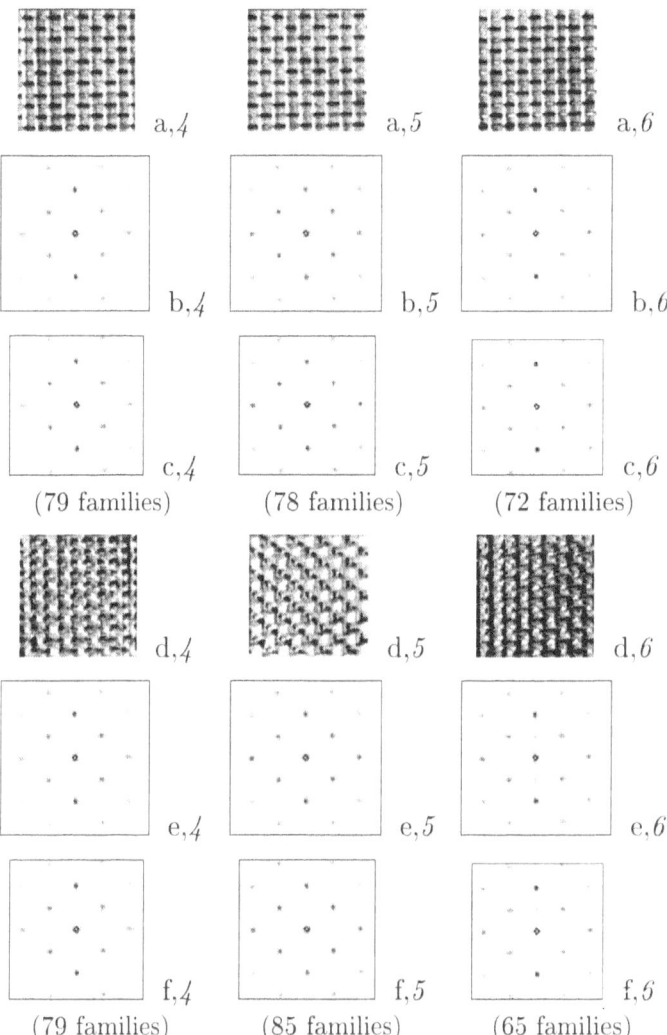

Figure 5.4.   Fragments *4 – 6* of the natural image texture D14; the same notation as in Figure 5.1.

overall visual appearance of these two texture types is preserved by the model.

All these simulation experiments were started from the same IRF "salt-and-pepper" sample. Each final sample is generated by 200 CSA macrosteps with the control parameters $c_0 = 0.$, $c_1 = 1$, and $c_2 = 0.001$. To see how close are the simulated and training sample with relation to the Gibbs model of Eq. (2.10), the chi-square distance between the normalized GLH and GLDHs of the training and simulated samples is used. The distance values are reduced from $601,000\ldots 1,329,000$ at the first macrostep ($t = 0$) to $1425\ldots 6125$ at the last macrostep ($t = 200$) for the samples in Figures 5.1 and 5.2 and from $965,000\ldots 2,712,000$ to $2300\ldots 7500$ for the samples in Figures 5.3 and 5.4. Therefore, both the training and simulated samples possess almost the same GLHs and GLDHs for all the chosen clique families. As a result, the interaction map and structure as well as potentials learnt for each simulated sample closely match the initial ones for the training sample. By this is meant that all the natural and simulated samples represent the same texture type as regarding the chosen Gibbs model of Eq. (2.10).

Figure 5.5 presents two more texture samples 256×256 simulated with the model parameters learnt for the samples D3-*1* and D14-*1* in Figures 5.1 and 5.4, respectively. The simulation uses the same control parameters and 200 CSA macrosteps. Once again, the interaction maps and recovered structures for the larger generated samples have almost no differences relative to the smaller samples in Figures 5.1 – 5.4.

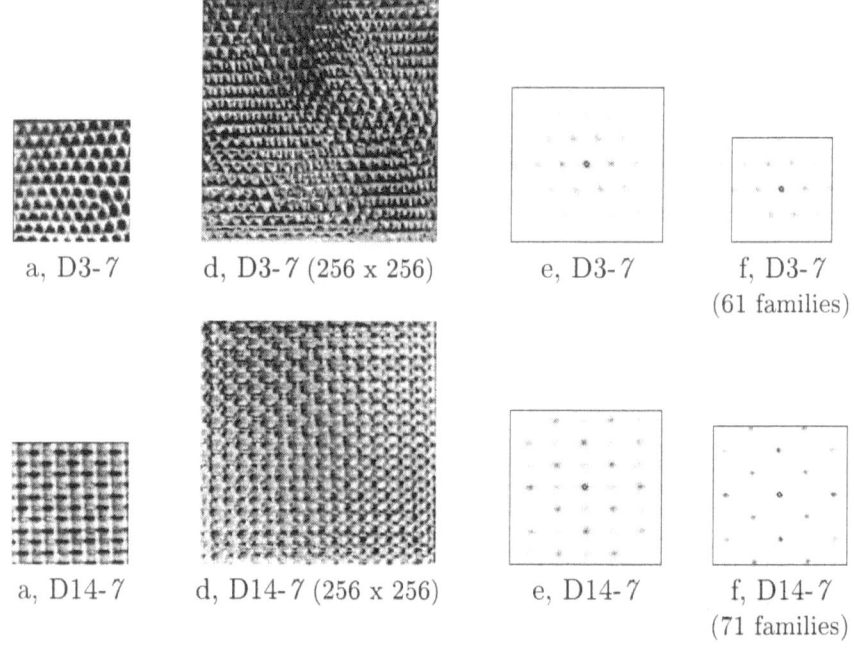

a, D3-*7*   d, D3-*7* (256 x 256)   e, D3-*7*   f, D3-*7*
(61 families)

a, D14-*7*   d, D14-*7* (256 x 256)   e, D14-*7*   f, D14-*7*
(71 families)

Figure 5.5.   Training 128×128 (a) and generated (d) 256×256 samples of the textures D3 and D14, the interaction maps (e), and recovered structures (f) for the generated samples.

It is easily seen that the visual similarity between the natural and simulated samples is rather weak although the basic spatial structure of the training patterns is reflected by simulation. The model of Eq. (2.10) is simplified by taking account only of the gray level differences. But the GLDHs for these training samples possess an approximate mirror symmetry and hence are almost invariant to the inversion of the image gray ranges that replaces each gray level $q \in \mathbf{Q}$ with the inverse value $Q - q$. As a result, the simulated image may demonstrate continuous transitions between the initial and inverted patterns of the texture represented by the training sample. To avoid such transitions, a more general image model of Eq. (2.3) exploiting the pairwise gray level co-occurrences should be used.

The above training and simulated samples are related to the same image textures D3 and D14 by the Gibbs model of Eq. (2.10) but have substantial visually perceived differences. Therefore the pairwise pixel interactions learnt by a simple thresholding of the interaction map do not approximate all the characteristic pixel interactions, and the textures D3 and D14 are unlikely to be considered as the purely stochastic ones. Notice that this holds only under a particular choice of most characteristic interaction structure, namely, for our simple thresholding of the interaction map. More complicated learning technique of Zalesny (1996) permits to considerably extend a subset of the stochastic textures which are efficiently modelled by the Gibbs models with multiple pairwise pixel interaction. At each step of this sequential learning, the current interaction structure is appended by a new clique family which is found to be the most characteristic one by comparing the interaction map for a sample generated under the current structure to the interaction map for a given training sample.

5.2. "Brodatz" image database

Table 5.1 lists abridged names of the 49 image textures selected from the album of Brodatz (1966) for our experiments. Some of these textures are used only in Chapters 6 and 7 for the experiments in texture retrieval and segmentation. The image samples are obtained by digitizing the pictures and transforming the resulting gray ranges into the reference range $[0, \ldots, 15]$.

To demonstrate experimentally which natural image textures can be assigned to stochastic textures under the Gibbs model of Eq. (2.10), we consider the 36 samples presented in Figures 5.6 – 5.8. The training samples and corresponding simulated samples generated by the CSA are marked by "T" and "G", respectively. The interaction structures are learnt by thresholding the interaction maps shown in Eqs. (3.15) and (3.16). Here, once more we use the 200 CSA macrosteps and control parameters $c_0 = 0.$, $c_1 = 1$, and $c_2 = 0.001$ for simulating all the samples.

The experiments show that some of the textures, namely, D4, D5,

TABLE 5.1. Selected Brodatz textures.

Texture	Name	Texture	Name
D1	Woven aluminum wire	D57	Handmade paper
D3	Reptile skin	D65	Handwoven rattan
D4	Pressed cork	D66	Plastic pellets
D5	Expanded mica	D68	Wood grain
D6	Woven aluminum wire	D69	Wood grain
D9	Grass lawn	D74	Coffee beans
D11	Homespun woolen cloth	D75	Coffee beans
D12	Bark of tree	D76	Grass fiber cloth
D14	Woven aluminum wire	D77	Cotton canvas
D16	Herringbone weave	D79	Grass fiber cloth
D17	Herringbone weave	D80	Straw cloth
D19	Woolen cloth	D82	Straw cloth
D20	French canvas	D83	Woven matting
D21	French canvas	D84	Raffia
D23	Beach pebbles	D85	Straw matting
D24	Calf leather	D87	Sea fan
D29	Beach sand	D92	Pigskin
D34	Netting	D93	Fur
D35	Lizard skin	D94	Brick wall
D36	Lizard skin	D95	Brick wall
D50	Raffia woven	D101	Cane
D52	Straw cloth	D102	Cane
D53	Straw cloth	D103	Loose burlap
D55	Straw matting	D105	Cheesecloth
D55	Straw matting		

D9, D29, D50, D57, D68, D69, D76, D77, D79, D80, D92, and D93 do really belong to the class of stochastic textures. In these cases, the natural and simulated patterns possess both the good visual resemblance and high proximity between the GLH and GLDHs chosen for the Gibbs model of Eq. (2.10).

In other cases such a proximity does not ensure the visual similarity. For example, there can be only a limited similarity as for the textures D24, D65, D74, d82, D83, D84, and D105 or the similarity can be absent. In the latter case either the texture is built from some regular texels which are not approximated by "star-like" subsets of pixel pairs (D6, D11, D17, D20, D34, D55, D82, D85, D101) or the training sample has substantial local

or global inhomogeneities (D23, D36, D66, D75, D95, D103). The chosen
sizes of the training sample (128×128) and search window (81×81) do not
permit to recover characteristic long-range "horizontal" interactions in the
textures such as D11, D17, or D85. Also, our models cannot mimic a 3D
visual appearance of the textured surfaces such as D5, D12, D23, or D74.

Here, we only restrict our consideration to a rough qualititative discrimination between homogeneous, weakly homogeneous, and inhomogeneous
textures. The textures are discriminated by direct visual comparisons of
the training and simulated samples. Table 5.2 reviews such a marking for
the textures in Figures 5.1–5.4 and 5.6–5.8. There are 14 stochastic (36.8%),
7 weakly homogeneous (18.4%), and 17 inhomogeneous textures (44.8%).
Therefore the Gibbs model of Eq. (2.10) describes adequately more than
half of the selected 36 textures even when the interaction structure is recovered by the simple thresholding of the interaction maps.

TABLE 5.2. Homogeneity of the 36 Brodatz textures (notation: **h** – homogeneous; **w**
– weakly homogeneous; **i** – inhomogeneous texture).

Name:	D3	D4	D5	D6	D9	D11	D12	D14	D17	D20
Mark:	i	h	h	i	h	i	i	i	i	i

Name:	D23	D24	D29	D34	D36	D50	D55	D57	D65	D66
Mark:	i	w	h	i	i	h	i	h	w	i

Name:	D68	D69	D74	D75	D76	D77	D79	D80	D82	D83
Mark:	h	h	w	i	h	h	h	h	w	w

Name:	D84	D85	D92	D93	D95	D101	D103	D105		
Mark:	w	i	h	h	i	i	i	w		

5.3. Interaction maps and texture features

Our experiments show that the model-based interaction maps can recover
most characteristic pairwise pixel interactions for representing a given texture only if the texture is spatially homogeneous, or translation invariant
in terms of the local conditional probabilities of signal cooccurrences or
differences. Also, the sample size should be sufficiently large to ensure that
the estimates of these conditional probability distributions by the sample
relative frequency distributions are consistent in a statistical sense.

The interaction map cannot represent the structural features perceived
easily by human vision if the local probability distribution of signal cooccurrences or differences vary substantially over the image. In such a case,

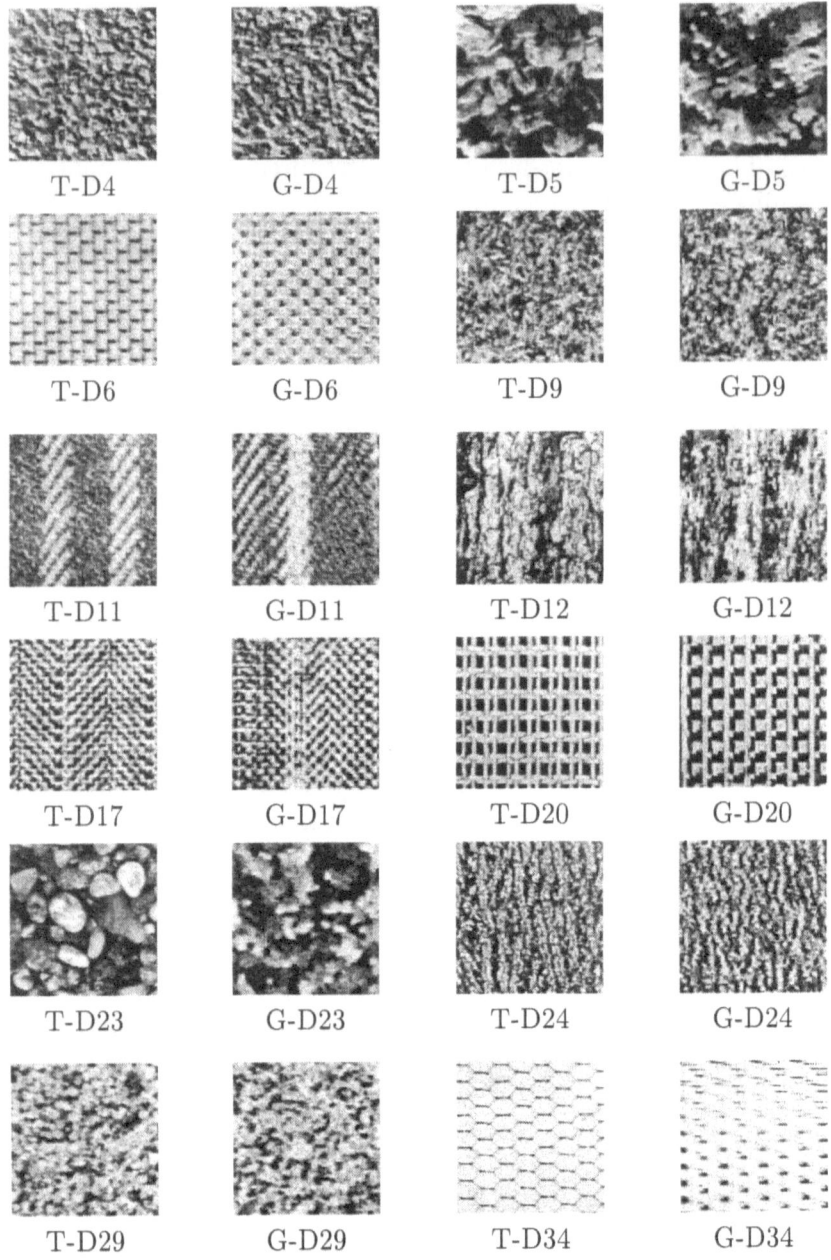

Figure 5.6.   Training (T) and simulated (G) 128 × 128 samples of different natural textures.

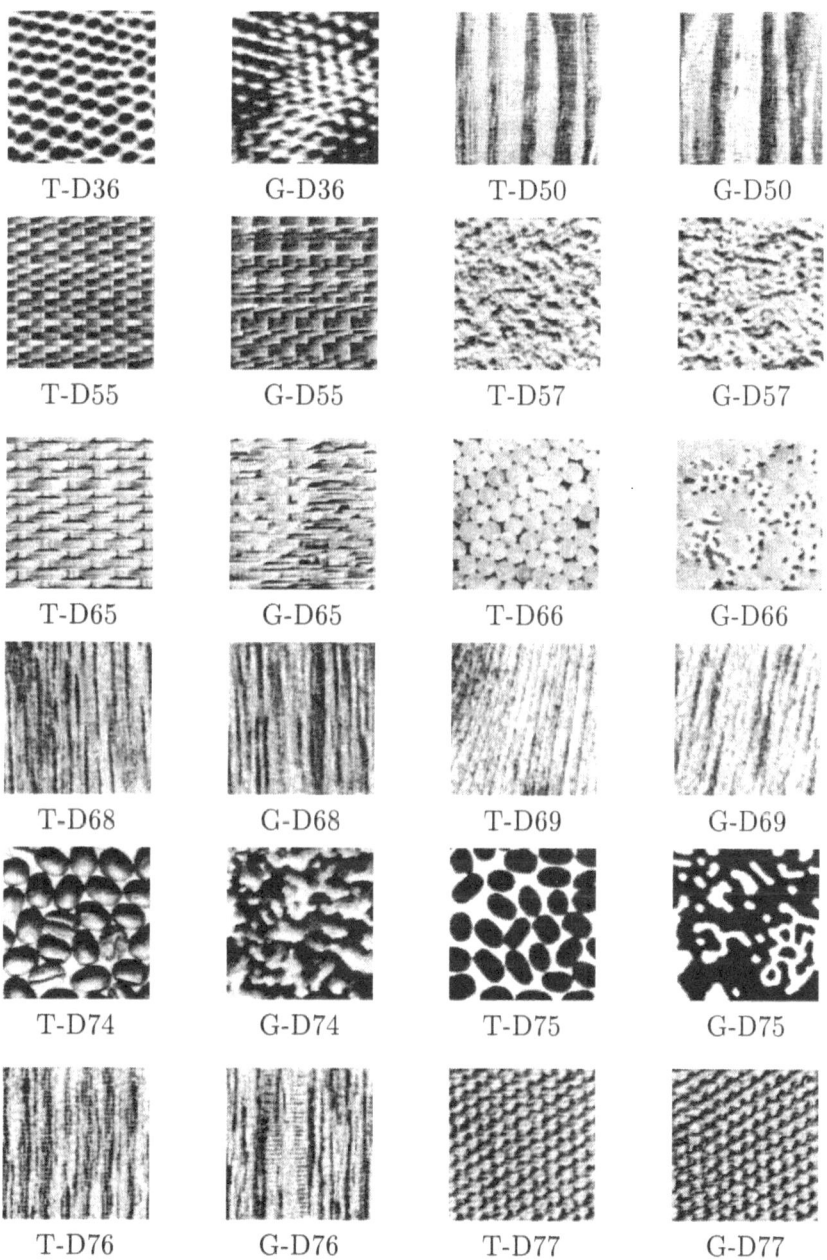

Figure 5.7. Training (T) and simulated (G) 128 × 128 samples of different natural textures.

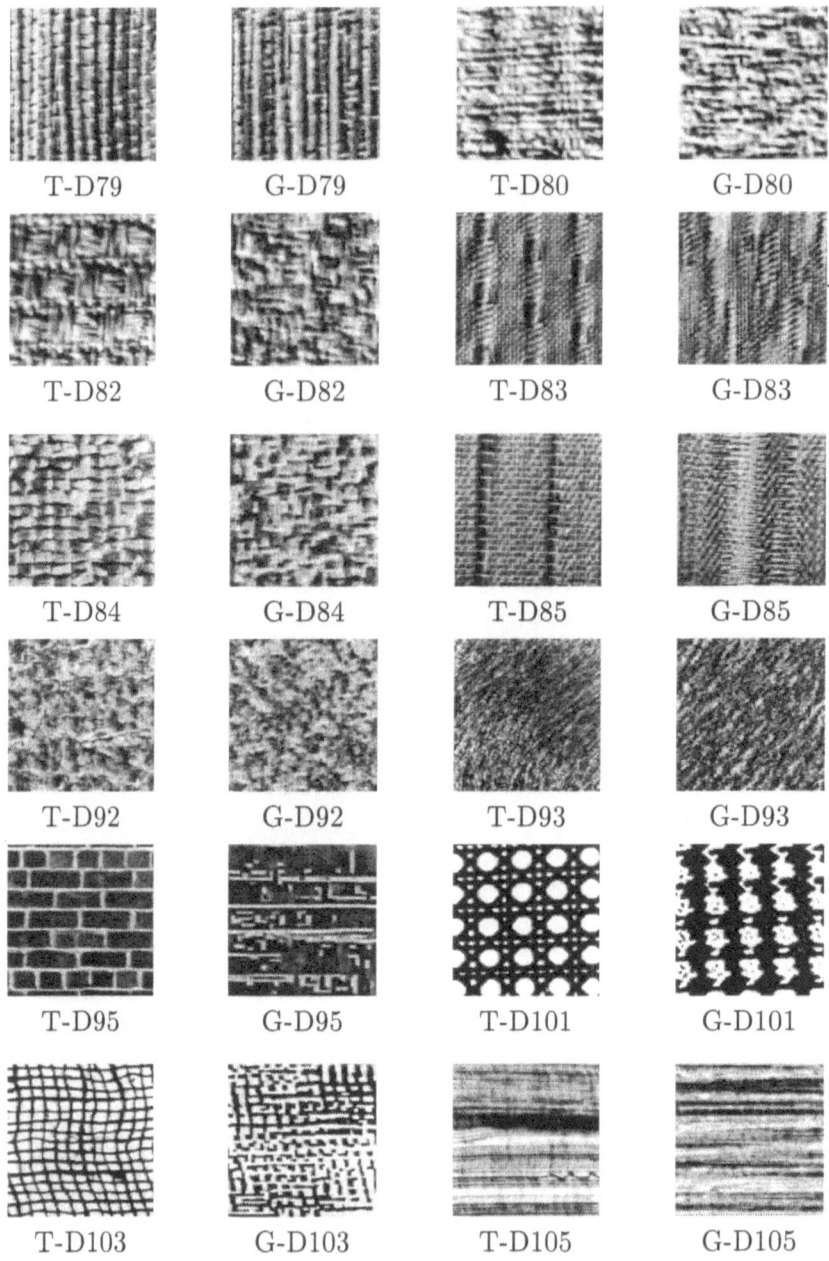

T-D79   G-D79   T-D80   G-D80

T-D82   G-D82   T-D83   G-D83

T-D84   G-D84   T-D85   G-D85

T-D92   G-D92   T-D93   G-D93

T-D95   G-D95   T-D101   G-D101

T-D103   G-D103   T-D105   G-D105

Figure 5.8.   Training (T) and simulated (G) 128 × 128 samples of different natural textures.

the different non-homogeneous pairwise pixel interactions are averaged in the GLDHs, GLCHs, or other signal histograms, so that the resulting inter-action maps and learnt interaction structures do not reflect essential local features of the interactions.

As is shown in this chapter and Chapters 6 and 7, the learning tech-nique for recovering the interaction structure (proposed in Chapter 3) holds much promise in texture simulation, retrieval, and segmentation in spite of its drawbacks. This learning technique was intoduced first in (Gimel'farb, 1996a)–(Gimel'farb, 1996c), (Jain and Gimel'farb, 1995), and (Gimel'farb and Jain, 1996). On the basis of this technique, the *feature-based interac-tion maps*, derived from the extended GLDHs, are proposed recently by Chetverikov and Haralick (1995) for analyzing such integral features of im-age textures as symmetry, anisotropy, and regularity. The similar analysis can be done using the model-based interaction maps, too.

For instance, if a homogeneous texture pattern possesses internal sym-metries, then the chosen structure or, at least, the initial interaction map reveals them. For instance, let us rotate or mirror the interaction structures in Figure 5.5 with respect to the shown initial ones. Let us compute, for each possible rotation or mirror angle, a relative number of the matching clique families. Then by a proper thresholding we can find three character-istic rotation and mirror angles for the interaction structure of the sample D3-7 or a single rotation and four mirror angles for the structure of the sample D14-7. Such a structural matching is successfully used in Chapter 6 for the scale- and rotation-invariant retrieval of the textures from an image data base; see also (Jain and Gimel'farb, 1995; Gimel'farb and Jain, 1996).

Also, isolated clusters of the long-range interactions that have rotation or mirror symmetries and almost the same inter-cluster distances count in favour of the texture granularity. Textures with a marked lineation have corresponding "lines" of the clique families in the chosen structures, whereas the randomness leads to connected "blobs" of the chosen families around the origin. Examples of corresponding imteraction structures are given in Chapter 6. But, to quantitatively describe such features one needs more theoretical and experimental effort in spite of a strong possibility to link these features with the interaction maps and structures.

5.4. CSA vs. traditional modelling scenario

The traditional modelling scenario exploits the following two processing stages (Chellappa and Jain, 1993; Li, 1995; Winkler, 1995):

(*i*) estimation of the Gibbs model parameters from a given training sample (in our case, it is done by analytic and stochastic approximation of the MLE of Gibbs potentials) and

(*ii*) generation of images by the pixel-wise stochastic relaxation using the
Gibbs model with the fixed estimated parameters.

As was discussed earlier, this scenario is obviously impracticable for texture
modelling. In theory, the stochastic approximation converges to the desired
potential MLEs only when a generated inhomogeneous Markov chain of
images reaches an equilibrium where it becomes a homogeneous one (that
is, when the stepwise changes of the model parameters tend to zero). There
exist no practicable statistical tests for the equilibrium to be used as a
stopping rule, and the theoretically justified schedules of stochastic approximation are impractically long because they are designed for the worst
cases (Younes, 1988). Unfortunately, only under such a schedule do the
obtained parameters possess the desired features of a training sample.

Thus, in the traditional scenario we have first to wait for an equilibrium
of an inhomogeneous Markov chain of images generated to find the MLEs
of potentials by stochastic approximation. Secondly, we have to wait for an
equilibrium of a homogeneous Markov chain generated under the obtained
model parameters to be sure that the images in the chain have the desired
GPD. But, in both the cases, there are no formal rules for testing whether
the equilibrium has actually been reached.

Our alternative scenario that combines the computation of the MLE and
image generation into a single CSA process pursues a more practical goal
of generating the samples with probabilities "around" the probability of a
given training sample. As is shown in Chapter 3, the CSA tends to minimize,
on an average, the chi-square distance between the signal histograms for the
generated and training samples. The histograms are sufficient statistics for
the Gibbs models under consideration, so the samples distributed "around"
a training sample are of the main practical interest.

The main practical aspect of the alternative scenario is that it results
in both faster and better modelling. Here, the term "better" means the
smaller chi-square distances between the histograms, but the visual similarity between the training and simulated samples is usually higher, too.
The alternative scenario has an explicit and natural stopping rule which
is absent in the traditional scenario. The stopping rule, based on the chi-
square distance between the histograms or on the difference between the
total energies, may guide, if necessary, the modelling process.

Figures 5.9 and 5.10 show results of generating samples of the Brodatz
texture D29 by the traditional and alternative scenarios. Plots under the
simulated samples show how the average chi-square distance between the
GLDHs for the simulated and training samples per clique family is changing
at each macrostep of stochastic relaxation. The training sample is shown
in Figure 5.6.

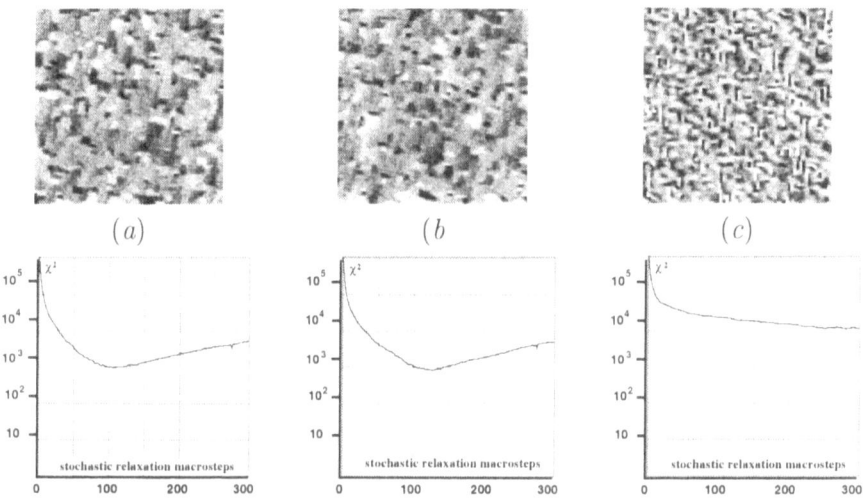

Figure 5.9.   Traditional modelling of the texture D29.

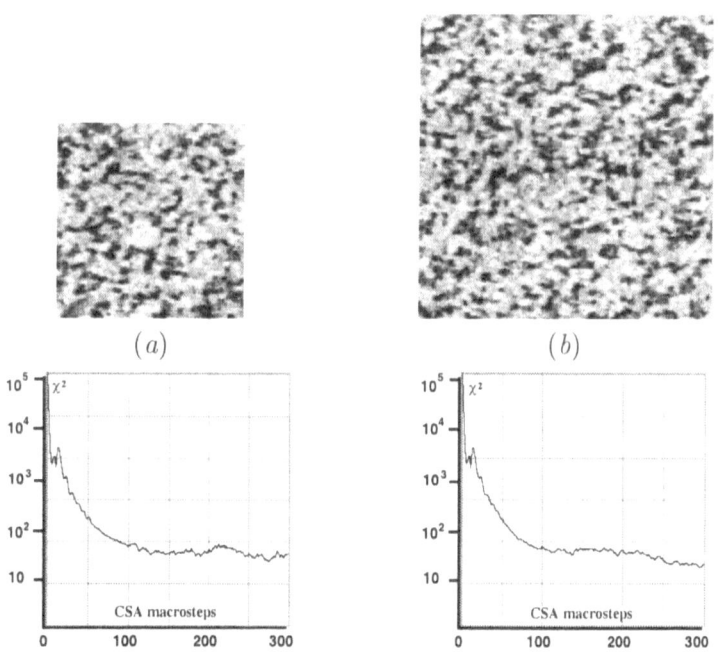

Figure 5.10.   CSA modelling of the texture D29.

Gibbs potentials to generate the samples in Figure 5.9,a and 5.9,b are estimated by 300 steps of stochastic approximation, and the samples themselves are obtained by 300 steps of pixel-wise stochastic relaxation with the fixed model parameters. The sample in Figure 5.9,c is obtained in a similar way but by 2000 and 300 steps, respectively.

In three examples the convergence to the training GLDHs is too slow; the final chi-square distance per clique family is still about $7,000 - 10,000$ after 300 steps of generation. Moreover, the first two samples even diverge from the training sample after the first 100 converging macrosteps that gave a better distance of $1,000$. Also, it is easily seen that the simulated samples possess almost no visual resemblance to the training sample.

The third example in Figure 5.9,c indicates that even $2,000$ steps of stochastic approximation for refining the potentials do not guarantee that the created Markov chain of samples has really reached an equilibrium and we may expect the better convergence at the subsequent generation stage.

The desired simulation results are obtained much easier and faster by the alternative CSA-based scenario (it should be noted that all the steps in both scenarios have the same computational complexity). For example, only 300 CSA steps result in the considerably better convergence of the generated samples to the training ones in Figure 5.10,a and 5.10,b. The final chi-square distances between the simulated and training GLDHs for these two samples are about two orders smaller $(30 - 70)$ than for the traditional scenario that performs in total $600 - 2300$ steps. As a result, these samples do visually resemble the training sample.

Figures 5.11, 5.12, 5.13, and 5.14 show four more textures from (Brodatz, 1966) simulated by the traditional and alternative scenarios. In all these cases, except for the samples D55,c and D93,b–d, we used in the traditional scenario 500 steps of stochastic approximation to refine the potentials and then 300 stochastic relaxation steps, under the final Gibbs model of Eq. (2.10), to simulate the sample (1000 and 500 steps for D55,a and 10,000 and 500 steps for D93,b–d, respectively). The alternative scenario uses only 300 CSA steps, except that the samples D93,h–j are obtained by 600, 1000, and 10,000 CSA steps, respectively. These latter samples are similar to the sample D93,g obtained by 300 CSA steps, suggesting that the CSA gives quite stable results.

These and other similar experiments suggest that the CSA-based scenario is more practicable in texture simulation than the traditional one.

5.5. "MIT VisTex" image database

The similar experimental results are obtained for 165 grayscale texture samples from the MIT Media Lab. "VisTex" 128×128 image database

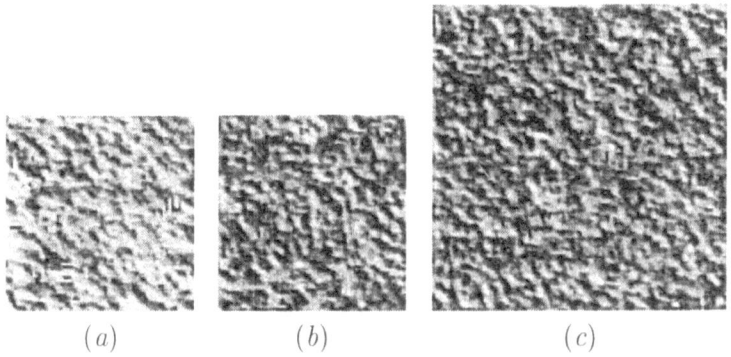

Figure 5.11.   Traditional (*a*) and CSA (*b,c*) modelling of the texture D4; the training sample is found in Figure 5.6.

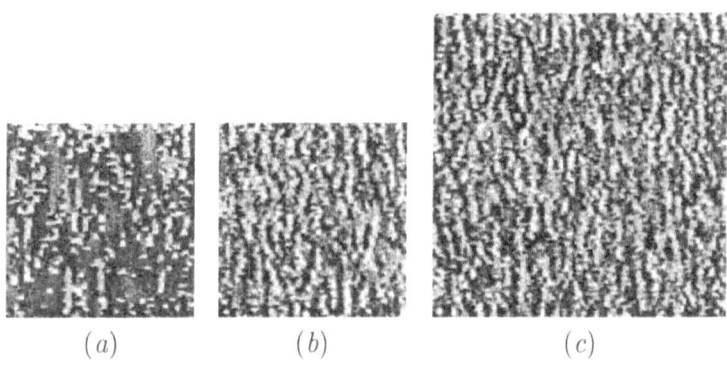

Figure 5.12.   Traditional (*a*) and CSA (*b,c*) modelling of the texture D24; the training sample is found in Figure 5.6.

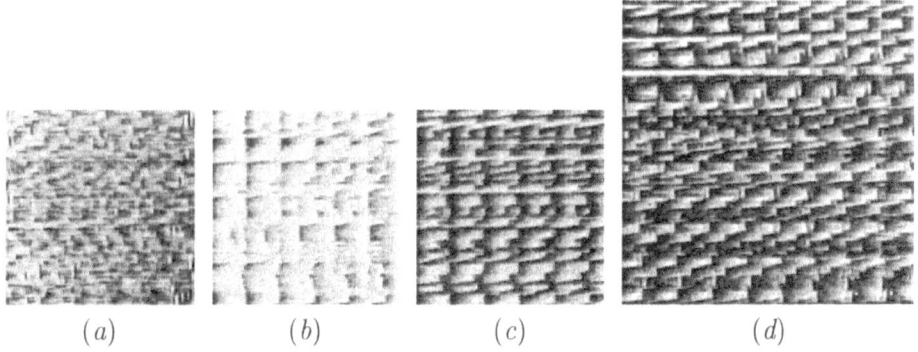

Figure 5.13.   Traditional (*a,b*) and CSA (*c,d*) modelling of the texture D55; the training sample is found in Figure 5.7.

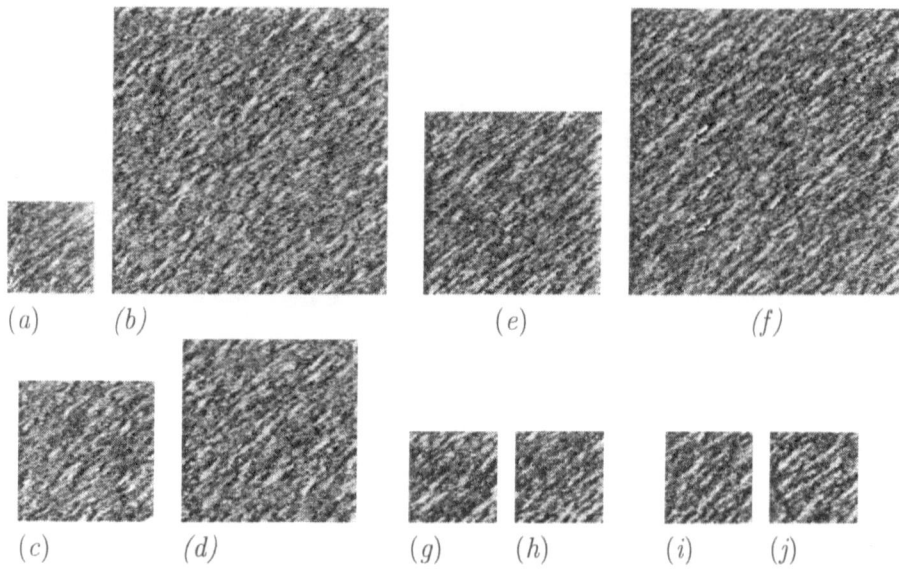

Figure 5.14. Training sample 64 × 64 (*a*) and traditional (*b,c,d*) and CSA (*e,f,g,h,i,j*) modelling of the texture D93.

in (Pickard et al., 1995). The training and simulated samples are presented in Figures 5.15 – 5.25.

There are 55 stochastic (33.3%), 53 weakly homogeneous (32.1%), and 57 inhomogeneous textures (34.6%) listed in Table 5.3.

As expected, our models result in spatially homogeneous versions of the scenes under consideration and do not approximate complex 3D scenes with only partially textured surfaces such as various buildings or terrains, or mimic spatially inhomogeneous paintings and mosaics. But, different types of fabrics, flat metallic surfaces, uniform fields of flowers, some water and wooden surfaces, in total more than 65% of all the textures are adequately described, to a greater or lesser extent.

Figure 5.26 indicates once more that the 300 macrosteps of the CSA-based modelling scenario that was used in Figures 5.15 – 5.25 are considerably superior to the 2000 macrosteps of stochastic approximation to refine the potentials and the subsequent 600 macrosteps of stochastic relaxation with the fixed potentials to generate the samples. In the latter case there is almost no convergence of the generated GLDHs to the training ones, and hence the visual resemblance between the training and simulated images is very poor. The CSA gives faster and much better results. We use the simplified model of Eq. (2.10), and the GLDHs in most of our training samples are almost symmetric with respect to inversion $q \to q_{max} - q$ of the gray

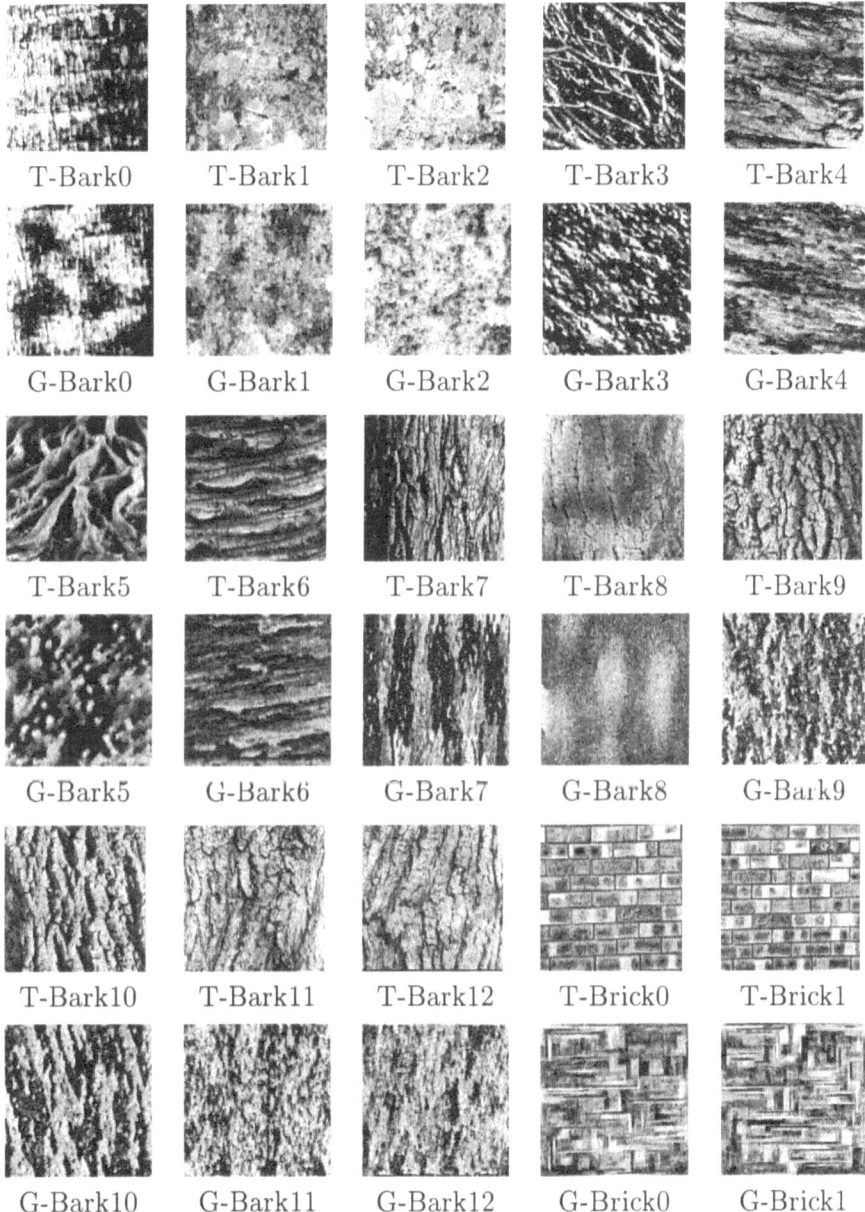

Figure 5.15.   Training (T) and simulated (G) 128 × 128 samples of different natural textures from MIT "VisTex" database.

Figure 5.16. Training (T) and simulated (G) 128 × 128 samples of different natural textures from MIT "VisTex" database.

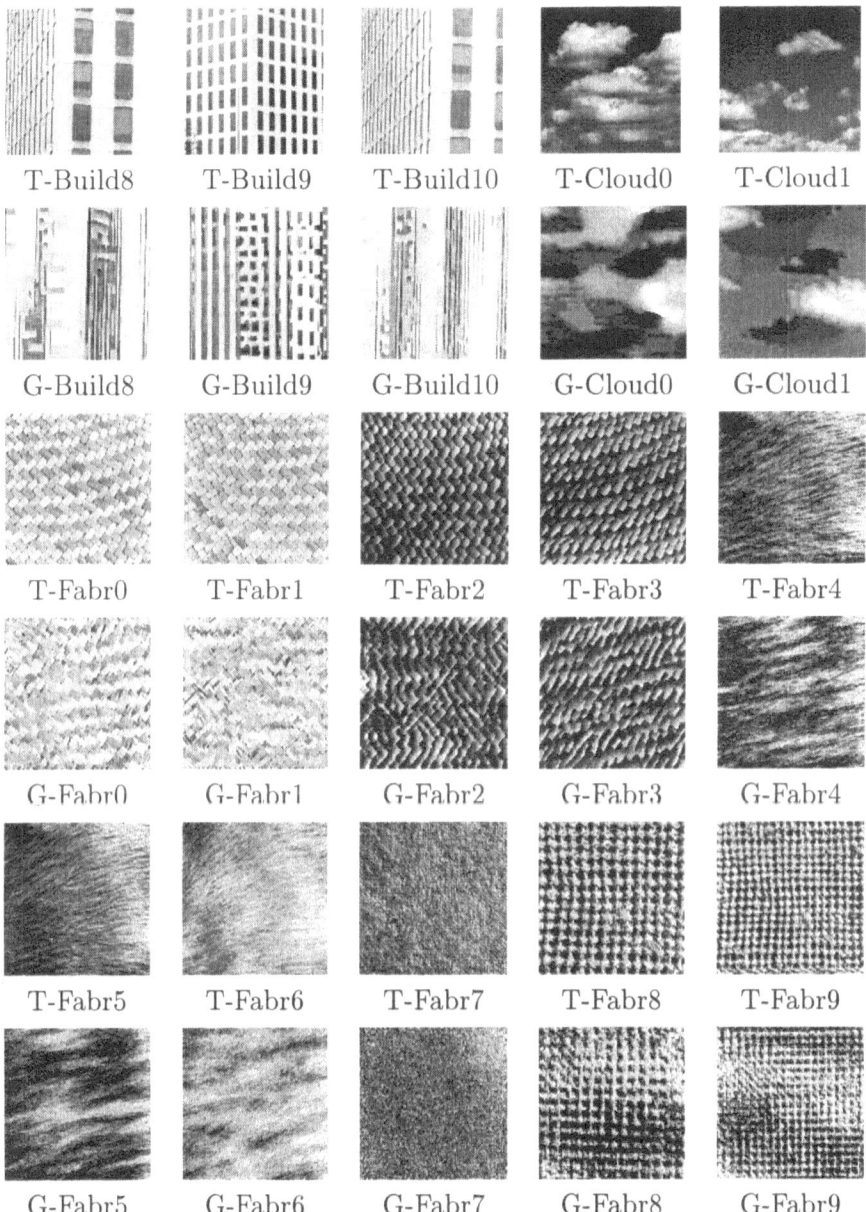

Figure 5.17.   Training (T) and simulated (G) 128 × 128 samples of different natural textures from MIT "VisTex" database.

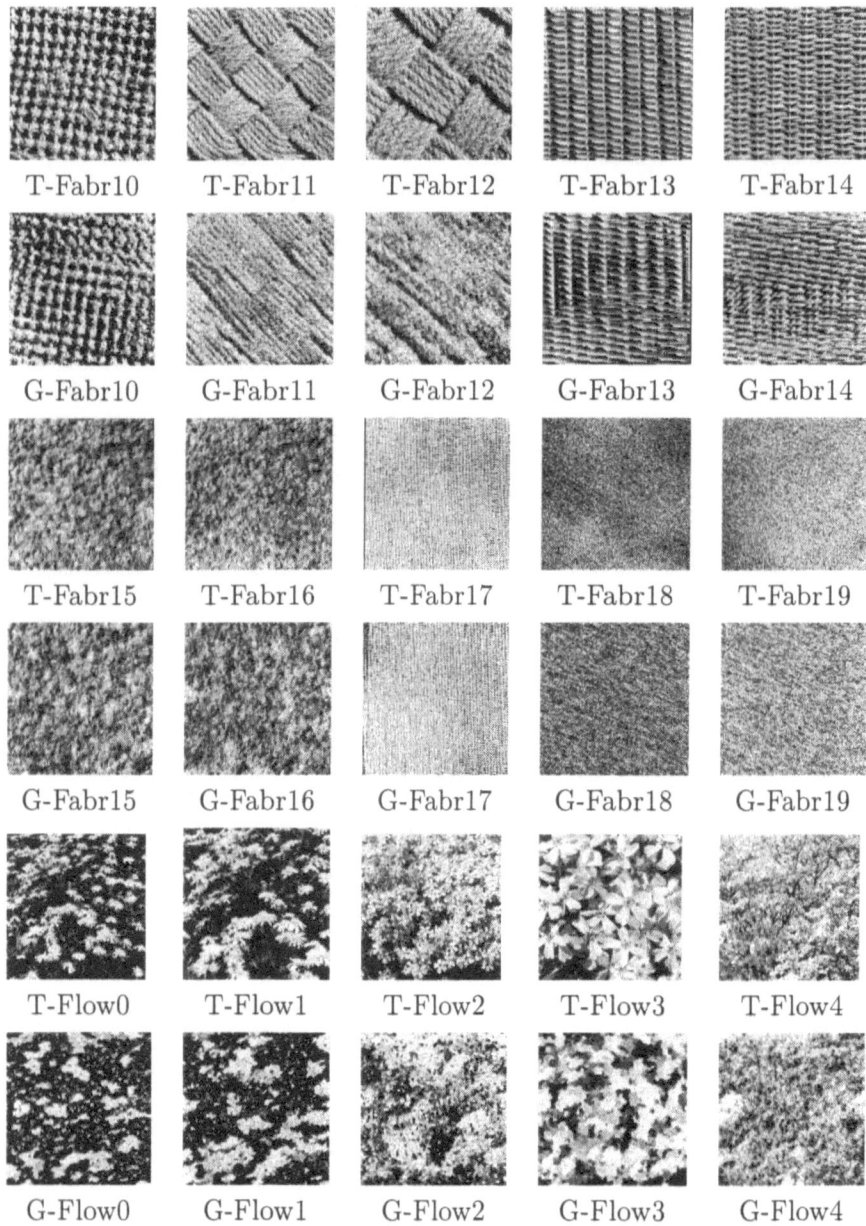

T-Fabr10	T-Fabr11	T-Fabr12	T-Fabr13	T-Fabr14
G-Fabr10	G-Fabr11	G-Fabr12	G-Fabr13	G-Fabr14
T-Fabr15	T-Fabr16	T-Fabr17	T-Fabr18	T-Fabr19
G-Fabr15	G-Fabr16	G-Fabr17	G-Fabr18	G-Fabr19
T-Flow0	T-Flow1	T-Flow2	T-Flow3	T-Flow4
G-Flow0	G-Flow1	G-Flow2	G-Flow3	G-Flow4

Figure 5.18. Training (T) and simulated (G) 128 × 128 samples of different natural textures from MIT "VisTex" database.

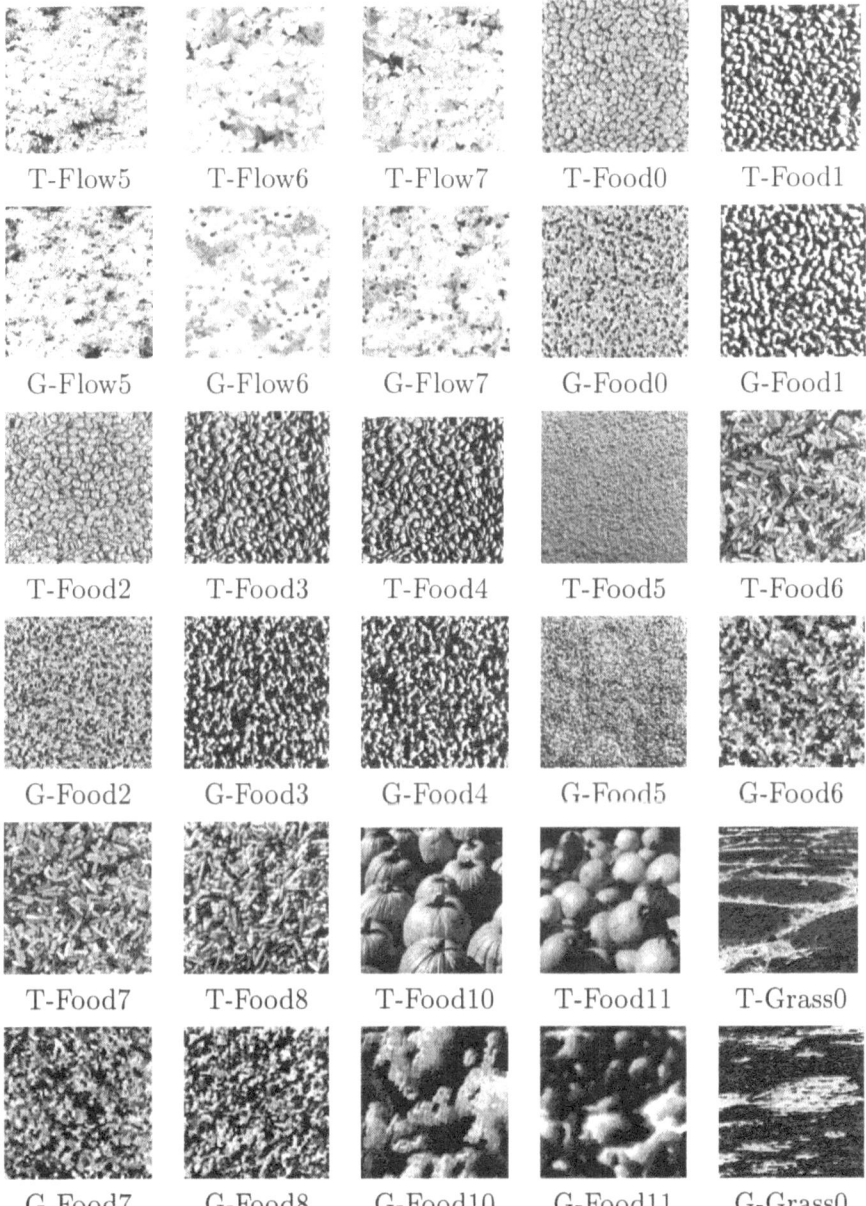

Figure 5.19.   Training (T) and simulated (G) 128 × 128 samples of different natural textures from MIT "VisTex" database.

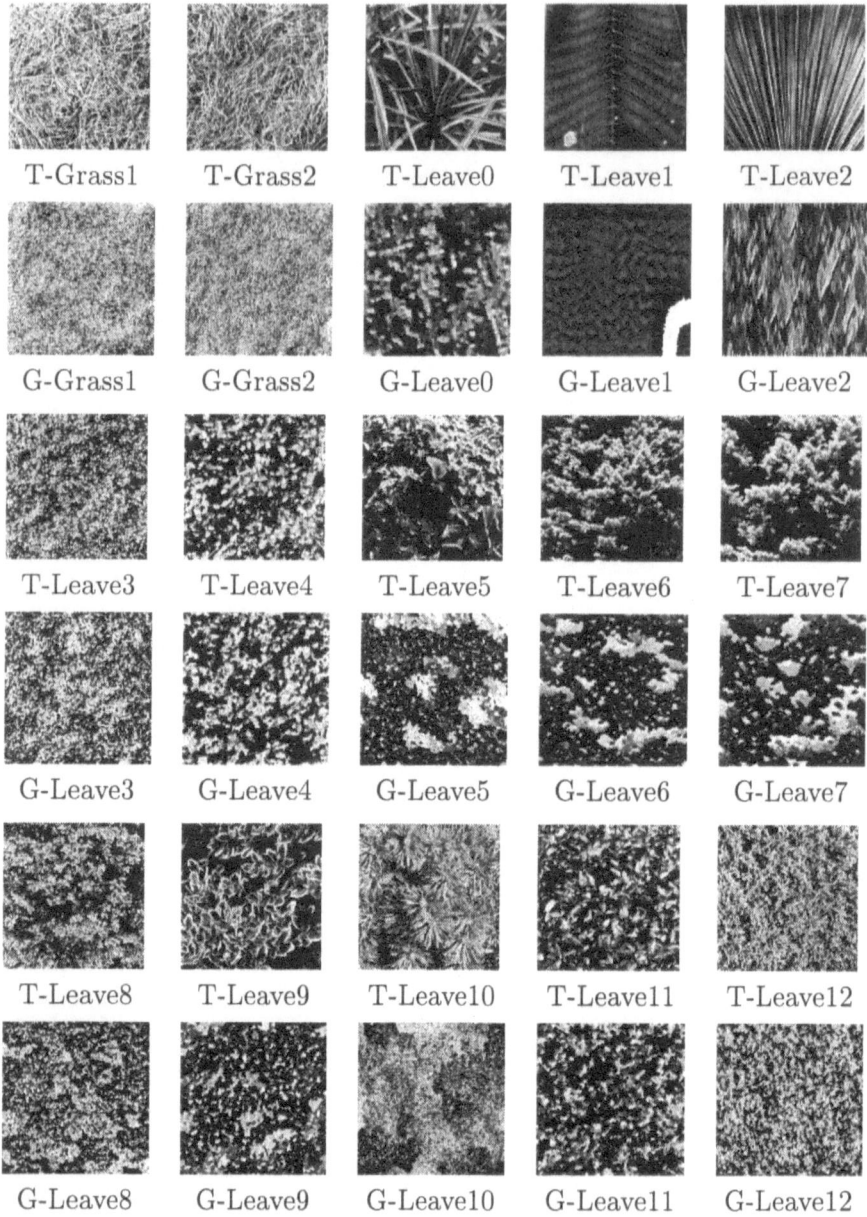

Figure 5.20.   Training (T) and simulated (G) 128 × 128 samples of different natural textures from MIT "VisTex" database.

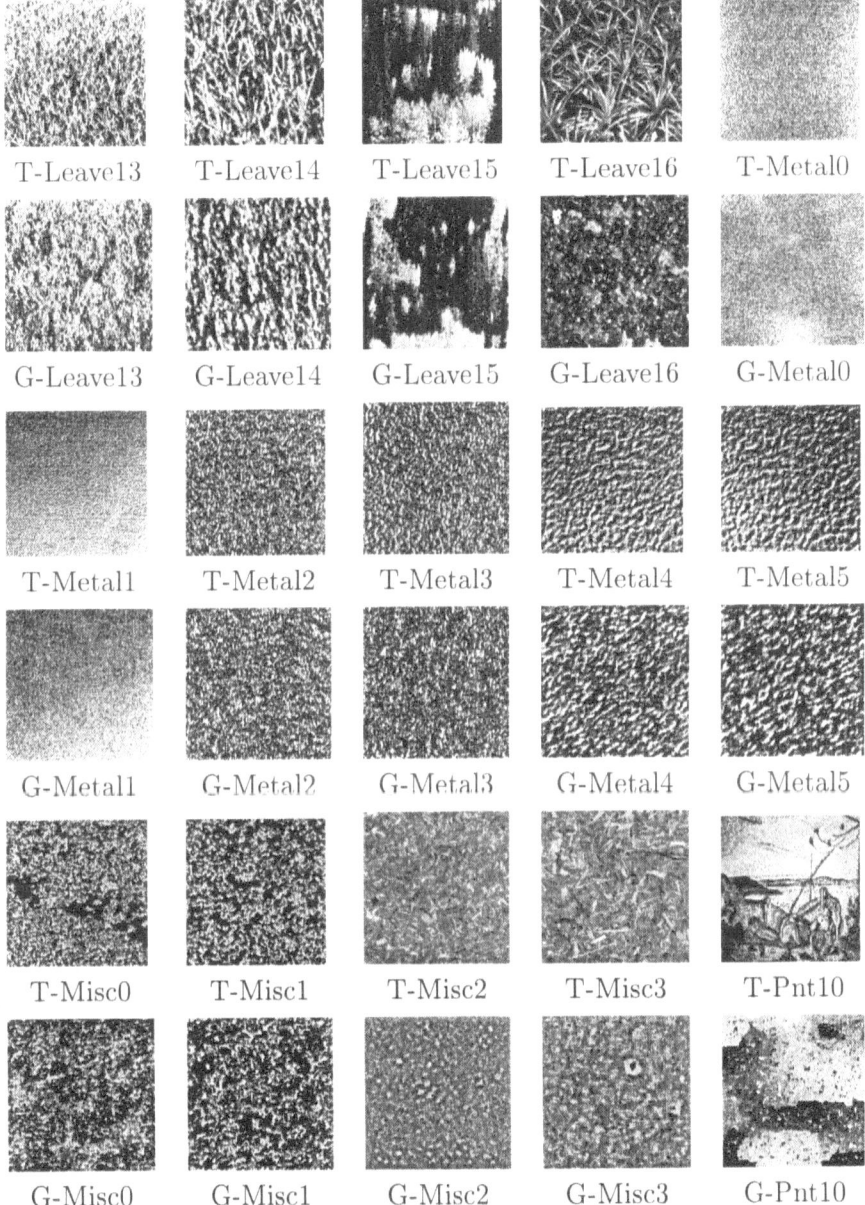

Figure 5.21. Training (T) and simulated (G) 128 × 128 samples of different natural textures from MIT "VisTex" database.

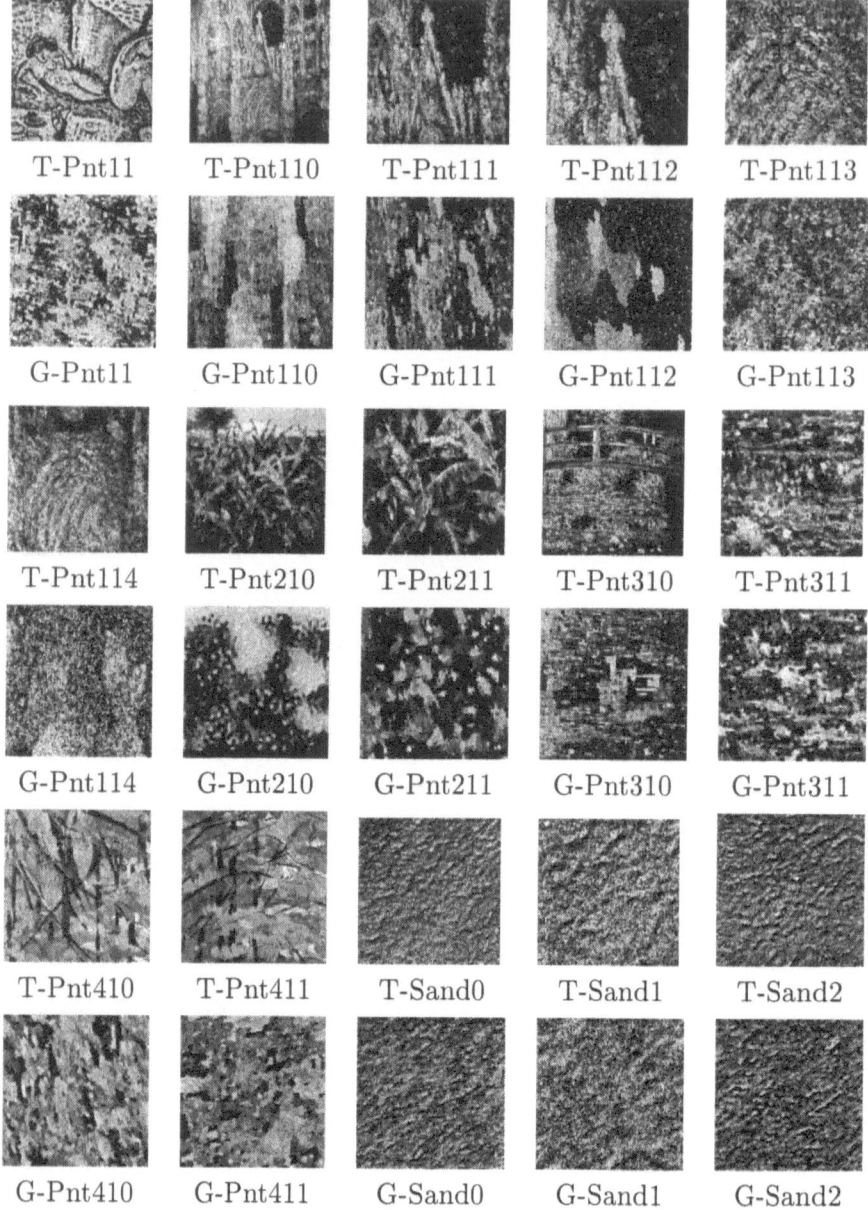

Figure 5.22.   Training (T) and simulated (G) 128 × 128 samples of different natural textures from MIT "VisTex" database.

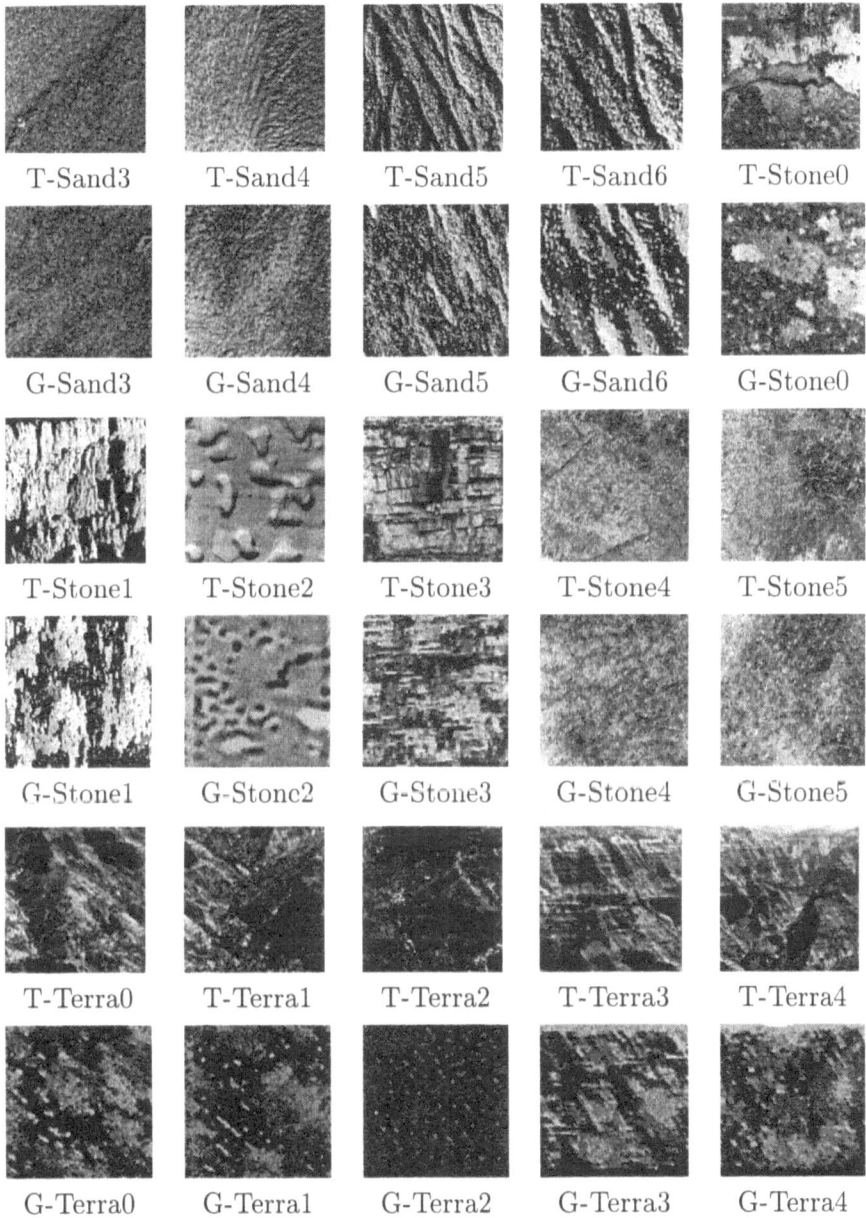

Figure 5.23.   Training (T) and simulated (G) 128 × 128 samples of different natural textures from MIT "VisTex" database.

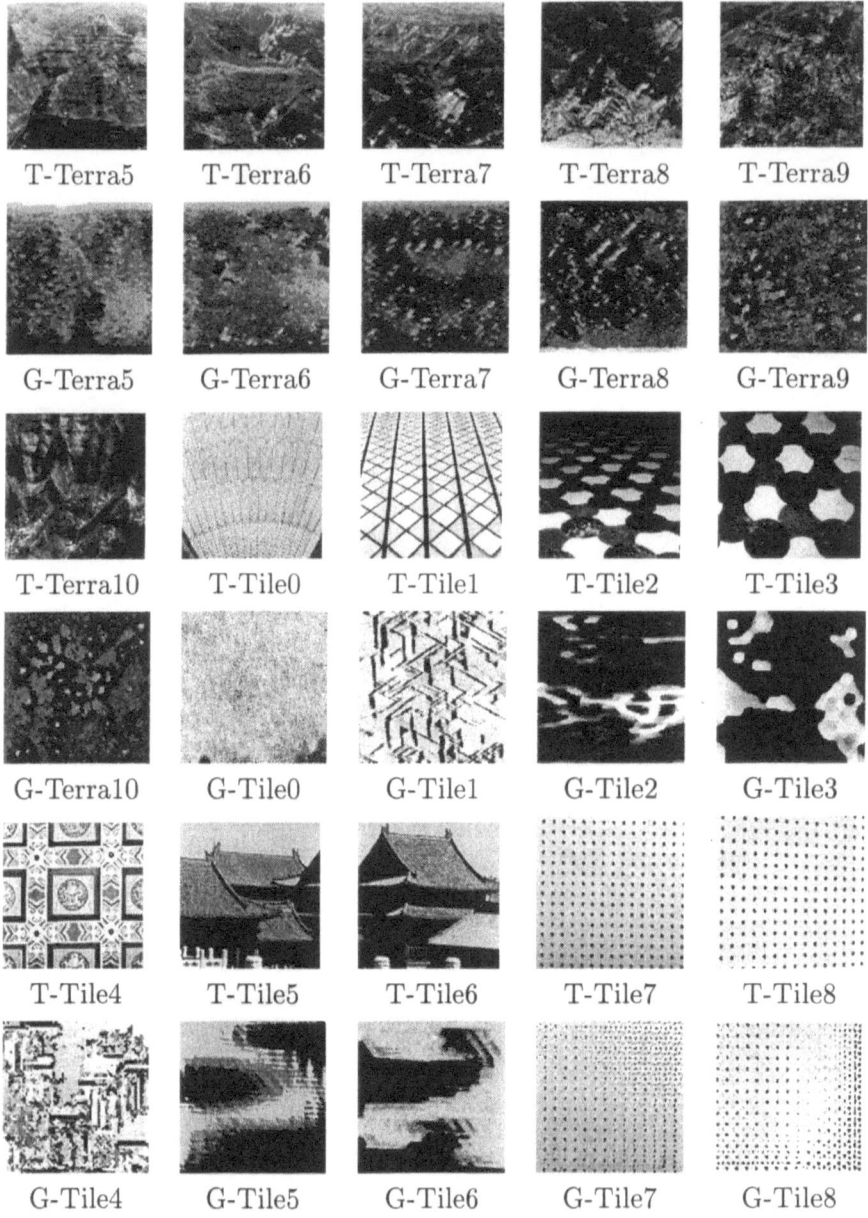

Figure 5.24. Training (T) and simulated (G) 128 × 128 samples of different natural textures from MIT "VisTex" database.

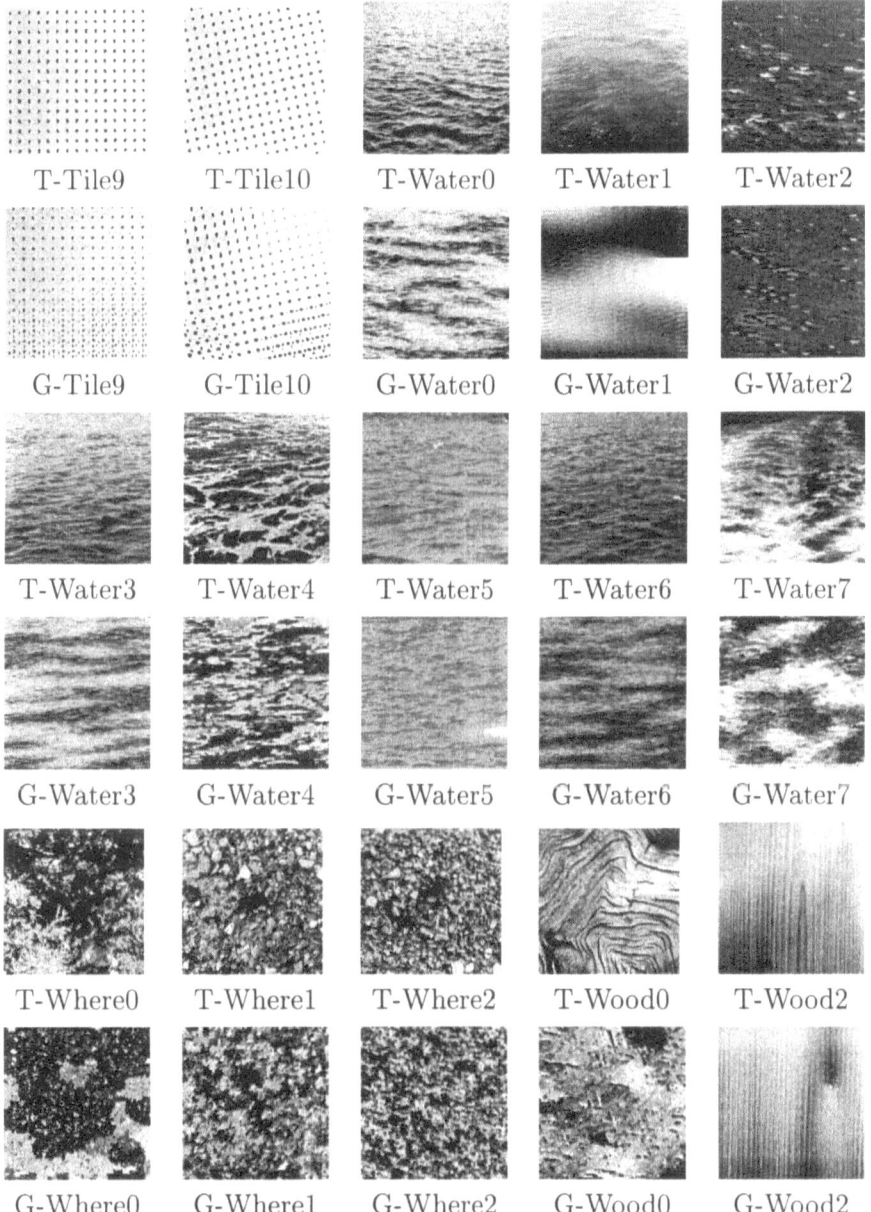

Figure 5.25. Training (T) and simulated (G) 128 × 128 samples of different natural textures from MIT "VisTex" database.

TABLE 5.3. Homogeneity of the 165 MIT "VisTex" textures (notation: **h**
– homogeneous; **w** – weakly homogeneous; **i** – inhomogeneous texture).

Name	Type: mark
Bark	0:i 1:w 2:w 3:i 4:i 5:i 6:w 7:w 8:w 9:i 10:w 11:w 12:i
Brick	0:i 1:i 2:i 3:i 4:i 5:w 6:i 7:i 8:i
Buildings	0:i 1:i 2:i 3:i 4:i 5:i 6:i 7:i 8:i 9:w 10:i
Clouds	0:w 1:w
Fabrics	0:w 1:w 2:h 3:h 4:h 5:w 6:w 7:h 8:h 9:h 10:h 11:i 12:i 13:w 14:w 15:h 16:h 17:h 18:h 19:h
Flowers	0:h 1:h 2:h 3:h 4:h 5:h 6:h 7:h
Food	0:w 1:h 2:w 3:h 4:h 5:h 6:w 7:w 8:h 10:i 11:w
Grass	0:w 1:i 2:i
Leaves	0:i 1:i 2:i 3:w 4:h 5:i 6:w 7:w 8:h 9:w 10:i 11:h 12:h 13:h 14:w 15:i 16:i
Metal	0:h 1:h 2:h 3:h 4:h 5:h
Miscellaneous	0:h 1:h 2:h 3:w
Paintings	10:i 11:i 110:i 111:i 112:i 113:w 114:w 210:i 211:i 310:w 311:w 410:w 411:w
Sand	0:h 1:h 2:h 3:h 4:h 5:w 6:w
Stone	0:w 1:w 2:i 3:w 4:h 5:h
Terrains	0:w 1:w 2:w 3:w 4:w 5:w 6:w 7:i 8:w 9:i 10:i
Tiles	0:w 1:i 2:i 3:i 4:i 5:i 6:i 7:w 8:w 9:h 10:h
Water	0h 1:i 2:i 3:h 4:h 5:h 6:w 7:w
WheresWaldo	0:i 1:h 2:h
Wood	0:i 2:h

range, Therefore, just as in Figure 5.5, the simulated images demonstrate continuous transitions between the initial and inverted patterns of the same texture learnt from the training sample.

Figure 5.27 shows the convergence of the CSA in terms of the above chi-square distance per clique family for the MIT "VisTex" texture Bark 0. This texture is not translation-invariant so that our modelling approximates it with translation-invariant textures that have very similar, in terms of the chi-square distance, GLDHs.

Both the modelling scenarios start from an IRF sample, and usually the chi-square distance for the alternative scenario obtained after a few hundred CSA macrosteps is about two orders smaller than for the traditional scenario that performs much more simulation macrosteps. Moreover, if the

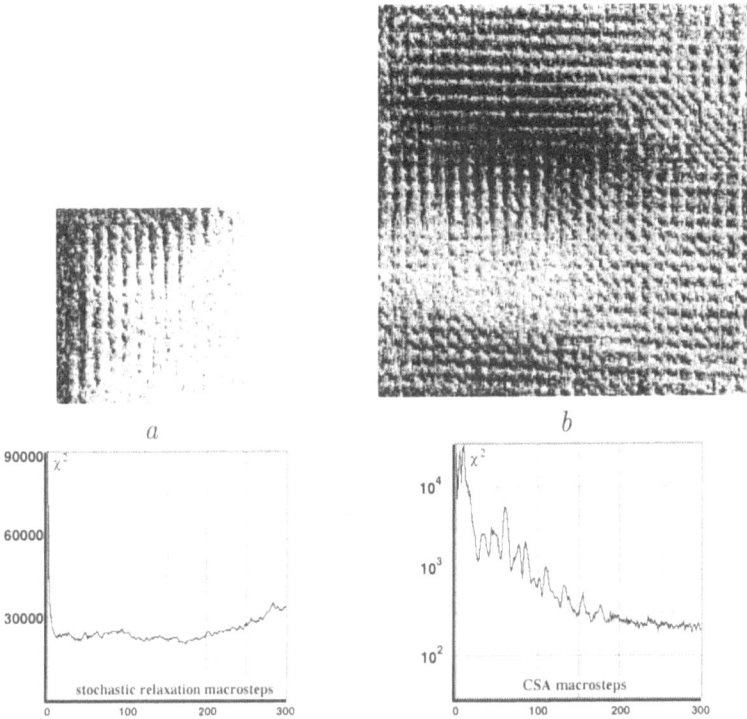

Figure 5.26.   Traditional and CSA modelling of the texture Fabrics 8 ("T-Fabr8" in Figure 5.17): (*a*) the sample 128 × 128 generated by the traditional modelling scenario (2,000 macrosteps of potential refinement and 300 macrosteps of stochastic relaxation) and (*b*) the sample 256 × 256 obtained by the CSA-based scenario (300 macrosteps).

potential refinement by stochastic approximation is terminated too early, then in many cases the traditional scenario demonstrates even a divergence of this distance after a relatively short converging stage.

That is why all the textures in this book are simulated by using only the CSA-based alternative scenario.

As shown in Figure 5.27, even inhomogeneous textures sometimes produce the model parameters that still reflect some global features of the training samples. In particular, such overall regularities are present in the simulated samples of tree barks, bricks, and buildings in Figures 5.15 – 5.17, samples Food 10 and Food 11 in Figure 5.19, and so forth. It is interesting to note that the simulated paintings Painting 10 – Painting 411 in Figures 5.21 and 5.22 bear some general resemblance to modern abstract artworks (more 256 × 256 examples are given in Figures 5.28 – 5.30). But we should not jump to a conclusion that Gibbs image models with multiple pairwise pixel interaction form the computational basis for the abstract art.

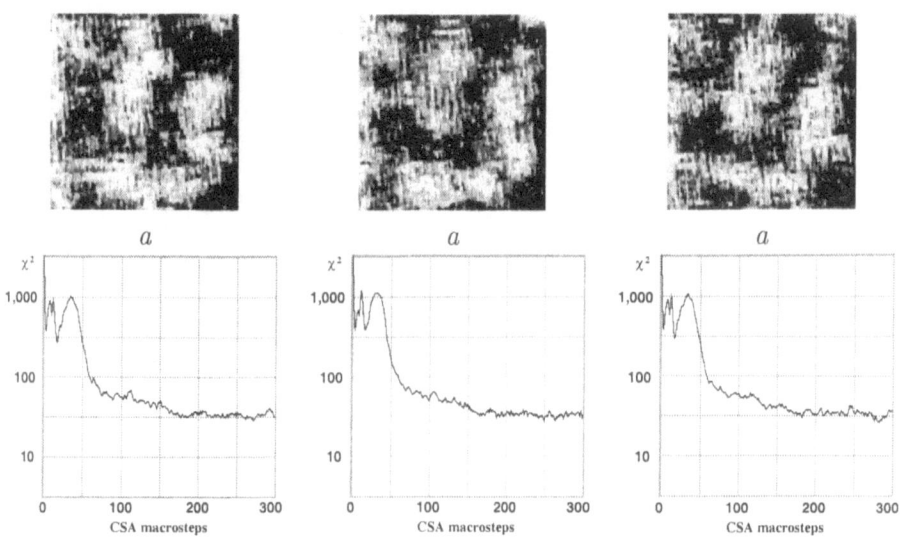

Figure 5.27. CSA modelling of the texture Bark 0 ("T-Bark0" in Figure 5.15): ($a1,b1,c1$) the samples 128×128 generated by 300 CSA macrosteps, ($a2,b2,c2$) the convergence plots for these samples.

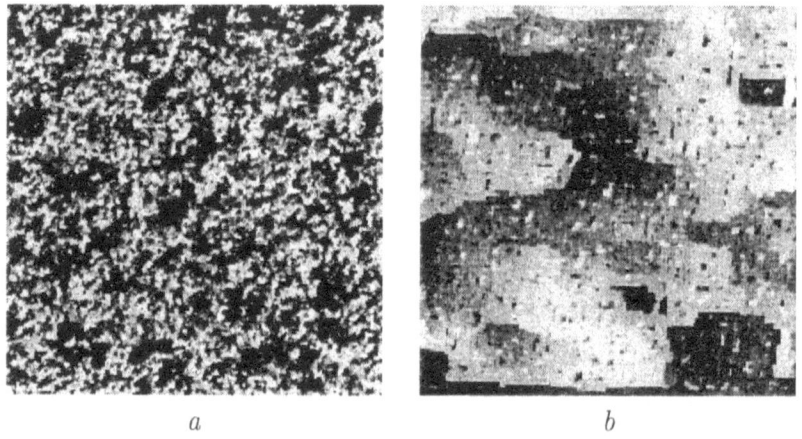

Figure 5.28. 256×256 samples simulated with the model parameters learnt from the homogeneous texture MIT "VisTex" Leaves 4 (a) and from the inhomogeneous texture Painting 10 (b).

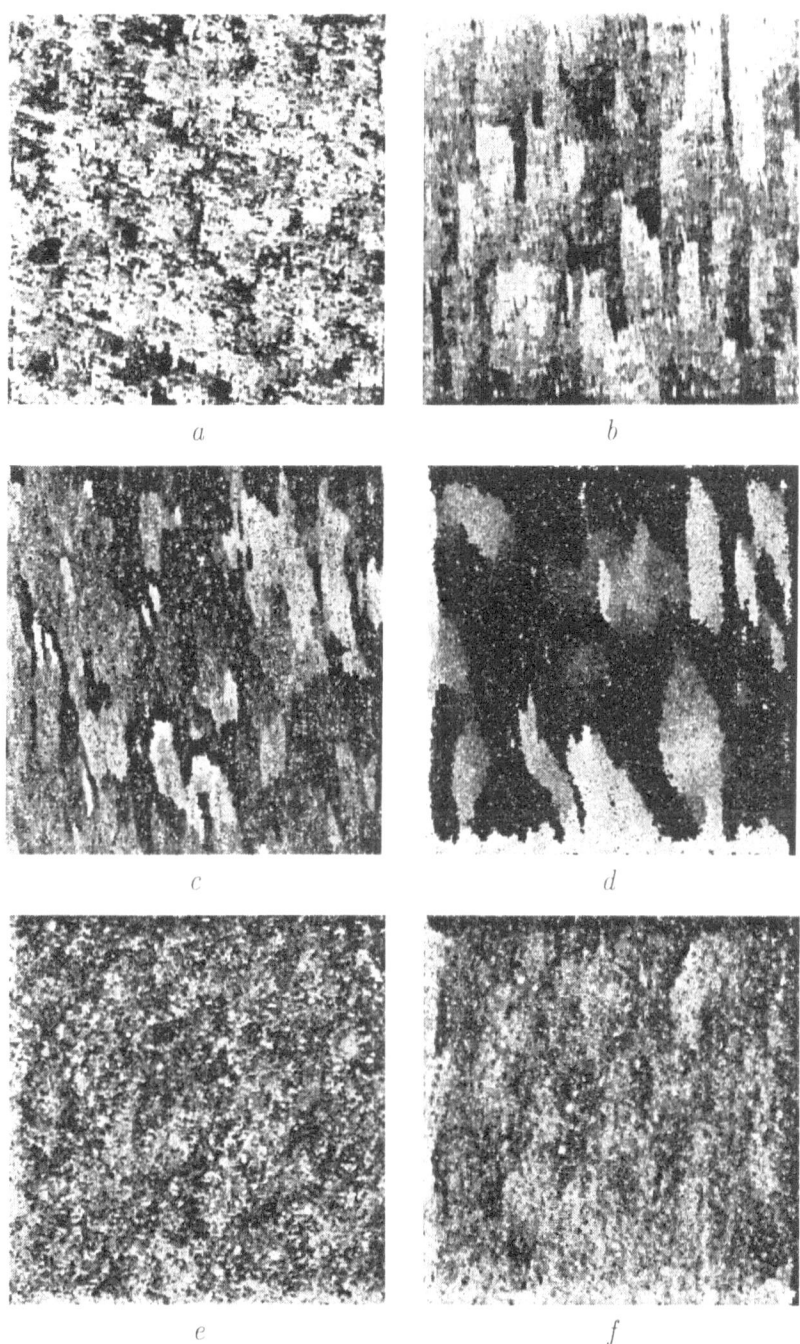

Figure 5.29.   256 × 256 samples simulated with the model parameters learnt from the in-homogeneous textures MIT "VisTex" Painting 11 (*a*), Painting 110 (*b*), Painting 111 (*c*), Painting 112 (*d*), Painting 113 (*e*), and Painting 114 (*f*).

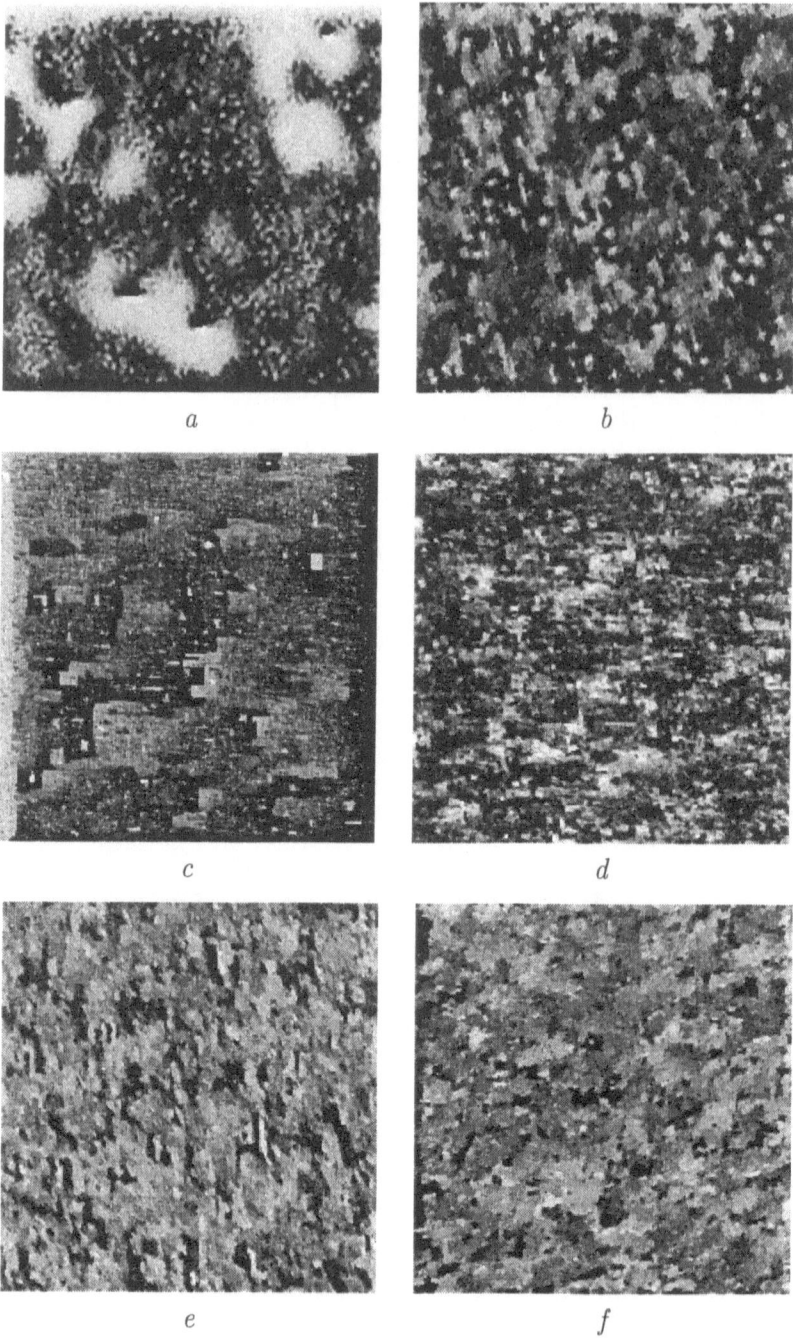

a

b

c

d

e

f

Figure 5.30.   256 × 256 samples simulated with the model parameters learnt from the in-homogeneous textures MIT "VisTex" Painting 210 (*a*), Painting 211 (*b*), Painting 310 (*c*), Painting 311 (*d*), Painting 410 (*e*), and Painting 411 (*f*).

CHAPTER 6

Experiments in Retrieving Natural Textures

In this chapter we introduce a particular measure of similarity between two textures and use it for a query-by-image retrieval of image textures. The measure is deduced from the Gibbs models of Eqs. (2.3) and (2.10) and is based on the model parameters that are learnt as described in Chapters 3 and 4. Most of these results were initially presented in (Jain and Gimel'farb, 1995; Gimel'farb and Jain, 1996).

When comparing a query image to a given image data base (IDB), the learnt interaction structure permits us to first match the query image to the IDB entries with due account of the admissible changes in its relative orientation and scale. Then, the total similarity between the two images is computed using the GLCHs or GLDHs, these being the sufficient statistics for the Gibbs models under consideration.

We discuss results in retrieving images from two experimental IDBs and outline the computational complexity of the proposed retrieval scheme. The first IDB, called below IDB-1, contains 30 fragments 170×170 of spatially homogeneous and weakly homogeneous digitized textures from (Brodatz, 1966), and we use 8 other non-overlapping fragments of each texture as the query images. The second IDB, called IDB-2, contains 120 fragments 64×64 of homogeneous, weakly homogeneous, and inhomogeneous textured images taken from the MIT Media Laboratory "VisTex" 128×128 collection of Pickard et al. (1995), and 3 other fragments of each texture are used as the query images.

Our approach presumes that the regions of both the queries and IDB images to be matched contain stochastic or, at least, weakly stochastic textures that are described by their characteristic interaction structures of multiple pairwise pixel interactions. The interaction structure of different subimages of the same texture is assumed to vary in orientation and scale to within given limits. The set of images having, within a given range of possible orientations and scales, high probabilities under the GPD with the learnt parameters is considered to belong to the same texture type as a given training sample used for learning.

6.1. Query-by-image texture retrieval

Query-by image retrieval from a large IDB has an obvious practical value in modern information storage and retrieval systems (Faloutsos et al., 1994; Holt and Hartwick, 1994; Flickner et al., 1995). The retrieval problems are rather simple for human vision, but very difficult for computational vision. As indicated in Chapter 1, this difficulty arises because we need to formally define the meaning of fuzzy terms such as texture, texture homogeneity, image similarity, and so forth which a human learns mostly by visual examples.

Human vision matches and retrieves textures quite easily at a subconscious level by using predominantly the "pattern-based" half of the brain (Marr, 1982). In computational vision, we try to model and implement the vision processes at the conscious level by transferring them to the other, "deduction-based", half of the brain. We will not go into the details of the psychology of human vision. But the long history of computational vision and image processing shows that almost any problem that is easy for human vision is, in reality, a very complex problem for computational vision. On the other hand, vision problems that are hard for human vision, for instance, an exact matching of or counting dots in random-dot pictures, are simple to implement in the computational framework.

In the paradigm of Marr (1982) we need to formulate a particular "pattern-based" vision problem as a "deduction-based" information processing task and deduce (if possible) its algorithmic solution. At the same time, the resulting computational models should also be of some practical value. Generally, these requirements contradict each other since the abstract computational models often do not effectively capture the entire reality of image formation and description. A vast majority of the known approaches of this type are rather successful in the first step of formulating the problem, but fail in the second one of solving it in a practicable way (our own experience is not an exception). In spite of this, the practical necessity of having computational algorithms whose performances are similar to that of human vision justifies the continuous attempts along this approach.

It is very difficult (if not impossible) to describe a variety of natural and artificial image textures by a single computational model. As a result, a number of different approaches have been used to model particular types of textures, to compute quantitative features associated with perceived textural properties, and to solve different application problems by using these models and features. These approaches implement almost all known image processing and analysis techniques, in particular, structural methods based on primitive elements and formal rules of their spatial placements, probabilistic methods investigating the textured images as samples of certain random fields, and spectral methods revealing specific features of the spatial frequency content in the image signals (Tuceryan and Jain, 1993).

In this chapter we use the probabilistic approach to state and solve the query-by-image texture retrieval problems. This approach can be implemented at least for a limited number of the artificial and natural grayscale image textures which belong to stochastic textures. The retrieval problem can be posed, in particular, in the following two ways.

(i) Retrieve all the IDB images containing regions with the same type of texture as in a given query image. This is a pattern recognition problem because we need to compute the similarity between the given image and the entries in the IDB and retrieve the most similar entry or a fixed number of the top-rank entries.

(ii) Different regions in the query image have to be labelled according to the IDB content. This is a more complex problem similar to segmentation problems discussed in Chapter 7.

We restrict our consideration to only the first problem of directly matching the query and IDB textures.

To simplify the problem, we assume that the IDB entries contain only stochastic or weakly stochastic textures. In this case any reasonably sized subimage of the IDB entry is of the same texture type as the whole image. Of course, in practice the purely stochastic textures are rarely observed. So, we restrict ourselves to those real images that can be represented in the IDB by specifying one or several homogeneous textured regions and to only homogeneous queries. The IDB images are retrieved by matching their prespecified regions to the query image. But we assume that the query image can differ, within certain limits of relative scales and orientations, from the IDB entries of the same texture type.

6.2. Similarity under scale and orientation variations

Gibbs models of Eqs. (2.3) and (2.10) describe a stochastic texture by a particular interaction structure and a subset of the GLCHs or GLDHs which are collected over each clique family and form the sufficient statistics of a model. This permits us to easily specify the concepts of self-similarity of a texture or of similarity between two textures by using one or another distance between these histograms. Also, the histograms as sufficient statistics allow to quantitatively define some basic features of image textures to be matched, for instance, the minimum size of the image patch that can be considered as a homogeneous texture.

6.2.1. SIZE OF A TEXTURE

If a given texture is spatially homogeneous then marginal probabilities of signal combinations in the cliques are assumed to be translationally invariant. In the case of the Gibbs image models of Eqs. (2.3) or (2.10), the marginal probabilities of gray level co-occurrences or differences are approximated by the sample relative frequency distributions obtained from the GLCHs or GLDHs for a given normalized training sample $\mathbf{g}^{\circ,\mathrm{rf}}$. Therefore the minimum size of an image representing a particular texture is dictated by a desired absolute or relative precision of such approximation.

Actually, the gray level co-occurrences or differences in the different cliques of the same family are statistically dependent. Although these dependencies are not clearly understood, they are typically rather weak. We will exploit the Chebyshev's inequality to estimate, with a large margin, the minimum size of an image that permits us to collect the GLCHs or GLDHs of the desired size.

The histogram size $|\mathbf{C}_a|$ which yields the maximum absolute error ε of the probability estimates with a confidence level $1 - \alpha$ is as follows:

$$|\mathbf{C}_a| \geq \frac{0.25}{\alpha \cdot \varepsilon^2}.$$

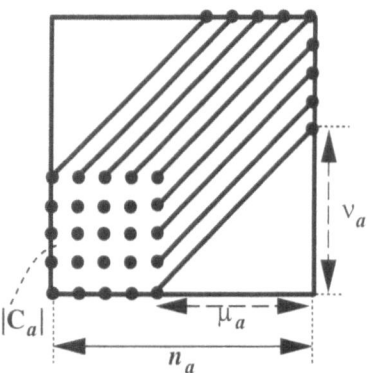

Figure 6.1. Clique family in a search window.

Let cliques $(i, j) \in \mathbf{C}_a$ occupy a square region of size $n_a \times n_a$ shown in Figure 6.1 where $n_a = \max\{\mu_a, \nu_a\} + \sqrt{|\mathbf{C}_a|}$. Then the upper bound for the minimal linear size n of a training sample is of the form:

$$n = \max\{\mu_{\max}, \nu_{\max}\} + \frac{0.5}{\varepsilon \cdot \sqrt{\alpha}}.$$

Table 6.1 presents examples of the minimal linear size n for the five confidence levels and five error values. In these examples, the longest-range pixel interaction is assumed to be either $\mu_{max} = \nu_{max} = 4$ or 40 so that the search window $\mathbf{W}$ for creating the interaction map and recovering the interaction structure is 9×9 or 81×81.

TABLE 6.1. Upper bounds for the minimum linear size of a square image, given the search window to build the interaction map.

Error ε	Confidence level $1 - \alpha$				
	0.85	0.90	0.95	0.99	0.995
Search window of size 9×9					
$\varepsilon = 0.25$	15	16	18	29	38
$\varepsilon = 0.10$	22	25	32	59	80
$\varepsilon = 0.05$	35	41	54	109	151
$\varepsilon = 0.01$	139	168	233	509	717
$\varepsilon = 0.005$	268	326	457	1009	1424
Search window of size 81×81					
$\varepsilon = 0.25$	46	47	49	60	69
$\varepsilon = 0.10$	53	56	63	90	111
$\varepsilon = 0.05$	66	72	85	140	182
$\varepsilon = 0.01$	170	199	264	540	748
$\varepsilon = 0.005$	299	357	488	1040	1455

These examples show that the texture patches must be larger than $25 \times 25 - 56 \times 56$ to estimate the marginal probabilities with the precision 90% anf confidence level 90%. To obtain the precision 95% and confidence 95%, the images should be larger than $54 \times 54 - 85 \times 85$. Therefore to empirically test the self-similarity of a texture or compare it to other textures in terms of the Gibbs models at hand, the texture has to be represented by a sufficiently large image patch. This casts some doubt on the validity of using small patches of size 32×32 or less pixels to learn the Gibbs model parameters or relevant integral features of the natural textures.

6.2.2. TEXTURE SIMILARITY MEASURE

Gibbs models in Eqs. (2.3) or (2.10) are based on the natural assumptions of the spatial uniformity (or translation invariance) of each texture and of the invariance to the gray range changes. Under these assumptions, a selected subset of the GLCHs or GLDHs for a given training sample or, what is the same, the learnt clique families $\mathbf{C} = [\mathbf{C}_a : a \in \mathbf{A}]$ and the corresponding potential values $\mathbf{V} = [\mathbf{V}_a : a \in \mathbf{A}]$ defining the characteristic interaction structure and strengths, respectively, can quantitatively describe this particular type of textures in the IDB.

We will measure the spatial self-similarity between different parts of a grayscale stochastic texture or the similarity between two stochastic textures as the similarity between the GLCHs or GLDHs collected for these images. This similarity measure has to take into account a given set $\boldsymbol{\Xi}$ of admissible variations in image scales and orientations or, more generally, of projective transformations of these images.

The total distance

$$D(\mathbf{g}_1, \mathbf{g}_2) \;=\; \sum_{a \in \mathbf{A}} \chi_a^2(\mathbf{g}_1, \mathbf{g}_2) \tag{6.1}$$

represents a natural quantitative measure of dissimilarity between the sample relative frequency distributions $\mathbf{F}(\mathbf{g}) = [\mathbf{F}_a(\mathbf{g}) : a \in \mathbf{A}]$ for the reference images $\mathbf{g}_1$ and $\mathbf{g}_2$. Here, $\chi_a^2(\mathbf{g}_1, \mathbf{g}_2)$ is the chi-square distance[1] between the sample relative frequency distributions for the clique family $\mathbf{C}_a$. As was indicated in Chapter 2, the sample relative frequency distributions are obtained by normalizing the GLCHs or GLDHs for the corresponding clique families.

Under the admissible translations, the corresponding clique families have the same inter-pixel shifts $[\mu_a, \nu_a]$. To take into account the admissible scale and orientation variations between the two images, we can minimize the distance in Eq. (6.1) over a given set $\boldsymbol{\Xi}$ of such relative geometric transformations. Experiments in Chapter 5 showed that each change in scale and orientation of a texture pattern results in a similar scale–orientation change in the interaction map or more specifically, in the inter-pixel shifts that define the corresponding clique families.

Let λ denote a scale factor. Let ϕ be an orientation angle. Let $\boldsymbol{\Xi}$ be a set of the rotation angles and scales to be exhausted in computing the dissimilarity measure. Every relative transformation $\xi \equiv [\lambda, \phi] \in \boldsymbol{\Xi}$ of a texture with respect to another texture makes an impact on its interaction map. This impact can be approximately described by a simple rearrangement of the initial GLCHs or GLDHs $\mathbf{H}(\mathbf{g}) = [\mathbf{H}_a(\mathbf{g}); a \in \mathbf{A}]$, in line with

---

[1] The chi-square distance between the sample relative frequency distributions of signal combinations $\mathbf{u} \in \mathbf{U}$ in the cliques of the family $\mathbf{C}_a$ has the following form (Lloyd, 1984)

$$\chi^2\left(\mathbf{F}_a(\mathbf{s}_1), \mathbf{F}_a(\mathbf{s}_2)\right) = \sum_{u \in \mathbf{U}} \frac{\left(F_a(\mathbf{u}|\mathbf{s}_1) - F_a(\mathbf{u}|\mathbf{s}_2)\right)^2}{F_a(\mathbf{u}|\mathbf{s}_1)}.$$

the transformed inter-pixel shifts $[\mu_{a(\xi)}, \nu_{a(\xi)}]$. Here,

$$
\begin{aligned}
\mu_{a(\xi)} &= \text{Int} \{0.5 + \lambda \left(\mu_\xi \cos \phi + \nu_\xi \sin \phi\right)\} ; \\
\nu_{a(\xi)} &= \text{Int} \{0.5 + \lambda \left(-\mu_\xi \sin \phi + \nu_\xi \cos \phi\right)\} ,
\end{aligned}
\tag{6.2}
$$

where $\text{Int}\{z\}$ denotes the integer part of a real number z. It is evident that the longer the interaction range, the higher the precision of such a transformation.

To obtain more precise transformation of an interaction map, one may use the extended GLCHs or GLDHs proposed by Chetverikov (1994). In this case the integer and fractional parts of the transformed inter-pixel shifts take part in the bi-linear interpolation of the initial histograms, and the transformed interaction map is built by using the interpolated histograms.

In our Gibbs models each clique family which is absent from the interaction structure has, by definition, the zero-valued potential $V_a = 0$, and its GLCH or GLDH is replaced by the similar histogram for the IRF. This permits us to compare first the chosen interaction structures under the admissible geometric transformations of Eq. (6.2). Most dissimilar image patches can be rejected by only checking their structural similarity given by a relative number of the matching characteristic clique families with respect to their total number in both the images. Then we exploit the total dissimilarity measure of Eq. (6.1) under the admissible transformations Ξ:

$$
D(\mathbf{g}_1, \mathbf{g}_2) = \min_{\xi \in \Xi} \sum_{a \in \mathbf{A}_{\text{comb}}(\xi)} \chi^2 \left(\mathbf{F}_a(\mathbf{g}_1), \mathbf{F}_{a(\xi)}(\mathbf{g}_2)\right)
\tag{6.3}
$$

to rank the structurally admissible IDB images and retrieve those images which possess simultaneously the high structural similarity and low dissimilarity between the corresponding signal histograms. Here, $\mathbf{A}_{\text{comb}}(\xi)$ denotes the combined set

$$
\mathbf{A}_{\text{comb}}(\xi) = \mathbf{A}_{\mathbf{g}_1} \bigcup \mathbf{A}_{\mathbf{g}_2}(\xi)
$$

of the clique families for the rotated and scaled query image $\mathbf{g}_2$ and particular IDB image $\mathbf{g}_1$.

Below we use the dissimilarity measure of Eq. (6.3) for a query-by-image texture retrieval provided that each query image is sufficiently large to estimate its interaction structure and corresponding marginal probabilities from the collected GLCHs or GLDHs.

6.3. Matching two textures

We use the dissimilarity measure in Eq.(6.3) to quantitatively specify the correspondence between a given texture and different texture types stored in the IDB and solve the image retrieval problem. The IDB entries are represented by the characteristic clique families and associated GLCHs or GLDHs.

The interaction structure estimated from the model-based interaction map as was shown in Chapter 3 serves as a low-level textural feature to discriminate between the different texture types. Then the total distance of Eq. (6.3) is used to obtain the final ranking of the IDB by the similarity to the query image.

If the clique family is present only in one interaction structure, then we replace the sample relative frequency distribution for the corresponding clique family absent in the other interaction structure by the uniform marginal probability distribution for gray level co-occurrences or by the triangle marginal probability distribution $\mathbf{M}_{irf}$ for gray level differences. So, the clique families which are absent in both Gibbs models need not be taken into account because for them the corresponding chi-square distance in Eq. (6.1) and (6.3) is equal to zero.

The dissimilarity measure in Eq. (6.3) takes account of both the structure and the strengths of the pairwise pixel interactions because the GLCHs or GLDHs for the images also specify the first analytic approximations of the MLEs of the potentials $\mathbf{V}$ for each clique family. The image retrieval algorithm based on this dissimilarity measure is as follows:

(i) Input a query image $\mathbf{g}$.

(ii) Compute all the GLCHs or GLDHs for the query image in a given search window $\mathbf{W}$ and estimate the interaction structure for this image, for instance, by thresholding the relative partial energies as is specified in Eqs. (3.15) and (3.16).

(iii) Compute the dissimilarity measures $D(\mathbf{g}_t, \mathbf{g})$ of Eq.(6.3) between the query image $\mathbf{g}$ and images $[\mathbf{g}_t : t = 1, 2, \ldots]$ in the IDB using a given range Ξ of possible rotations and scales of the interaction structure of the query image. In so doing, the GLCHs or GLDHs and the associated interaction structures for the images in the IDB can be computed off-line.

(iv) Retrieve from the IDB the image with the smallest dissimilarity $D(\ldots)$ as the first choice; additional images can also be retrieved based on the ranked order of $D(\ldots)$ values.

The size of the search window $\mathbf{W}$, threshold to learn the characteristic clique families, and range Ξ of possible relative rotations and scales are the only user-specified parameters of the algorithm. The larger the window, the more precise the estimation of the interaction structure of spatially homogeneous textures with prevailing long-range interactions such as the Brodatz's texture D3 in Figure 3.1). However, a larger window size also means slower retrieval. In practice, the window sizes are usually chosen empirically but this problem also needs a theoretical analysis as was discussed in the previous section.

6.4. Experiments with natural textures

The retrieval abilities of the proposed measure in Eq. (6.3) are investigated by using two experimental IDBs. The first one, IDB-1, contains 30 homogeneous and weakly homogeneous natural textures of size 170×170 pixels, 8 bits/pixel, selected from the images of Brodatz (1966) that have already been used for our simulation experiments in Chapter 5. The second one, IDB-2, contains 120 images (64×64 pixels, 8 bits/pixel) of homogeneous, weakly homogeneous, and inhomogeneous textured images selected from the images of the collection "VisTex" of the MIT Media Laboratory (Pickard et al., 1995)) that have been used in Chapter 5, too.

In our retrieval experiments, we use the search window 101×101 for learning the interaction structures of the images in IDB-1 and 41×41 for the images in IDB-2. Therefore the longest (city block) interaction range is $\mu_{\max} = \nu_{\max} = 50$ in the first case and 20 in the second case.

All the 30 samples stored in the IDB-1 are taken from the 30 initial digitized and normalized 512×512 textures (one sample per texture type). The 8 other non-overlapping fragments of every such texture are used as the query images: in total, $8 \times 30 = 240$ different 170×170 queries to IDB-1. The 120 samples stored in IDB-2 are taken from the 120 initial 128×128 textures (one sample per texture type). The 3 other non-overlapping fragments of each such texture are used as the queries ($3 \times 120 = 360$ different 64×64 queries to IDB-2).

6.4.1. IMAGE DATA BASE IDB-1

The textures in IDB-1 are shown in Figures 6.2 and 6.3 (the full names are listed in Table 5.1). The learnt clique families are given in Figures 6.4 and 6.5. Jusat as in Figures 3.12 – 3.14, the interaction structures are shown with only the the coordinate axes (μ_a, ν_a) pointing out the origin $(0, 0)$. A few examples of the initial 512×512 images of the Brodatz textures showing the arrangement of the IDB sample and 8 query samples are presented in Figures 6.6–6.11 where the first choices of the retrieval obtained for each query are also indicated.

Experiments with these images as well as similar experiments with the same textures but using 128×128 IDB samples and 15 non-overlapping 128×128 queries per texture type confirm the feasibility of texture retrieval by image query based on the dissimilarity measure of Eq. (6.3).

Tables 6.2 and 6.3 summarize retrieval results obtained by exhausting different rotation angles in the range $(-18°, 18°)$ and different relative scales

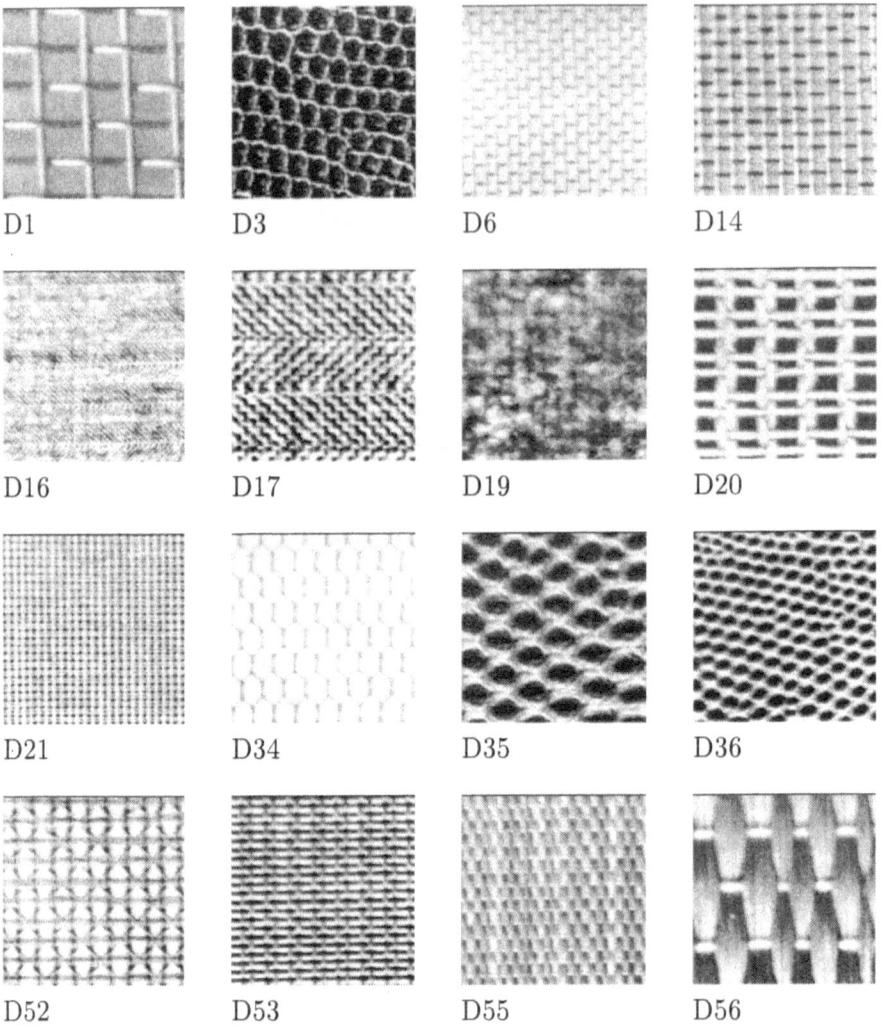

D1   D3   D6   D14

D16   D17   D19   D20

D21   D34   D35   D36

D52   D53   D55   D56

Figure 6.2.   Experimental image data base IDB-1 (30 samples 170 × 170 of digitized textures), see also Figure 6.3.

of the query interaction structure in the range (0.8, 1.25). For the angle step of 9° and scale step of 0.05 in Table 6.2, the first choice was mostly the right one, namely, 88.4% of the correct retrievals among the first choices only and 95.5% of them among the first and the second choices. If closely related texture pairs such as D3 (Reptile skin) and D36 (Lizard skin), D94 (Brick wall) and D95 (Brick wall), and D79 (Grass fiber cloth) and D105 (Cheesecloth) are placed into the same texture types, the retrieval results

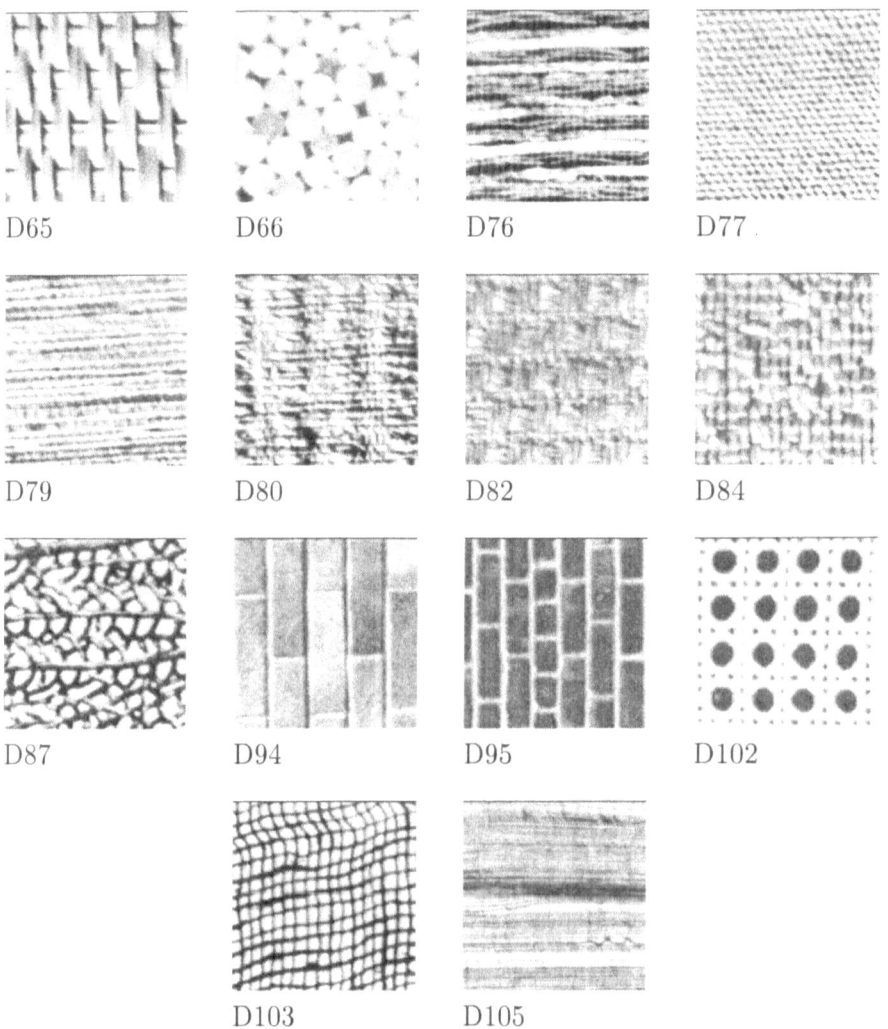

Figure 6.3.   Experimental image data base IDB-1 (30 samples 170 × 170 of digitized textures), see also Figure 6.2.

improve as follows: 97.1% for the correct first choice and 98.8% for the correct first and the second choices. The angle step of 4.5° and scale step of 0.025 in Table 6.3 gave very similar results: 88.8% of the correct first choices only and 96.7% of the correct retrievals among the first and second choices that are improved in the same way to 96.3% and 97.9%, respectively.

It should be noted that visually the two texture types D3 and D36 in Figures 6.2, 6.6, and 6.8 and D94 and D95 in Figure 6.3 are really of

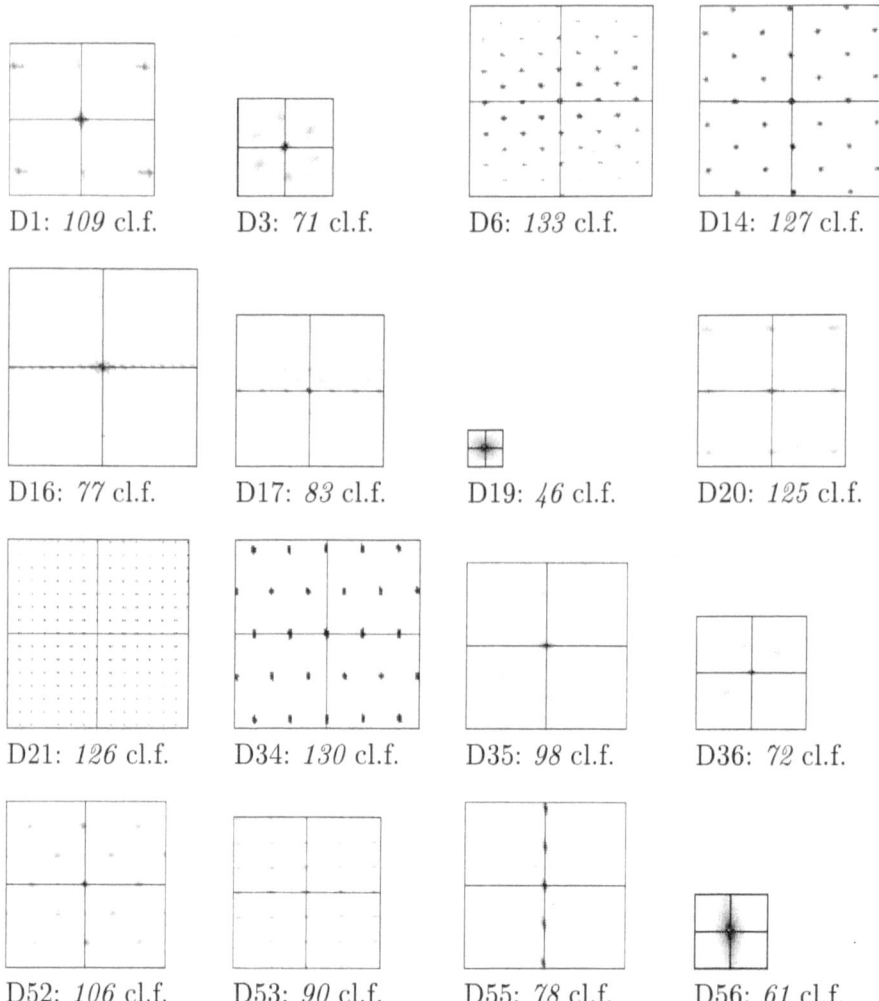

Figure 6.4. Learnt interaction structures for the IDB-1 images (the shorthand notation "cl.f." stands for "clique families"), see also Figure 6.5.

the same type while the texture pair D79 and D105 in Figures 6.3, 6.10, and 6.11 have noticeable visual differences. But, in the last case, due to a poor vertical homogeneity in these textures, both interaction structures are quite similar and mostly reflect the more or less homogeneous horizontal interactions as is shown in Figure 6.5.

Figures 6.6–6.11 indicate only the first choices for the retrieval experiment in Table 6.3. We do not show the lower-rank choices from the IDB because, in most cases, the top choice is the correct one. For the queries of the texture type D3 in Figure 6.6, all the first choices are correct but

TABLE 6.2.

Query type (number)	First choice from the IDB: IDB entry name (number of retrieved images of that type)	Second choice from the IDB IDB entry name (number of retrieved images of that type)
D1(*8*)	**D1**(*8*)	D66(*3*), D94(*5*)
D3(*8*)	**D3**(*7*), D36(*1*)	**D3**(*1*,D36(*5*), D94(*1*)
D6(*8*)	**D6**(*8*)	D34(*8*)
D14(*8*)	**D14**(*8*)	D55(*8*)
D16(*8*)	**D16**(*7*), D105(*1*)	**D16**(*1*), D105(*7*)
D17(*8*)	**D17**(*8*)	D76(*2*), D103(*1*), D105(*5*)
D19(*8*)	**D19**(*8*)	D66(*8*)
D20(*8*)	**D20**(*8*)	D102(*8*)
D21(*8*)	**D21**(*8*)	D17(*1*), D52(*5*), D65(*2*)
D34(*8*)	**D34**(*8*)	D6(*5*), D55(*1*), D102(*2*)
D35(*8*)	**D35**(*5*), D19(*3*)	D16(*1*), D65(*3*), D80(*3*), D94(*1*)
D36(*8*)	D3(*5*), **D36**(*1*), D80(*1*), D87(*1*)	D3(*1*), **D36**(*4*), D80(*2*), D87(*1*)
D52(*8*)	**D52**(*8*)	D34(*1*), D102(*7*)
D53(*8*)	**D53**(*8*)	D6(*3*), D20(*1*), D52(*2*), D102(*2*)
D55(*8*)	**D55**(*8*)	D94(*8*)
D56(*8*)	**D56**(*8*)	D65(*2*), D94(*6*)
D65(*8*)	**D65**(*8*)	D56(*8*)
D66(*8*)	D19(*1*), **D66**(*7*)	D19(*7*), **D66**(*1*)
D76(*8*)	**D76**(*8*)	D105(*8*)
D77(*8*)	**D77**(*8*)	D79(*8*)
D79(*8*)	**D79**(*2*), D105(*6*)	D76(*4*), **D79**(*3*), D105(*1*)
D80(*8*)	**D80**(*8*)	D19(*3*), D87(*5*)
D82(*8*)	**D82**(*8*)	D102(*8*)
D84(*8*)	**D84**(*8*)	D19(*1*), D65(*4*), D80(*2*), D87(*1*)
D87(*8*)	**D87**(*8*)	D84(*8*)
D94(*8*)	**D94**(*8*)	D56(*8*)
D95(*8*)	D94(*7*), **D95**(*1*)	D56(*2*), D94(*1*), **D95**(*5*)
D102(*8*)	D20(*2*), **D102**(*6*)	D20(*6*), **D102**(*2*)
D103(*8*)	**D103**(*8*)	D3(*3*), D76(*4*), D94(*1*)
D105(*8*)	**D105**(*8*)	D76(*7*), D79(*1*)

Retrieval from IDB-1; the rotation angles $(-18°, 18°)$; step $9°$, and scale factors $(0.8, 1.25)$; step 0.05, are exhausted for each query structure. Correct retrieval for 240 query images: **88.4**% among the first choices only and **95.5**% among the two top choices. After placing the similar pairs (*i*) D79 and D105; (*ii*) D3 and D36; (*iii*) D94 and D95 into the same texture types: **97.1**% among the first choices only and **98.8**% among the two top choices.

TABLE 6.3.

Query type (*number*)	First choice from the IDB: IDB entry name (*number of retrieved images of that type*)	Second choice from the IDB IDB entry name (*number of retrieved images of that type*)
D1(*8*)	**D1**(*8*)	D66(*3*), D94(*4*), D102(*1*)
D3(*8*)	**D3**(*8*)	D36(*5*), D94(*2*), D103(*1*)
D6(*8*)	**D6**(*8*)	D34(*8*)
D14(*8*)	**D14**(*8*)	D55(*8*)
D16(*8*)	**D16**(*8*)	D105(*8*)
D17(*8*)	**D17**(*8*)	D76(*3*), D105(*5*)
D19(*8*)	**D19**(*8*)	D66(*8*)
D20(*8*)	**D20**(*8*)	D102(*8*)
D21(*8*)	**D21**(*8*)	D17(*3*), D52(*2*), D95(*1*), D105(*2*)
D34(*8*)	**D34**(*8*)	D6(*8*)
D35(*8*)	**D35**(*5*), D19(*3*)	D16(*1*), D65(*3*), D80(*3*), D94(*1*)
D36(*8*)	D3(*5*), **D36**(*1*), D80(*1*), D87(*1*)	D3(*1*), **D36**(*4*), D80(*2*), D87(*1*)
D52(*8*)	**D52**(*8*)	D102(*8*)
D53(*8*)	**D53**(*8*)	D6(*3*), D20(*1*), D52(*2*), D102(*2*)
D55(*8*)	**D55**(*8*)	D94(*8*)
D56(*8*)	**D56**(*8*)	D65(*4*), D94(*4*)
D65(*8*)	**D65**(*8*)	D56(*8*)
D66(*8*)	D19(*1*), **D66**(*7*)	D19(*7*), **D66**(*1*)
D76(*8*)	**D76**(*7*), D79(*1*)	**D76**(*1*), D79(*5*), D105(*2*)
D77(*8*)	**D77**(*8*)	D79(*8*)
D79(*8*)	**D79**(*3*), D105(*5*)	**D79**(*5*), D105(*3*)
D80(*8*)	**D80**(*8*)	D19(*3*), D87(*5*)
D82(*8*)	**D82**(*8*)	D102(*8*)
D84(*8*)	**D84**(*8*)	D19(*1*), D65(*4*), D80(*2*), D87(*1*)
D87(*8*)	**D87**(*8*)	D84(*8*)
D94(*8*)	**D94**(*8*)	D56(*8*)
D95(*8*)	D94(*7*), **D95**(*1*)	D56(*1*), D65(*1*), D94(*1*), **D95**(*5*)
D102(*8*)	D20(*2*), **D102**(*6*)	D20(*6*), **D102**(*2*)
D103(*8*)	**D103**(*8*)	D3(*4*), D76(*4*)
D105(*8*)	D79(*3*), **D105**(*5*)	D76(*1*), D79(*4*), **D105**(*3*)

Retrieval from IDB-1; the rotation angles $(-18°, 18°)$; step $4.5°$, and scale factors $(0.8, 1.25)$; step 0.025, are exhausted for each query structure. Correct retrieval for 240 query images: **88.8%** among the first choices only and **96.7%** among the two top choices. After placing the similar pairs (*i*) D79 and D105; (*ii*) D3 and D36; (*iii*) D94 and D95 into the same texture types: **96.3%** among the first choices only and **97.9%** among the two top choices.

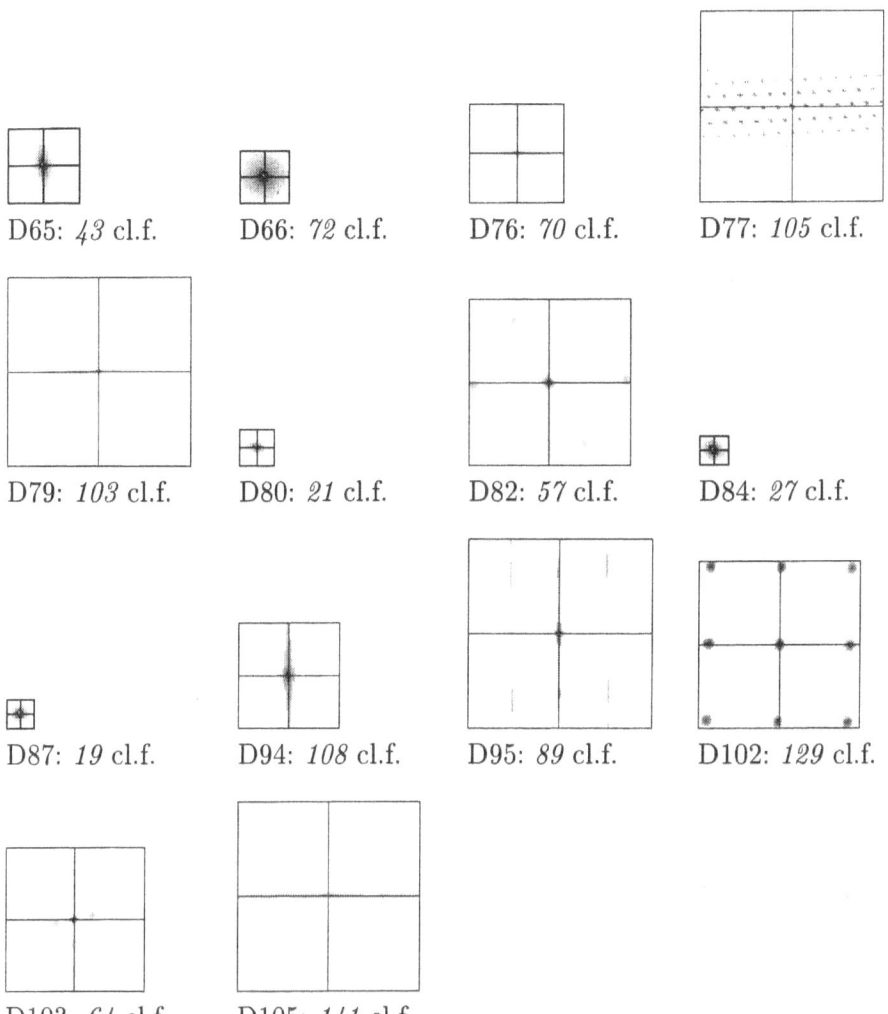

D65: *43* cl.f.   D66: *72* cl.f.   D76: *70* cl.f.   D77: *105* cl.f.

D79: *103* cl.f.   D80: *21* cl.f.   D82: *57* cl.f.   D84: *27* cl.f.

D87: *19* cl.f.   D94: *108* cl.f.   D95: *89* cl.f.   D102: *129* cl.f.

D103: *64* cl.f.   D105: *141* cl.f.

Figure 6.5.   Learnt interaction structures for the IDB-1 images (the shorthand notation "cl.f." stands for "clique families"), see also Figure 6.4.

obtained with different relative rotations and scales of the queries due to obvious differences between these images and the IDB image. For the queries 1–8 in Figure 6.6 these optimal rotation angles and scales are, respectively, as follows:

$$(-4.5°, 1.075), \ (-4.5°, 1.075), \ (-4.5°, 0.800), \ (-4.5°, 0.925),$$
$$(4.5°, 0.925), \ (-18.0°, 0.875), \ (-18.0°, 0.875), \ (-13.5°, 0.875).$$

The similar results, as shown in Tables 6.2 and 6.3, are obtained for all

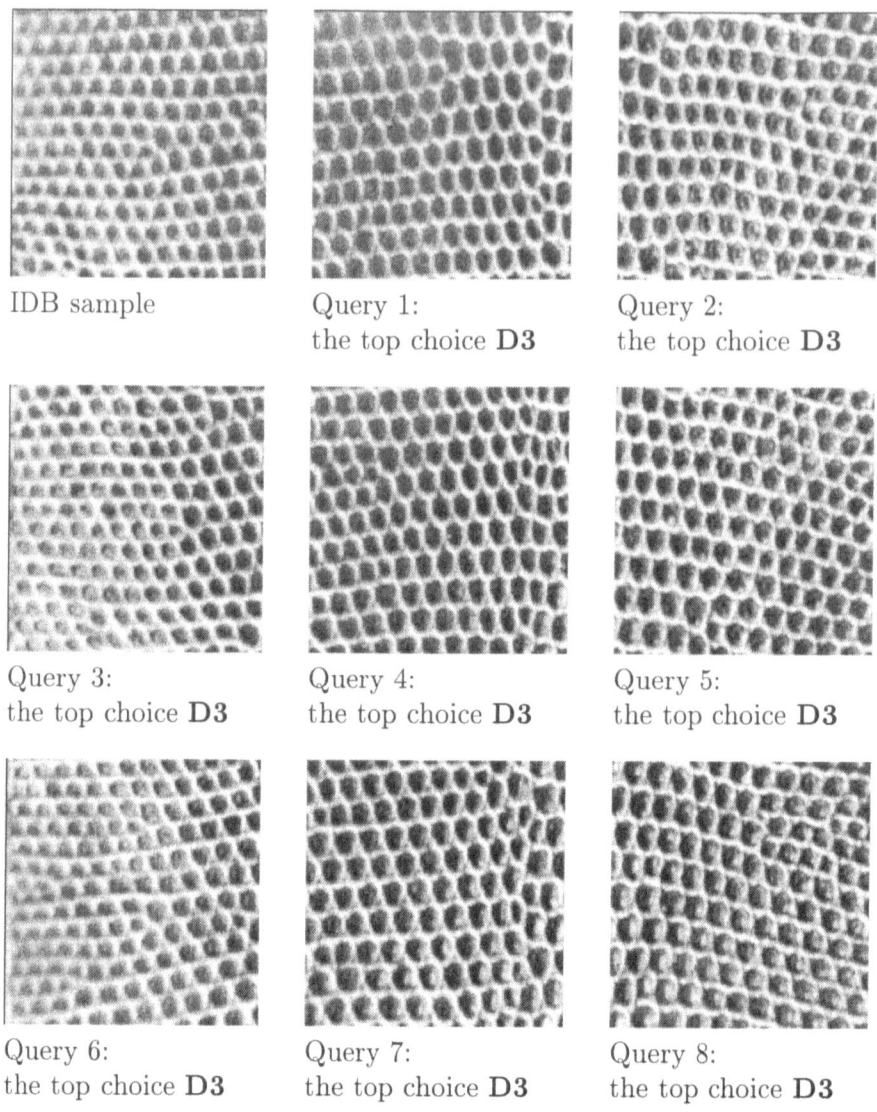

IDB sample   Query 1:   Query 2:
                        the top choice **D3**   the top choice **D3**

Query 3:   Query 4:   Query 5:
the top choice **D3**   the top choice **D3**   the top choice **D3**

Query 6:   Query 7:   Query 8:
the top choice **D3**   the top choice **D3**   the top choice **D3**

Figure 6.6.   IDB and query samples of the texture type D3 (Reptile skin).

other texture types. The correct first choices for the queries of the texture type D19 are with low confidence because its interaction structure, due to a weak homogeneity, is quite similar to the structure of D66 type as is indicated in Figures 6.7 and 6.9. Spatial non-uniformity in the query or IDB images leads to averaging of the different pairwise interactions, so in these situations our retrieval algorithm can give almost arbitrary results

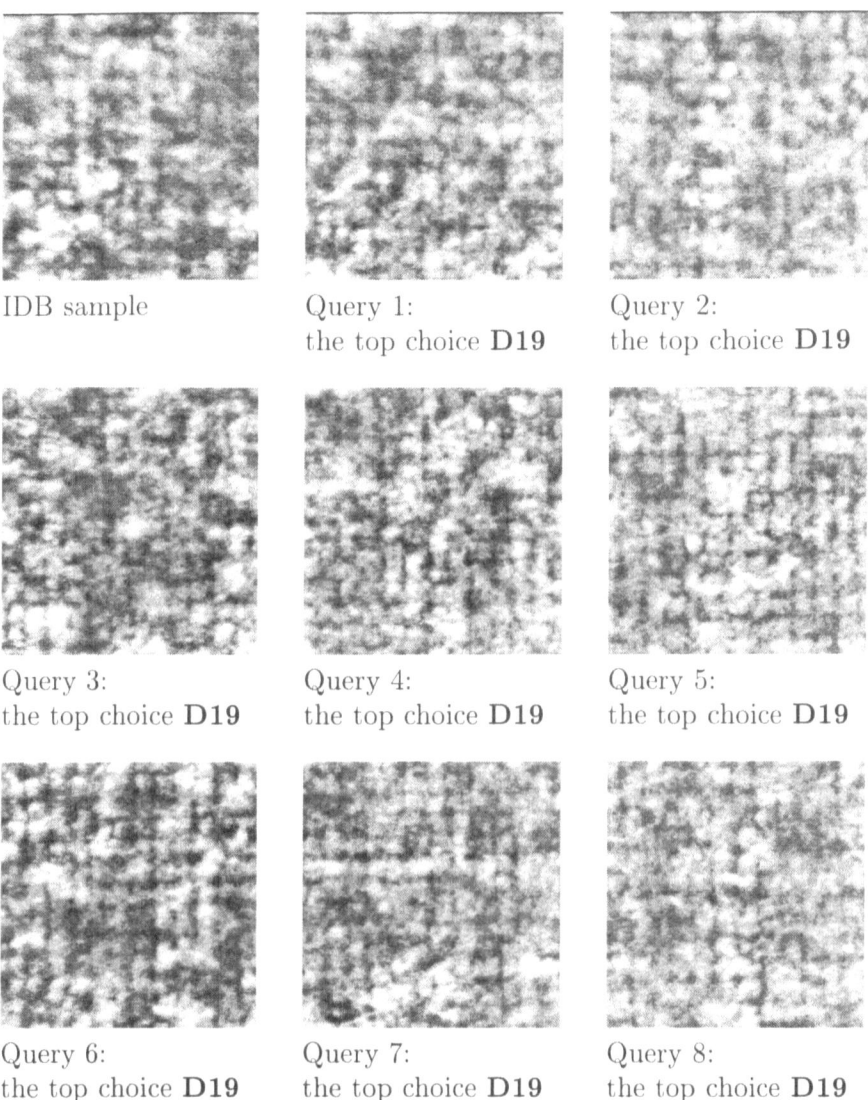

Figure 6.7. IDB and query samples of the texture type D19 (Woolen cloth).

(see, for instance, Figure 6.8).

The IDB-1 was intentionally constructed as to simultaneously include both highly homogeneous textures (in terms of the similarity between the local and global GLCHs or GLDHs) and weakly homogeneous textures. Examples of the weakly homogeneous textures are presented in Figures 6.8–6.11). Our strategy obtains rather stable and consistent retrieval results

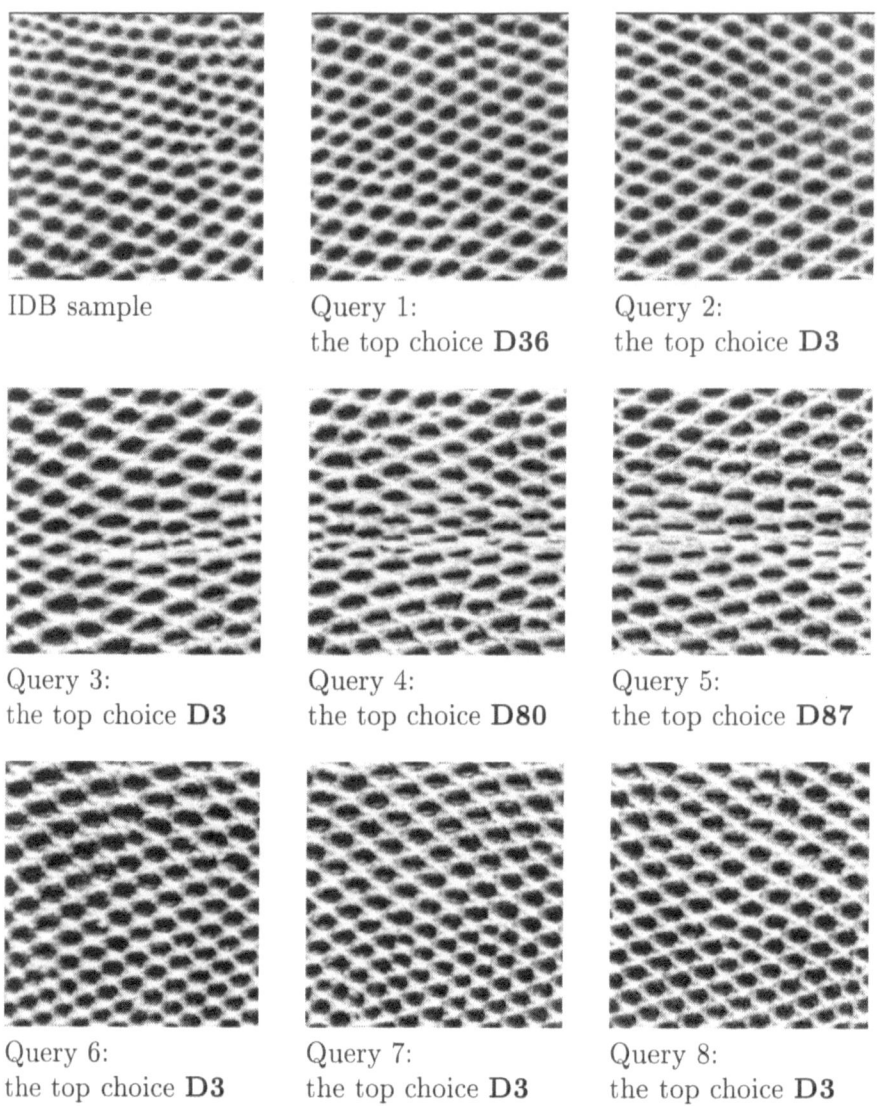

IDB sample	Query 1: the top choice **D36**	Query 2: the top choice **D3**
Query 3: the top choice **D3**	Query 4: the top choice **D80**	Query 5: the top choice **D87**
Query 6: the top choice **D3**	Query 7: the top choice **D3**	Query 8: the top choice **D3**

Figure 6.8. IDB and query samples of the texture type D36 (Lizard skin). The textures D36 and D3 shown in Figure 6.6 are strongly similar. But the inhomogeneities of the queries 4 and 5 result in the wrong first choice of the D80 or D87 texture types, and all the three top choices are of the types D80, D84, and D87.

for the homogeneous, or stochastic textures, but can fail for the weakly homogeneous or inhomogeneous ones. In these latter cases the GLCHs or GLDHs and the learnt interaction structures differ much from one fragment to other even in the same textured image and can become more similar to

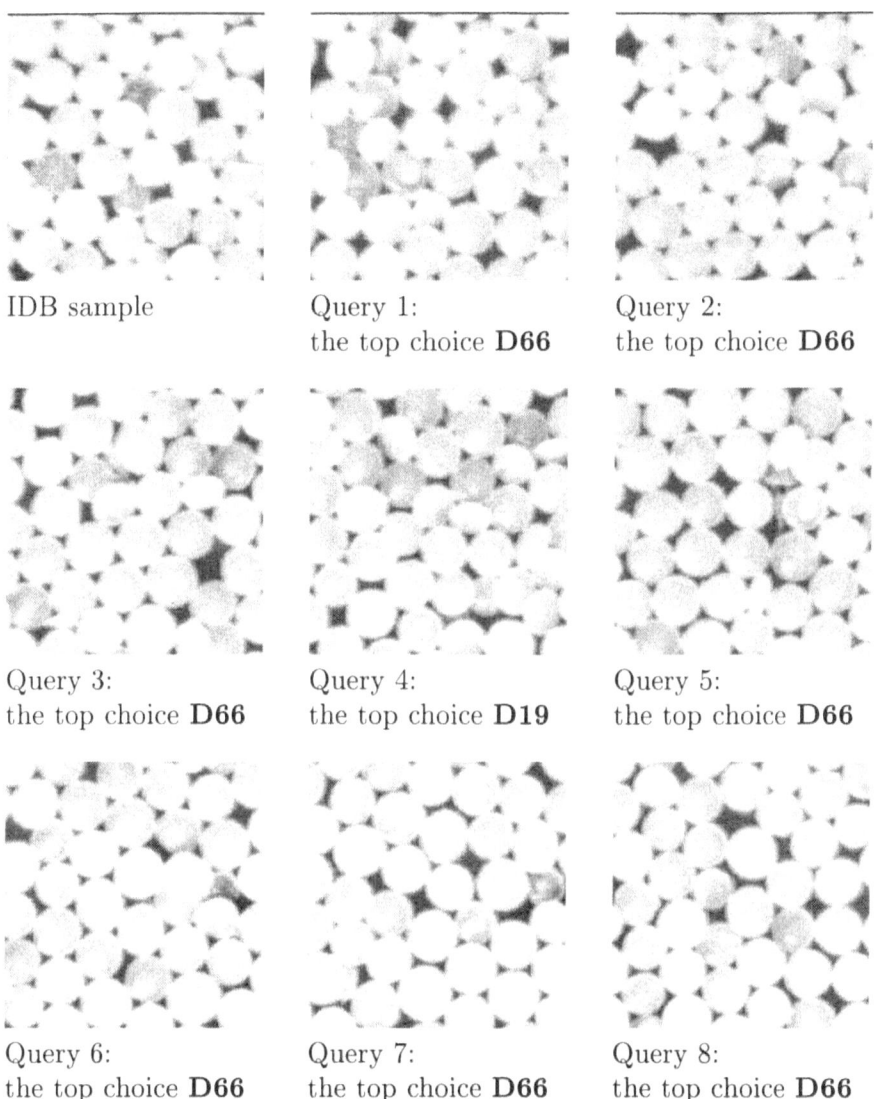

Figure 6.9. IDB and query samples of the texture type D66 (Plastic pellets). Here, the interaction structures in Figures 6.4 and 6.5 for the weakly homogeneous textures D66 and D19 are strongly similar.

some other texture types.

One more difficulty is inherent, for instance, in the texture types similar to the textures D56 and D65 in Figures 6.2 and 6.3 with the characteristic "very-long-range" pairwise pixel interactions. Because the image sizes

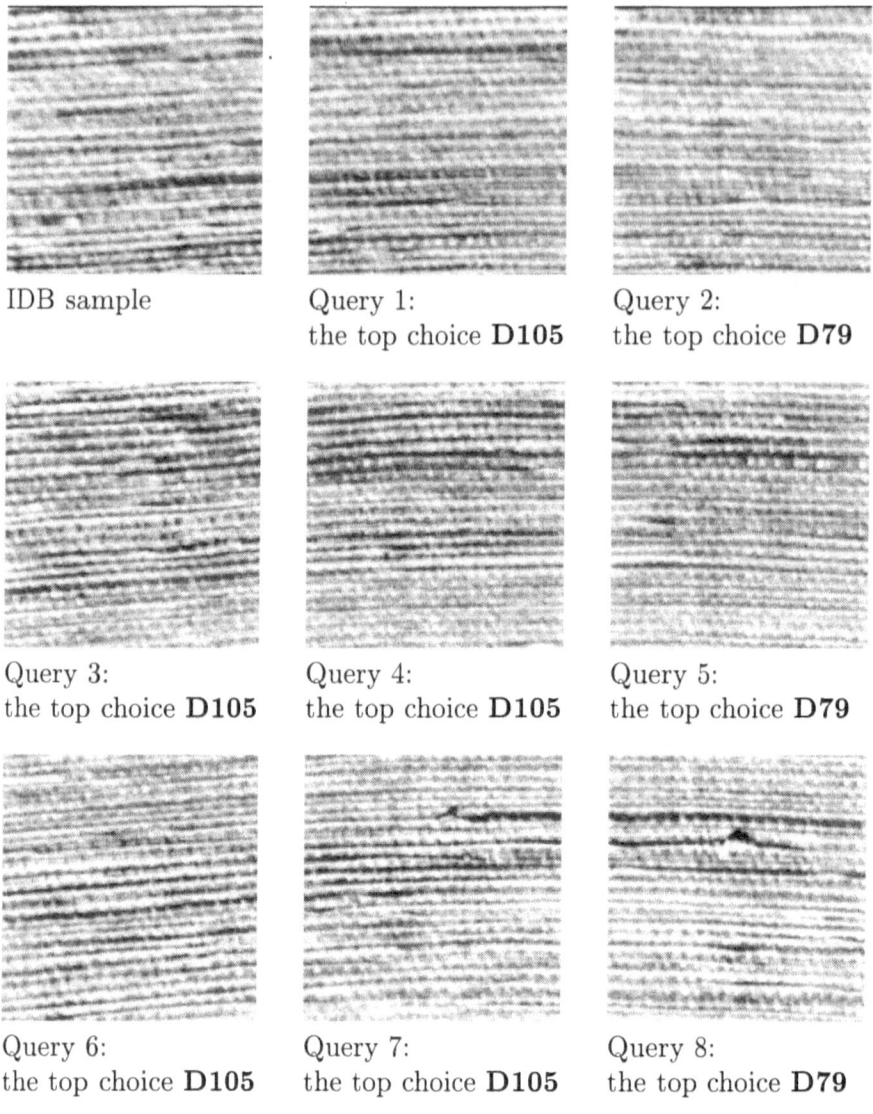

IDB sample   Query 1:   Query 2:
                           the top choice **D105**   the top choice **D79**

Query 3:   Query 4:   Query 5:
the top choice **D105**   the top choice **D105**   the top choice **D79**

Query 6:   Query 7:   Query 8:
the top choice **D105**   the top choice **D105**   the top choice **D79**

Figure 6.10. IDB and query samples of the texture type D79 (Grass fiber cloth). The interaction structures in Figure 6.5 of the textures D79 and D105 are strongly similar. Because both the textures are weakly homogeneous in the vertical direction, the corresponding periodicities in the pixel interactions have not been found by the proposed approach.

are relatively small compared with these patterns, the long-range vertical periodicity which is easily perceived and may be extrapolated by human vision cannot be recovered by the proposed approach. This is easily seen

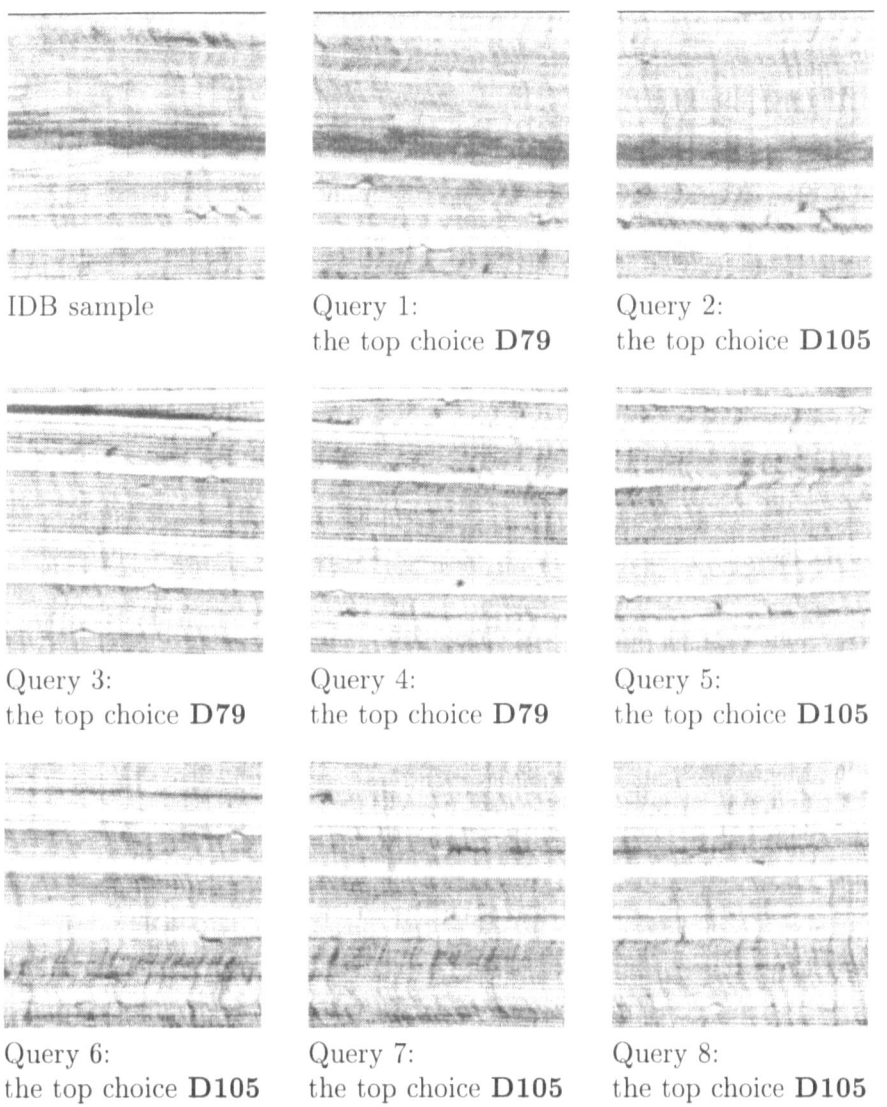

IDB sample

Query 1:
the top choice **D79**

Query 2:
the top choice **D105**

Query 3:
the top choice **D79**

Query 4:
the top choice **D79**

Query 5:
the top choice **D105**

Query 6:
the top choice **D105**

Query 7:
the top choice **D105**

Query 8:
the top choice **D105**

Figure 6.11.   IDB and query samples of the texture type D105 (Cheesecloth). The interaction structures in Figure 6.5 for the textures D79 and D105 are strongly similar (see, also, Figure 6.10).

in the interaction structures shown in Figures 6.4 and 6.5. In the above experiments, we did retrieve all the correct first choices for these textures but only due to the sufficiently distinct GLCHs and GLDHs for the learnt short-range structures.

In spite of these drawbacks, the overall performance of the proposed approach in our experiments with the data base IDB-1 is sufficiently good. Therefore one may conclude that the problem of searching in the IDB for all the images that belong to the same texture type as the input query image can be solved in part by the proposed approach.

The homogeneity of a texture type can be formally estimated by comparing the self-similarity of the interaction maps for the different subimages of the same type. This permits, in principle, to find for each IDB entry the representative uniform parts to be used in retrieving a given texture type. But, it should be noted that the images with very similar GLDHs over a given set of clique families can sometimes differ substantially in the visual appearances. Thus, the discriminating ability of the proposed computational approach can differ from that of the human vision even though, in most of our experiments, both the classifications were similar (that is, the first or, at least, the second retrieved images did correspond to the right texture type of the query image).

Retrieval with rejections. The proposed approach has no provisions to reject a query which does not visually match any of the images in the IDB if its interaction structure and GLDHs are similar in terms of our dissimilarity measure in Eq. (6.3) to a texture type represented in the IDB. But, it does allow us to reject the queries that differ substantially in their interaction structure and strength from the IDB entries. Table 6.4 summarizes results of the experiments that are similar to those presented in Table 6.3 but involve an auxiliary rejection criterion.

The rejection is based on a relative number of common clique families in the matched images. For any given IDB entry, the matched query image is rejected if the following condition holds:

$$\max_{\xi=[\varphi,s]\in\Xi} \frac{|\mathbf{A}_{\text{comm}}(\xi)|}{|\mathbf{A}_{\text{comb}}(\xi)|} \leq \theta_{\text{rej}}.$$

Here, $\mathbf{A}_{\text{comm}}(\xi)$ is a set of the clique families that are common to both the matched query image $\mathbf{g}$ and the IDB one $\mathbf{g}_t$ under a given relative transformation ξ of the query image:

$$\mathbf{A}_{\text{comm}}(\xi) = \mathbf{A}_{\mathbf{g}_t} \bigcap \mathbf{A}_{\mathbf{g}}(\xi)$$

and θ_{rej} denotes a particular rejection threshold. The experiments shown in Table 6.4 were conducted with the threshold $\theta_{\text{rej}} = 0.20$. In these experiments, the thresholds higher than 0.20 rejected even some valid queries of the type D77 (Cotton canvas). So, we selected the threshold 0.20 to reject, on the average, 183 "invalid" queries from 240 initial ones for any IDB-1 entry. In other cases, such a threshold has to be found empirically.

TABLE 6.4.

Query type (*number*)	First choice: IDB entry name (*number of retrieved images of that type*)	Second choice: IDB entry name (*number of retrieved images of that type*)	Number of rejected queries
D1(*8*)	**D1**(*8*)	D66(*4*), D94(*3*), D102(*1*)	150
D3(*8*)	**D3**(*8*)	D36(*5*), D94(*2*), D103(*1*)	210
D6(*8*)	**D6**(*8*)		232
D14(*8*)	**D14**(*8*)		232
D16(*8*)	**D16**(*8*)	D76(*1*), D105(*7*)	198
D17(*8*)	**D17**(*8*)	D76(*1*), D105(*3*)	223
D19(*8*)	**D19**(*8*)	D66(*8*)	112
D20(*8*)	**D20**(*8*)	D102(*8*)	207
D21(*8*)	**D21**(*8*)		232
D34(*8*)	**D34**(*8*)		232
D35(*8*)	**D35**(*5*), D19(*3*)	D16(*1*), D66(*3*), D80(*4*)	157
D36(*8*)	D3(*5*), **D36**(*1*), D80(*1*), D87(*1*)	D3(*1*), **D36**(*4*), D80(*2*), D87(*1*)	191
D52(*8*)	**D52**(*8*)	D102(*7*)	223
D53(*8*)	**D53**(*8*)		232
D55(*8*)	**D55**(*8*)	D94(*8*)	208
D56(*8*)	**D56**(*8*)	D65(*4*), D94(*4*)	127
D65(*8*)	**D65**(*8*)	D56(*8*)	131
D66(*8*)	D19(*1*), **D66**(*7*)	D19(*7*), **D66**(*1*)	113
D76(*8*)	**D76**(*7*), D79(*1*)	**D76**(*1*), D79(*5*), D105(*2*)	155
D77(*8*)	**D77**(*8*)		232
D79(*8*)	**D79**(*3*), D105(*5*)	**D79**(*5*), D105(*3*)	196
D80(*8*)	**D80**(*8*)	D19(*3*), D87(*5*)	144
D82(*8*)	**D82**(*8*)	D84(*4*), D102(*4*)	179
D84(*8*)	**D84**(*8*)	D19(*1*), D65(*4*), D80(*2*), D87(*1*)	121
D87(*8*)	**D87**(*8*)	D84(*8*)	121
D94(*8*)	**D94**(*8*)	D56(*8*)	144
D95(*8*)	D94(*7*), **D95**(*1*)	D56(*1*), D65(*1*), D94(*1*), **D95**(*5*)	169
D102(*8*)	D20(*2*), **D102**(*6*)	D20(*6*), **D102**(*2*)	208
D103(*8*)	**D103**(*8*)	D76(*4*)	201
D105(*8*)	D79(*3*), **D105**(*5*)	D76(*1*), D79(*4*), **D105**(*3*)	198

Retrieval with rejections from IDB-1 (the threshold $\theta_{\mathrm{rej}} = 0.20$): the rotation angles $(-18°, 18°)$; step $4.5°$, and scale factors $(0.8, 1.25)$; step 0.025, are exhausted for each query structure.

Crude segmentation.   If we assume that homogeneous textured regions in a given query image are sufficiently large, then it is possible to answer the question whether the query image contains a texture of a particular type which is present in the IDB (with the same distinctions from human vision as were mentioned above). It should be emphasized that the more representative and unique the local structure of the pairwise pixel interaction in a particular texture, the more reliable the retrieval and more stable the classification results.

For instance, the IDB samples from the homogeneous textures D6, D14, D17, D21, D34, D53, D77 representing different woven fabrics have very specific and pronounced structures shown in Figures 6.4 and 6.5. These structures differ substantially from all other texture types in IDB-1 so that all the incorrect IDB entries for them are rejected in Table 6.4. But, the inhomogeneity of the IDB and query samples leads to significant differences in the interaction structures and, thereby, to invalid retrieval choices and fewer rejected queries. For instance, the 4th and 5th query images for the texture type D36 in Figure 6.8 have very distinctive inhomogeneities which result in the first retrieved choice of the type D80 or D87 with fairly distinct visual appearances.

The interaction structure and GLCHs or GLDHs can be similar for very different spatially inhomogeneous images due to averaging of different interactions over the lattice. Such images can hardly be discriminated from each other during our retrieval. Nevertheless, if the resulting "averaged" structures have marked differences as, for instance, the IDB entries D19, D84, D87 in Figures 6.4 and 6.5 such a discrimination is still possible.

It should be noted that we stratify the textures by their spatial homogeneity only by a qualititative comparison of the visual and computational experimental results. Therefore such a stratification is only a matter of convention, and a texture which was inhomogeneous with respect to a particular simulation scheme in Chapter 5 may become the homogeneous one for a particular query-by-image retrieval scheme.

6.4.2. IMAGE DATA BASE IDB-2

Some typical 64×64 images from IDB-2 are shown in Figures 6.12 and 6.13 to be compared to the same 128×128 textures in Figures 5.15–5.25. The texture types were assigned to the images by the authors of the "VisTex" collection Pickard et al. (1995) and reflect a logical rather than visual similarity between them. Although there are several visually similar images with the assigned different texture types, we preserve this original classification in our experiments. Figures 6.14 and 6.15 present the typical arrangement of the IDB and query samples within the initial "VisTex" 128×128 images.

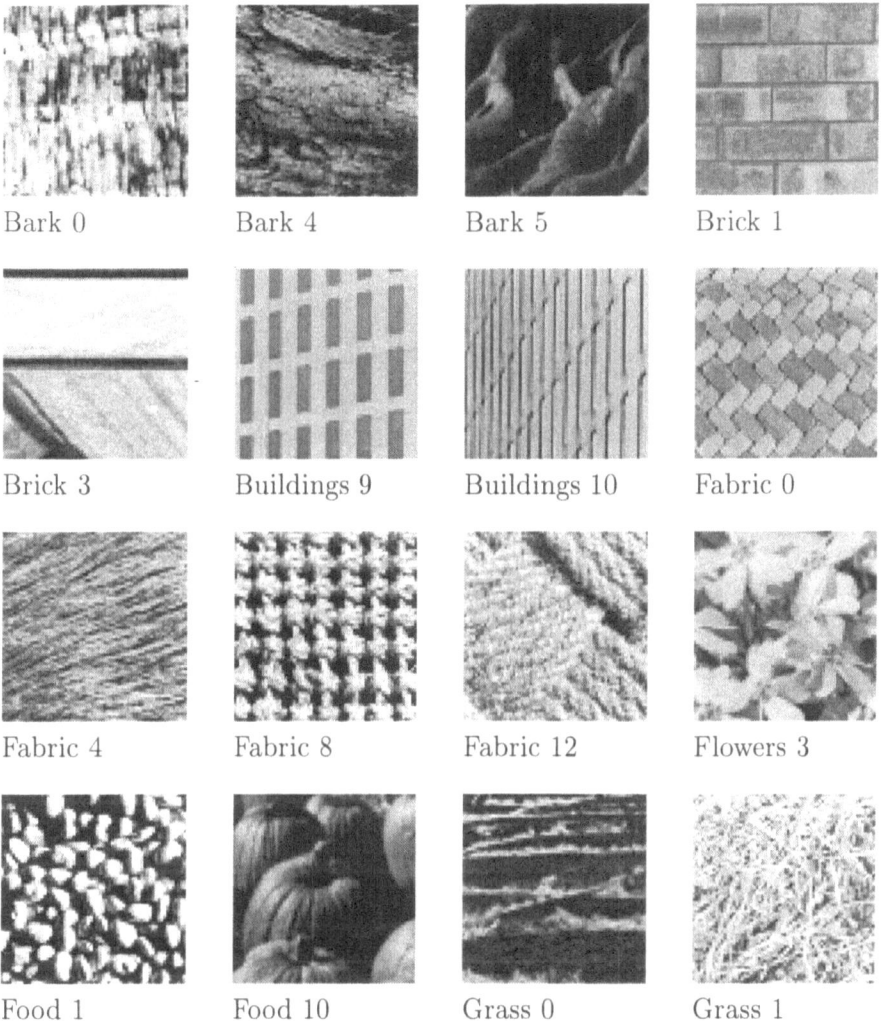

Figure 6.12.   Typical images from the experimental image data base IDB-2, see also Figure 6.13.

It should be noted that most of these images are too inhomogeneous to result in a correct match between the query image and the IDB images: see, for instance, the images "Bark 0", "Bark 5", "Brick 3", "Fabric 0", "Fabric 4", "Grass 0", "Grass 1", "Grass 2" in Figures 6.12 and 6.13. The interaction structures of the latter five images have marked non-linear transformations over the lattice. So, the experiments with this IDB had the following goals:

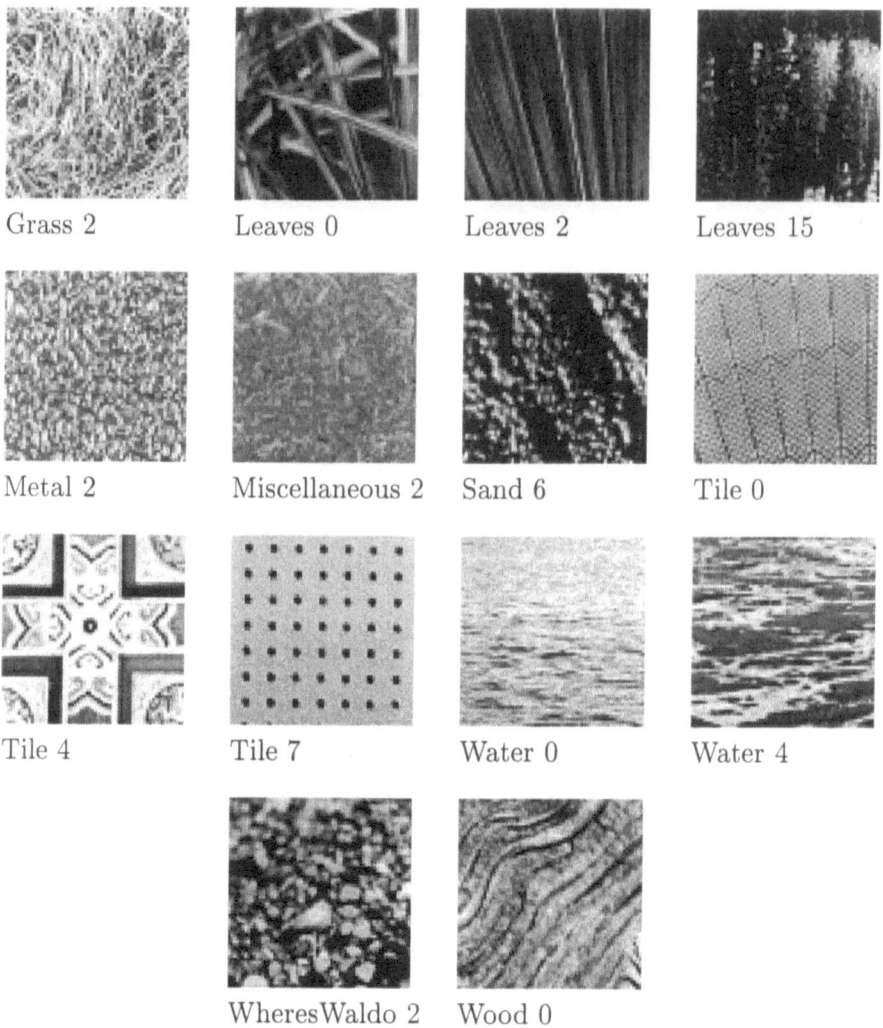

Grass 2   Leaves 0   Leaves 2   Leaves 15

Metal 2   Miscellaneous 2   Sand 6   Tile 0

Tile 4   Tile 7   Water 0   Water 4

WheresWaldo 2   Wood 0

Figure 6.13. Typical images from the experimental image data base IDB-2, see also Figure 6.12.

(i) to separate homogeneous and weakly homogeneous textures from all the inhomogeneous ones with respect to a particular retrieval scheme and

(ii) to evaluate the ability of the proposed approach to represent typical structural features of very different natural textures.

Table 6.5 summarizes the results of the retrieval experiment with the same ranges and steps of rotations and scales as in Table 6.2. As was expected,

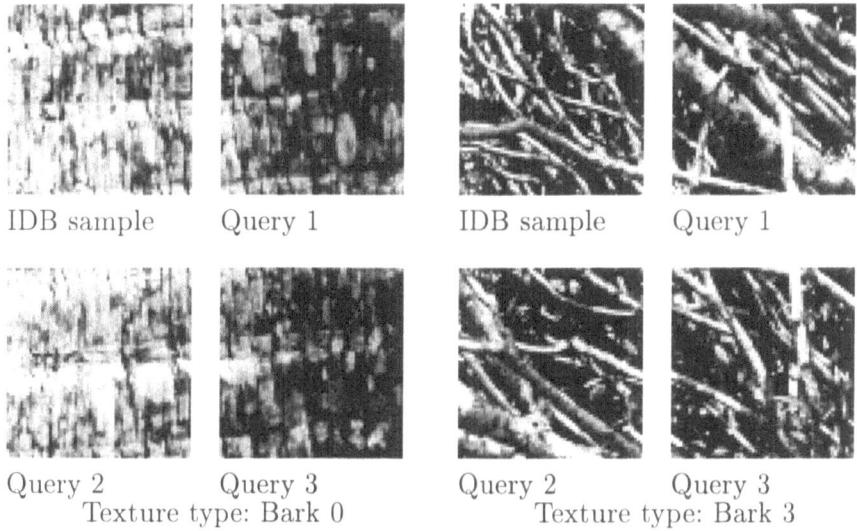

IDB sample   Query 1   IDB sample   Query 1

Query 2   Query 3   Query 2   Query 3
        Texture type: Bark 0   Texture type: Bark 3

Figure 6.14.   IDB and query samples for some texture types from IDB-2.

the retrieval results are in total lower than for IDB-1. The correct retrieval
of the assigned texture types is as follows: 60.6% among the first choices
only, 73.3% among the top two choices, and 77.8% among the top three
choices. But, in this experiment we did not take into account the obvious
inhomogeneities in the matched images. To estimate the homogeneity, we
use the following rough classification:

- Each texture taken from the "VisTex" 128×128 collection is homogeneous if the first retrieved choice for all the three 64×64 queries
  belongs to the same texture type as the corresponding IDB sample.
- It is weakly homogeneous if only two of the queries gave this texture
  type.
- It is considered inhomogeneous if the three different texture types are
  retrieved.
- It is similar to another texture type if at least two queries gave the
  first retrieved choice of the same texture type but this choice differs
  from the texture type of the IDB sample.

These classes, marked with **h**, **w**, **i**, and **s**, respectively, in Table 6.6, can
be compared to the similar stratification by the simulation experiments in
Table 5.3.

This allows us to exclude from IDB-2 all the inhomogeneous textures
as well as three texture types "Grass 0", "Leaves 2", and "Tile 0" which
are quite similar structurally to some other texture types due to their high
non-uniformity. In this case, only 95 homogeneous and weakly homogeneous

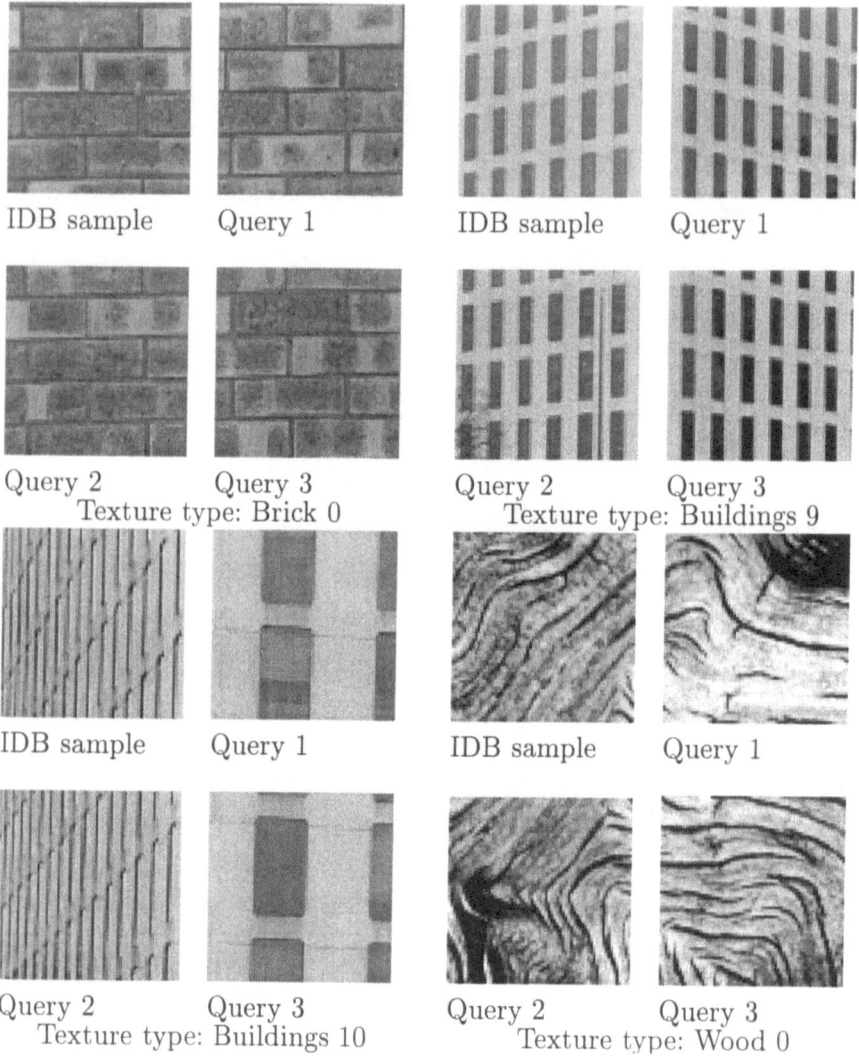

IDB sample   Query 1   IDB sample   Query 1

Query 2   Query 3   Query 2   Query 3
      Texture type: Brick 0   Texture type: Buildings 9

IDB sample   Query 1   IDB sample   Query 1

Query 2   Query 3   Query 2   Query 3
      Texture type: Buildings 10   Texture type: Wood 0

Figure 6.15. IDB and query samples for some texture types from IDB-2.

entries remain in IDB-2 (comparing to 108 such entries for the total MIT "VisTex" database in Table 5.3). Also, we join the texture types "Flowers" and "Leaves" in the same texture type "Plants" because of high similarities between some of their samples. In such a case, the retrieval performance is given in Table 6.7: there are 76.8% correct answers among the first choices, 84.9% among the top two choices, and 89.1% among the top three choices.

Figures 6.16 – 6.21 show the top choices obtained in these experiments for the queries of the subtypes 0–4,7 of the texture type "Bark of tree".

TABLE 6.5. Retrieval from IDB-2; the rotation angles $(-18°, 18°)$; step $9°$, and scale factors $(0.8, 1.25)$; step 0.05, are exhausted for each query structure. Correct retrieval for 360 query images: **66.6%** among the first choices only, **73.3%** among the top two choices, and **77.8%** among the top three choices.

Texture type	Number of the IDB entries	Number of the queries	Own type among the first choices	Own type among the top two choices	Own type among the top three choices
Bark	13	39	26	33	34
Brick	7	21	13	15	15
Buildings	3	9	3	3	3
Fabric	20	60	44	45	47
Flowers	8	24	8	17	18
Food	11	33	28	31	32
Grass	3	9	0	0	2
Leaves	17	51	29	36	41
Metal	6	18	13	16	16
Miscellaneous	4	12	6	8	8
Sand	7	21	10	17	17
Tile	8	24	18	19	22
Water	8	24	18	19	22
WheresWaldo	3	9	4	4	4
Wood	2	6	3	3	3

As it was noted earlier, the inhomogeneities in such weakly homogeneous images cause noticeable variations in the interaction structures. In more detail, it is shown for the above images in Figures 6.22 – 6.25. Therefore, in such cases the proposed retrieval can only coarsely discriminate between the natural texture types. It sometimes contradicts the visual perception of these textures. For instance, the interaction structure for the "Water 7" texture is quite similar, under possible rotation and scale transformations, to the structure for the "Bark 4" texture. The same holds for the samples "Leaves 0" and "Bark 0", "Miscellaneous 3" and "Bark 1" or "Bark 2", "Sand 5" and "Bark 7", and so forth.

Nevertheless, even for such natural textured images as collected in IDB-2 the proposed approach gives rather reasonable retrieval results in the absence of any "high-level" semantic or contextual knowledge.

TABLE 6.6. Rough estimates of the texture homogeneity for IDB-2; **h** – homogeneous, **w** – weakly homogeneous, **i** – inhomogeneous texture, and **s** – similar to another texture type.

Texture type	Texture subtypes and their estimated homogeneity												
Bark	0	1	2	3	4	5	6	7	8	9	10	11	12
	i	w	w	w	s	h	w	w	w	h	w	w	h
Brick	0	1	2	3	4	5	7						
	h	h	s	h	s	i	i						
Buildings	8	9	10										
	i	h	i										
Fabric	0	1	2	3	4	5	6	7	8	9	10	11	12
	i	h	i	s	w	w	w	s	h	h	h	h	h
	13	14	15	16	17	18	19						
	s	w	h	h	h	h	s						
Flowers	0	1	2	3	4	5	6	7					
	w	w	s	w	s	s	w	s					
Food	0	1	2	3	4	5	6	7	8	10	11		
	h	h	h	h	s	s	h	h	h	w	h		
Grass	0	1	2										
	s	i	i										
Leaves	0	1	2	3	4	5	6	7	8	9	10	11	12
	s	i	s	h	h	i	i	h	w	w	i	h	w
	13	14	15	16									
	w	h	w	i									
Metal	0	1	2	3	4	5							
	i	w	w	w	h	h							
Miscellaneous	0	1	2	3									
	i	w	i	h									
Sand	0	1	2	3	4	5	6						
	w	w	h	w	i	s	w						
Tile	0	1	3	4	7	8	9	10					
	s	s	w	w	h	s	h	h					
Water	0	1	2	3	4	5	6	7					
	w	w	h	h	s	w	h	i					
WheresWaldo	0	1	2										
	i	i	h										
Wood	0	2											
	i	h											

6.5. Complexity and practicality

This chapter shows that the texture attributes derived from the Gibbs image models with multiple pairwise pixel interactions can be used for solving

TABLE 6.7. Retrieval from IDB-2 after excluding 22 non-uniform entries and 3 entries similar to other texture types.

Type	IDB entries	Queries	Top choice	%	2 top choices	%	3 top choices	%
Bark	12	36	26	**72.2**	32	**88.9**	33	**91.7**
Brick	5	15	11	**73.3**	14	**93.3**	14	**93.3**
Buildings	1	3	3	**100.0**	3	**100.0**	3	**100.0**
Fabric	18	54	43	**79.6**	46	**85.2**	47	**87.0**
Flowers and Leaves	19	57	36	**63.2**	48	**84.2**	54	**94.7**
Food	11	33	28	**84.8**	31	**93.9**	32	**97.0**
Metal	5	15	12	**80.0**	13	**86.7**	13	**87.7**
Miscellaneous	2	6	5	**83.3**	6	**100.0**	6	**100.0**
Sand	6	18	12	**66.7**	15	**83.3**	15	**83.3**
Tile	7	21	13	**61.9**	16	**76.2**	18	**85.7**
Water	7	21	16	**76.2**	16	**76.2**	19	**90.5**
WheresWaldo	1	3	3	**100.0**	3	**100.0**	3	**100.0**
Wood	1	3	3	**100.0**	3	**100.0**	3	**100.0**
In total:	95	285	219	**76.8**	242	**84.9**	254	**89.1**

IDB sample   Query 1   Query 2   Query 3

Top choice:

Food 6   WhereWaldo 2   Leaves 0

Figure 6.16. Top retrieval choices for query samples of the inhomogeneous texture Bark 0 (the 5 rotation angles (−18°, 18°); step 9°, and 10 scale factors (0.8, 1.25); step 0.05, are exhausted for each query structure).

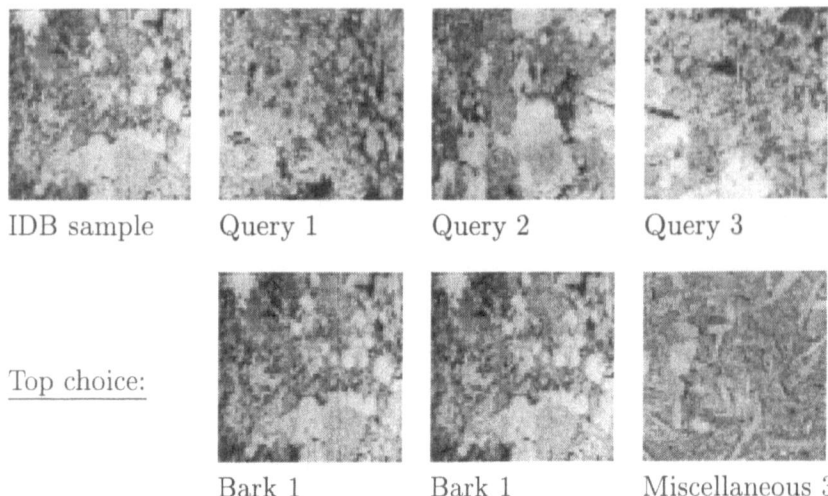

IDB sample   Query 1   Query 2   Query 3

Top choice:

Bark 1   Bark 1   Miscellaneous 3

Figure 6.17. Top retrieval choices for query samples of the weakly homogeneous textures Bark 1 (the 5 rotation angles $(-18°, 18°)$; step $9°$, and 10 scale factors $(0.8, 1.25)$; step 0.05, are exhausted for each query structure).

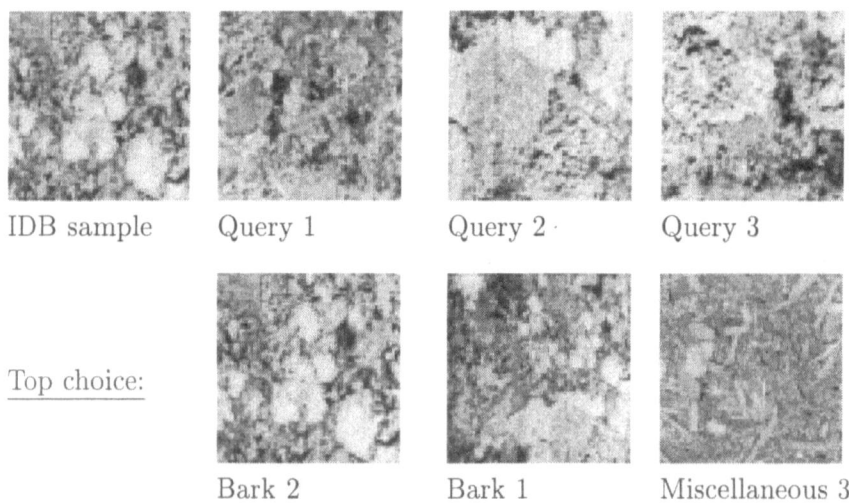

IDB sample   Query 1   Query 2   Query 3

Top choice:

Bark 2   Bark 1   Miscellaneous 3

Figure 6.18. Top retrieval choices for query samples of the weakly homogeneous textures Bark 2 (the 5 rotation angles $(-18°, 18°)$; step $9°$, and 5 scale factors $(0.8, 1.25)$; step 0.05, are exhausted for each query structure).

the problem of image retrieval from a database. The texture similarity measure of Eq. (6.3) exploits the characteristic interaction structure and the associated GLCHs or GLDHs to describe every given texture sample and takes into account possible rotation and scale differences between the tex-

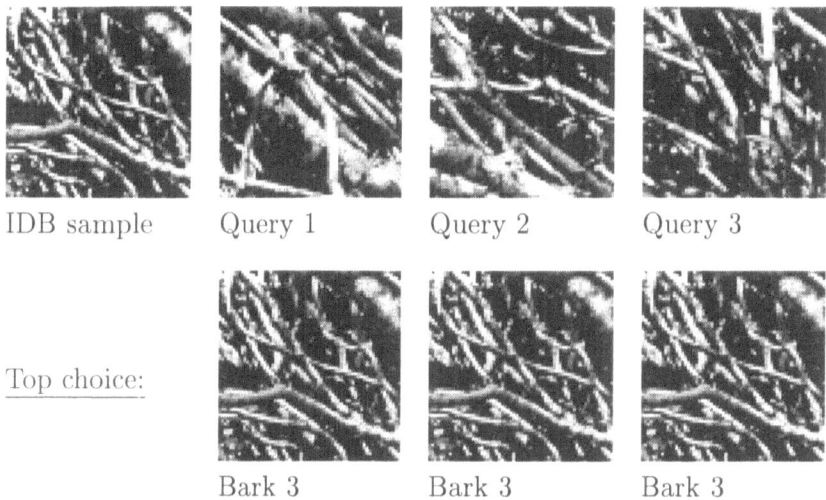

IDB sample   Query 1   Query 2   Query 3

Top choice:

Bark 3   Bark 3   Bark 3

Figure 6.19. Top retrieval choices for query samples of the homogeneous texture Bark 3 (the 5 rotation angles $(-18°, 18°)$; step $9°$, and 10 scale factors $(0.8, 1.25)$; step 0.05, are exhausted for each query structure).

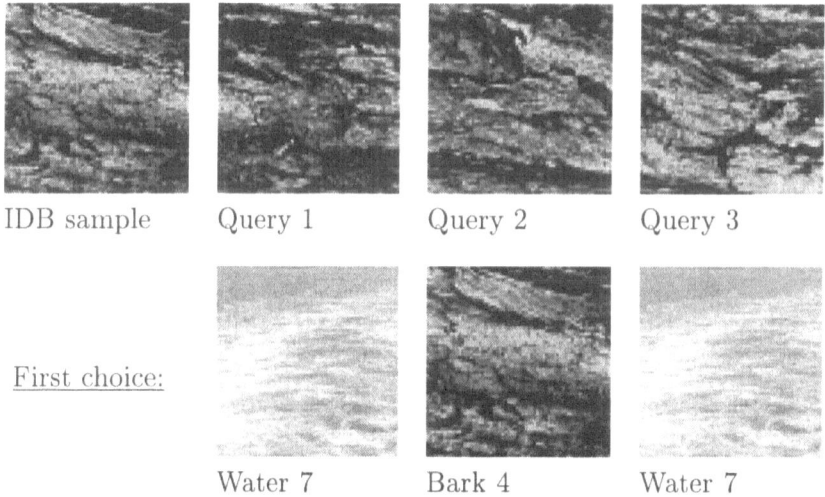

IDB sample   Query 1   Query 2   Query 3

First choice:

Water 7   Bark 4   Water 7

Figure 6.20. Top retrieval choices for query samples of the weakly homogeneous texture Bark 4 (the 5 rotation angles $(-18°, 18°)$; step $9°$, and 10 scale factors $(0.8, 1.25)$; step 0.05, are exhausted for each query structure).

tures to be matched.

The computational complexity of this approach is rather moderate: for instance, less than *1 sec* per a single IDB entry was spent on a HP 9000 Model 715/50 workstation for a single 170×170 query image if a 101×101

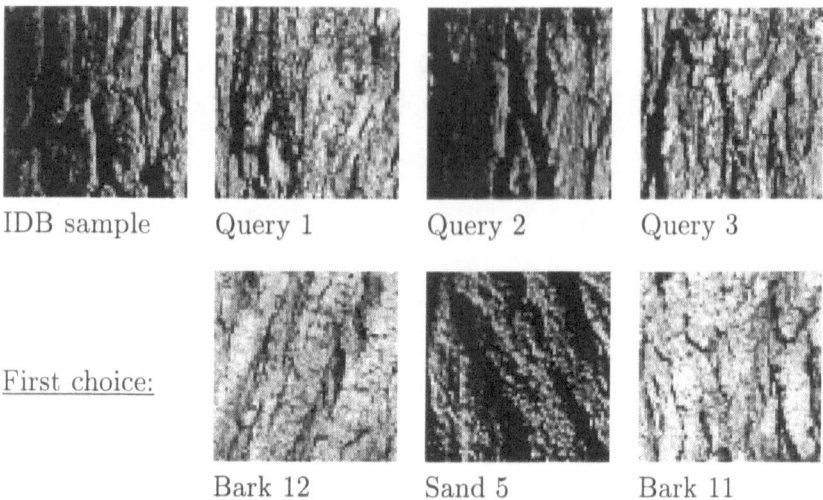

IDB sample   Query 1   Query 2   Query 3

First choice:

                   Bark 12   Sand 5   Bark 11

Figure 6.21. Top retrieval choices for query samples of the weakly homogeneous texture Bark 7 (the 5 rotation angles ($-18°$, $18°$); step $9°$, and 10 scale factors (0.8, 1.0); step 0.05, are exhausted for each query structure).

window is used to compute all the GLDHs and recover the interaction structure and if 9 different rotation angles and 17 different scales are examined for finding the matching score. The above experimental results show the feasibility of this retrieval approach.

Generally, we assume that the IDB contains only homogeneous parts of every piecewise-homogeneous texture so that any IDB entry is represented by one or several homogeneous textures. In practice such an assumption may not always hold. Our experiments show that some weakly homogeneous and inhomogeneous textures can still be retrieved, but with much lower confidence. Another limitation of the proposed retrieval method is the use of only pairwise interactions in the Gibbs model. While this facilitate the model parameter estimation as was shown in Chapters 3 and 4 and simplify the matching computations, it restricts the scope of the model with respect to the retrieval problems.

In spite of these drawbacks, the dissimilarity measure of Eq. (6.3) results in a reliable texture retrieval and can potentially be used in practice.

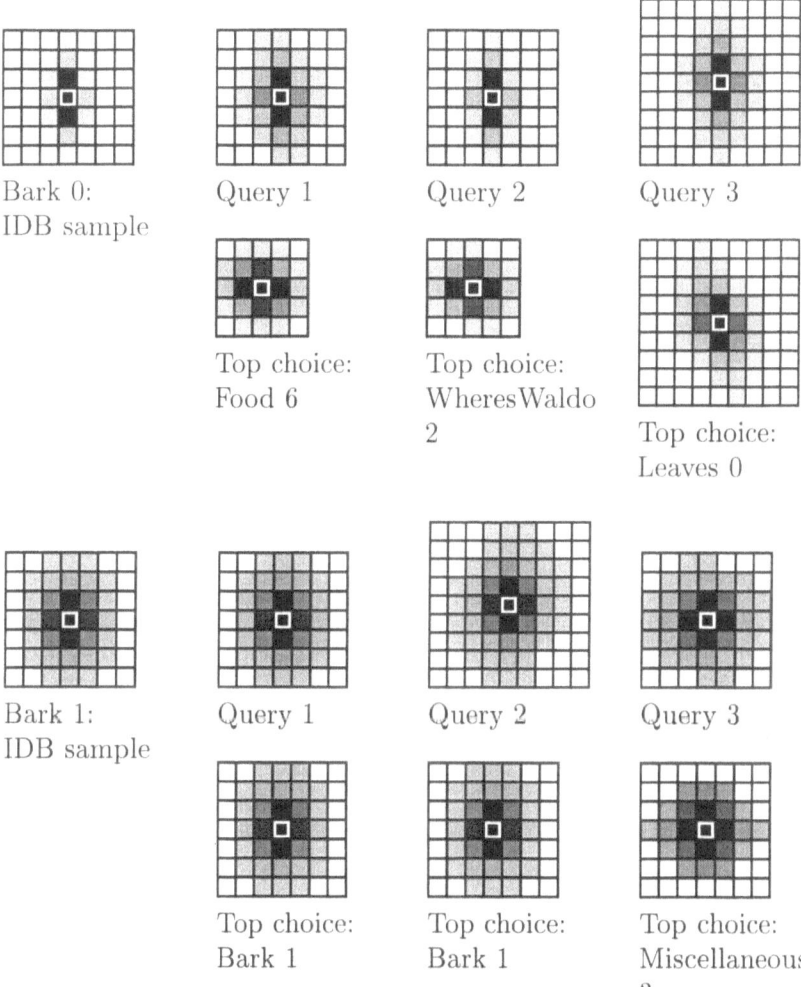

Figure 6.22. Interaction structures for the IDB and query samples and for the top-choice retrieved IDB samples in Figures 6.16 and 6.17.

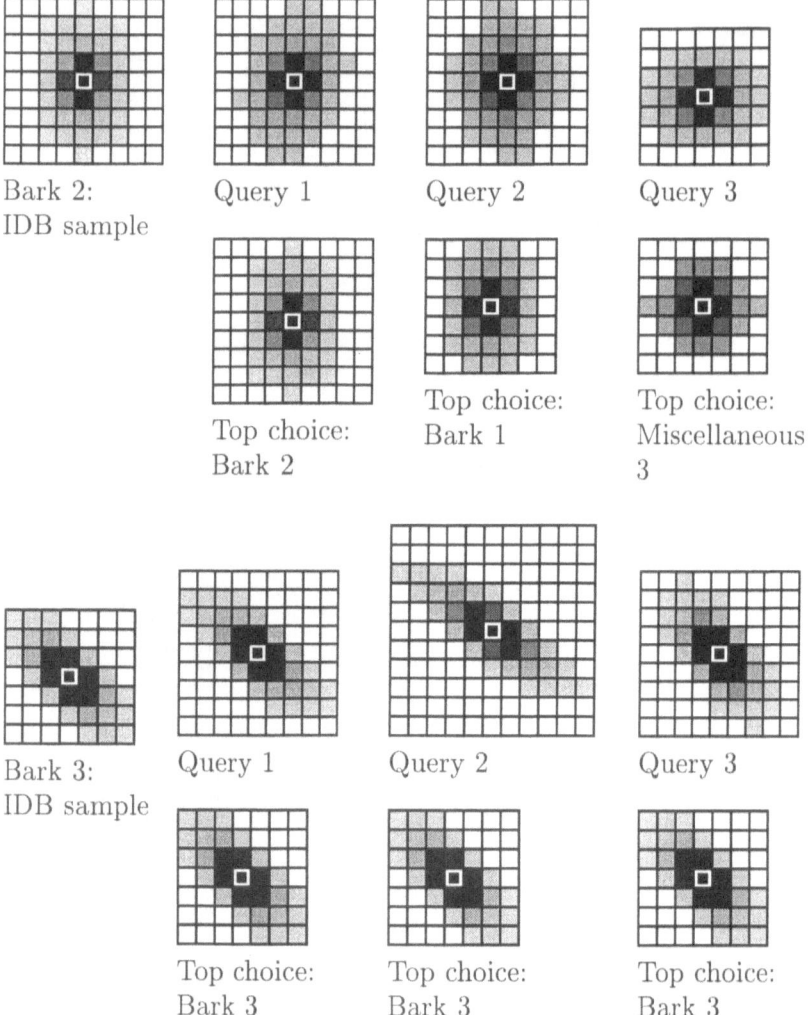

Bark 2:   Query 1   Query 2   Query 3
IDB sample

                Top choice: Top choice: Top choice:
                Bark 2   Bark 1   Miscellaneous
                                                3

Bark 3:   Query 1   Query 2   Query 3
IDB sample

                Top choice: Top choice: Top choice:
                Bark 3   Bark 3   Bark 3

Figure 6.23. Interaction structures for the IDB and query samples and for the top-choice retrieved IDB samples in Figures 6.18 and 6.19.

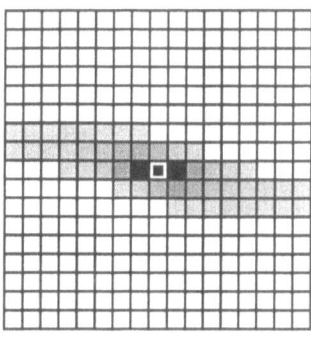

Bark 4:
IDB sample

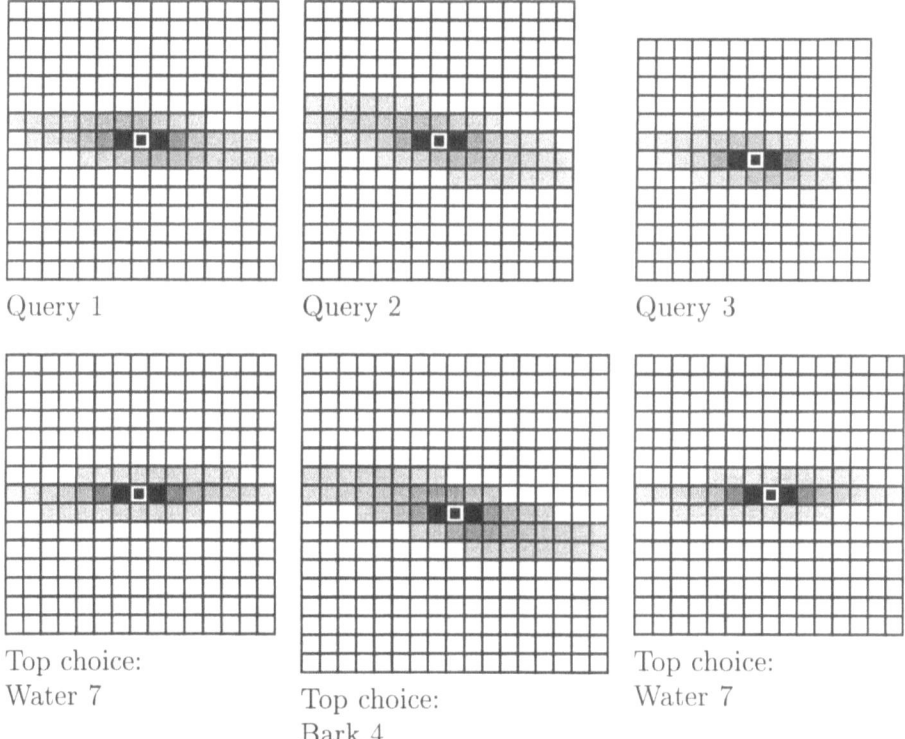

Query 1

Query 2

Query 3

Top choice:
Water 7

Top choice:
Bark 4

Top choice:
Water 7

Figure 6.24. Interaction structures for the IDB and query samples and for the top-choice retrieved IDB samples in Figure 6.20.

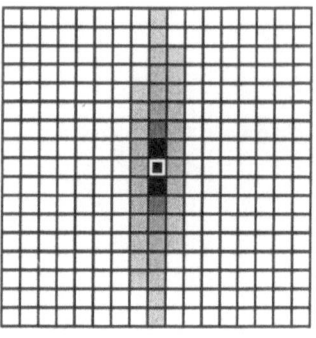

Bark 7:
IDB sample

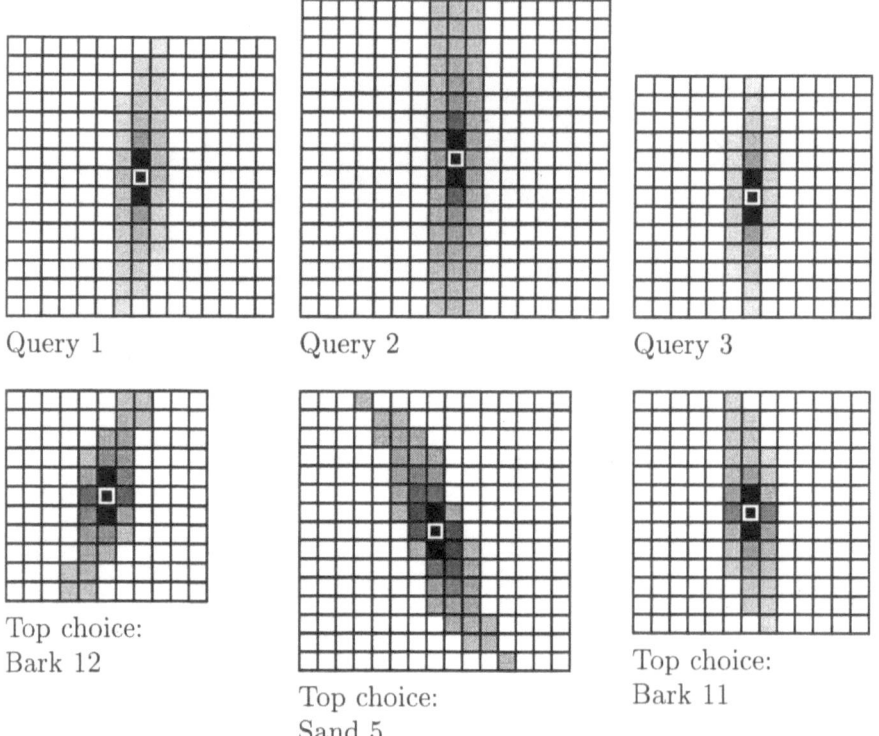

Query 1   Query 2   Query 3

Top choice: Top choice:
Bark 12   Bark 11

                         Top choice:
                         Sand 5

Figure 6.25. Interaction structures for the IDB and query samples and for the top-choice retrieved IDB samples in Figure 6.21.

CHAPTER 7

Experiments in Segmenting Natural Textures

This chapter presents results of the supervised segmentation of different natural textured images and artificial collages of natural textures. The conditional Gibbs model of region maps in Eq. (2.26) suggests that to segment a piecewise-homogeneous grayscale texture we have to simply simulate the region maps relating to this texture. Such a segmentation is also based on the CSA that adapts the model potentials as to approximate the GL/RLH and GLC/RLCHs or GLD/RLCHs for a given training sample by the similar histograms for the image to be segmented and its simulated region map.

7.1. Initial and final segmentation

Generally, only the intra-region interactions between the region labels and gray levels, that is, the interactions of gray levels within the corresponding homogeneous regions, are closely similar in a given training pair of the grayscale texture and region map used for learning the model parameters and in the test images of the same type which have to be segmented. Only the intra-region interactions are described by the variant of the conditional model in Eq. (2.26) with zero-valued inter-region potentials and resulting potential centering of Eq. (2.30).

The inter-region interactions may differ much for the piecewise-homogeneous grayscale textures of the same type because of the quite distinct relative arrangement and sizes of the homogeneous regions with the same texture. Moreover, the subregions where the cliques of a particular family cross the region borders have usually very small sizes in comparison with the regions themselves. Therefore, the collected GLD/RLCHs may inadequately describe the true marginal probabilities of the inter-region interactions in the training sample so that the test image to be segmented and its segmentation map, simulated under the general Gibbs model of Eq. (2.26), may possess very different inter-region statistics.

In our experiments below, the characteristic interaction structure and potentials are learnt separately under the centering conditions of Eq. (2.29) and of Eq. (2.30) from the GL/RLH and GLD/RLCHs for the same training sample. Then the following two-stage procedure is applied to segment a given piecewise-homogeneous test image.

1. *Initial segmentation.* At the first stage, the model of Eq. (2.26) with the potential centering of Eq. (2.30) is used to obtain the initial segmentation map for a given image. The CSA is starting from a sample of the IRF of region labels.

2. *Final segmentation.* At the second stage, the model of Eq. (2.26) with the potential centering of Eq. (2.29) is applied to form the final region map. The CSA is starting from the initial map obtained at the first stage.

7.2. Artificial collages of Brodatz textures

7.2.1. FIVE-REGION COLLAGE

Figure 7.1 shows the five-region collage 256×256 of the natural textures and its ideal region map. The textures are selected arbitrarily from the album (Brodatz, 1966). The training collage 256×256 and its region map are presented in Figure 7.2. This training pair is formed by a random arrangement of five patches 64×64 of these texture types. There are four randomly chosen positions for the patch of type 0 and three such positions for the patches of each other type $(1, \ldots, 4)$.

The initial segmentation map, shown in Figure 7.3, a, is obtained after 300 macrosteps of the CSA with the control parameters $c_0 = 0$; $c_1 = 1$; $c_2 = 0.001$. Figure 7.3, b, demonstrates the final segmentation map obtained also after 300 macrosteps of the CSA with the same parameters which was started from the initial map. The final map is sufficiently close to the ideal one in spite of small deviations of the region borders. The deviations are mostly due to local similarities between small patches of these textures. Also, the inter-region interactions in the training sample differ slightly from the same interactions in the test collage because of the distinct region borders and arrangement. Figure 7.3, c displays the segmentation errors (here, the black pixels indicate all the positions with different labels in the ideal and segmentation maps). The relative error rate is equal to 3.08%.

Figure 7.4 displays separately the texture regions obtained by the final segmentation. It is evident that all the found regions are equivalent to the ideal ones as regarding the intra-region texture homogeneity.

7.2.2. DIFFERENT FOUR-REGION COLLAGES

This section presents results of segmenting the 8 four-region collages 256×256 of different homogeneous, weakly homogeneous, and inhomogeneous textures from the album (Brodatz, 1966). All the training collages have

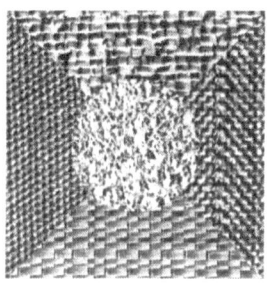 

Figure 7.1.   Collage 256×256 to be segmented and its ideal region map.

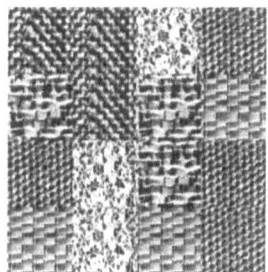 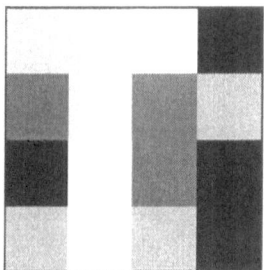

Figure 7.2.   Training collage 256×256 and its region map.

the same region map with four regions. Each region $0, \ldots, 3$ contains 16384 pixels and is formed by random arrangement of four square patches 64×64.

The training map is used for cutting the corresponding patches from the initial Brodatz textures 256×256. Each training collage is formed by the textures ranked by their serial numbers in (Brodatz, 1966), in particular,

1. D3, D4, D5, and D9 in Figure 7.5,

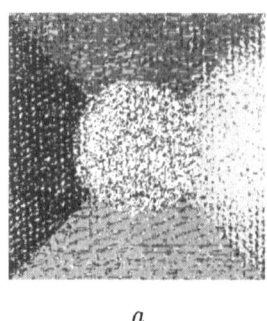 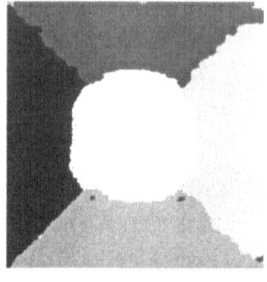 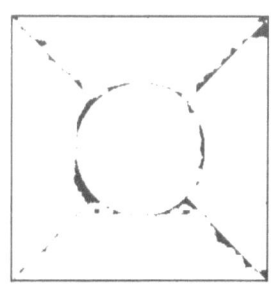

a $\qquad\qquad\qquad\qquad$ b $\qquad\qquad\qquad\qquad$ c

Figure 7.3.   Initial and final segmentation maps and positions of the segmentation errors.

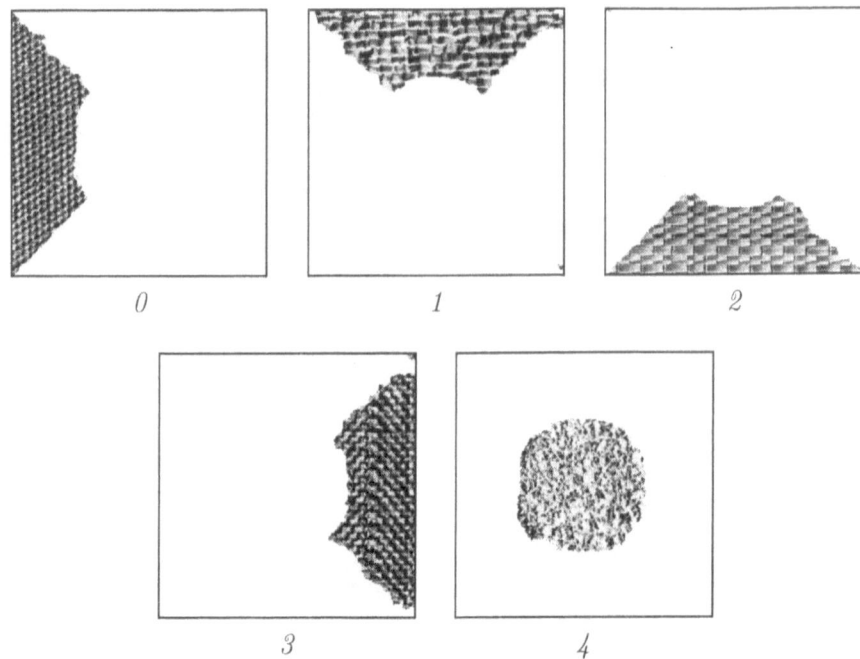

Figure 7.4.   Texture regions after the segmentation.

2. D11, D12, D17, D20 in Figure 7.6,
3. D23, D24, D29, D34 in Figure 7.7,
4. D50, D55, D57, D65 in Figure 7.8,
5. D66, D68, D69, D74 in Figure 7.9,
6. D75, D76, D77, D79 in Figure 7.10,
7. D83, D84, D85, D92 in Figure 7.11, and
8. D93, D95, D101, D103 in Figure 7.12.

All the test collages also share a single region map generated by using the Markov/Gibbs map model of Eq. (2.17). The model parameters were learnt from the training map. The regions 0,1,2,3 in the test map have different areas: 23382, 17502, 17604, and 7048 pixels, respectively. The test collages are also formed by cutting the corresponding patches from the same initial textures 236×256 as the trining ones. But because the test region map is distinct from the training one, each test collage contains the texture patches which are much different from the training ones.

Figures 7.5 – 7.12 display the training and test grayscale collages, their region maps, and initial and final segmentation maps obtained by the CSA for the training and test collages. Both the initial and final segmentation maps are obtained by 300 macrosteps of the CSA with the control parameters $c_0 = 0$, $c_1 = 1$, and $c_2 = 0.001$. The Gibbs model parameters are learnt from the training sample. In these experiments we take no account of the pixelwise interactions, and assume that the pairwise pixel interactions depend only on gray level differences in the cliques. Therefore, the image patches which differ by their average gray value may still represent the same texture type with respect to this particular Gibbs model.

Table 7.1 lists the total error rates for the final segmentation maps in terms of relative numbers of the misclassified pixels, that is, pixels having the different labels in the ideal and obtained maps. The segmentation results indicate that the training samples have notable distinctions from the test ones as regarding the statistics of the inter-region pairwise interactions. It is obvious also from Figure 7.13 that presents the results of segmenting one more test collage with the textures D23, D24, D29, and D34. Its region map is quite similar to the training one but has the different region labeling (namely, the region number $3 - k$ instead of the number $k = 0, \ldots, 3$ in the training map). In this case the total error rate becomes significantly lower than in Figure 7.7 (7.20% vs. 34.43 %). Thus, in the case of our conditional Gibbs model of Eq. (2.26), the closer the inter-region interactions in the test images to the training sample, the lower the expected error.

TABLE 7.1. Relative segmentation errors (in percents)

Figure	7.5	7.6	7.7	7.8	7.9	7.10	7.11	7.12
Training collage	2.46	0.31	3.44	0.01	0.05	0.02	2.67	0.42
Test	25.50	20.98	34.43	19.20	19.16	21.01	19.67	16.78

It is worthy to note that borders between the texture regions present main difficulties to the proposed segmentation because it is the inter-region statistics and, therefore, the potentials that most likely differ in the test images comparing to the training sample. It is easily seen in Figure 7.14 that indicates the differences between the ideal and segmentation maps for all the above-mentioned four-region test collages. The more the textures have to be separated in the test image, the less definite is the reconstruction of their borders. This holds true, especially, for our experiments because (i) we exploit the simplified Gibbs model of Eq. (2.26) with the same fixed interaction structure for all the textures and regions and (ii) the characteristic interaction structure is recovered by simply thresholding the relative partial energies in the interaction map with no account of resulting discrimination between the individual texture types. Thus, there exist potentialities to amplify the obtained experimental results to be investigated in future.

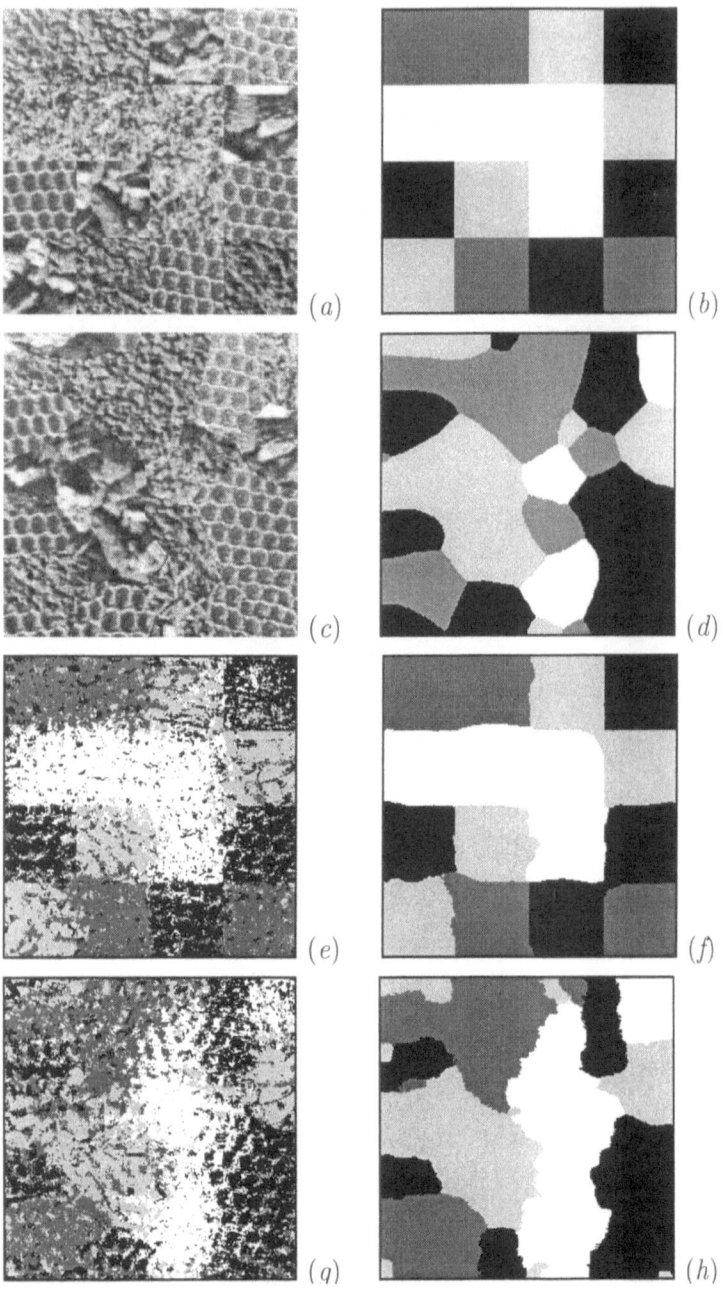

Figure 7.5. Segmentation of the four-region collage of the textures D3, D4, D5, D9 (the training collage (*a*) and its region map (*b*), test collage (*c*) and its ideal map (*d*), initial (*e*) and final (*f*) segmentation maps for the training collage, and initial (*g*) and final (*h*) segmentation maps for the test collage).

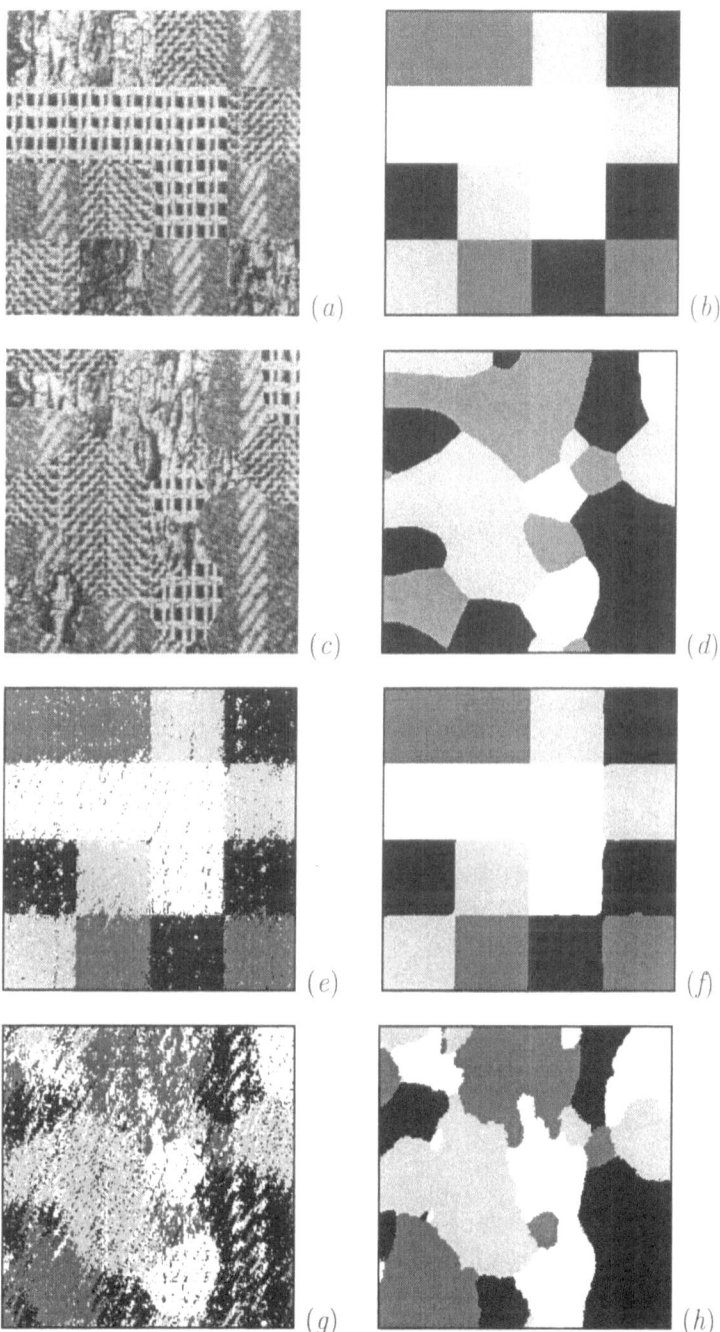

Figure 7.6. Segmentation of the four-region collage of the textures D11, D12, D17, D20 (the same notation as in Figure 7.5).

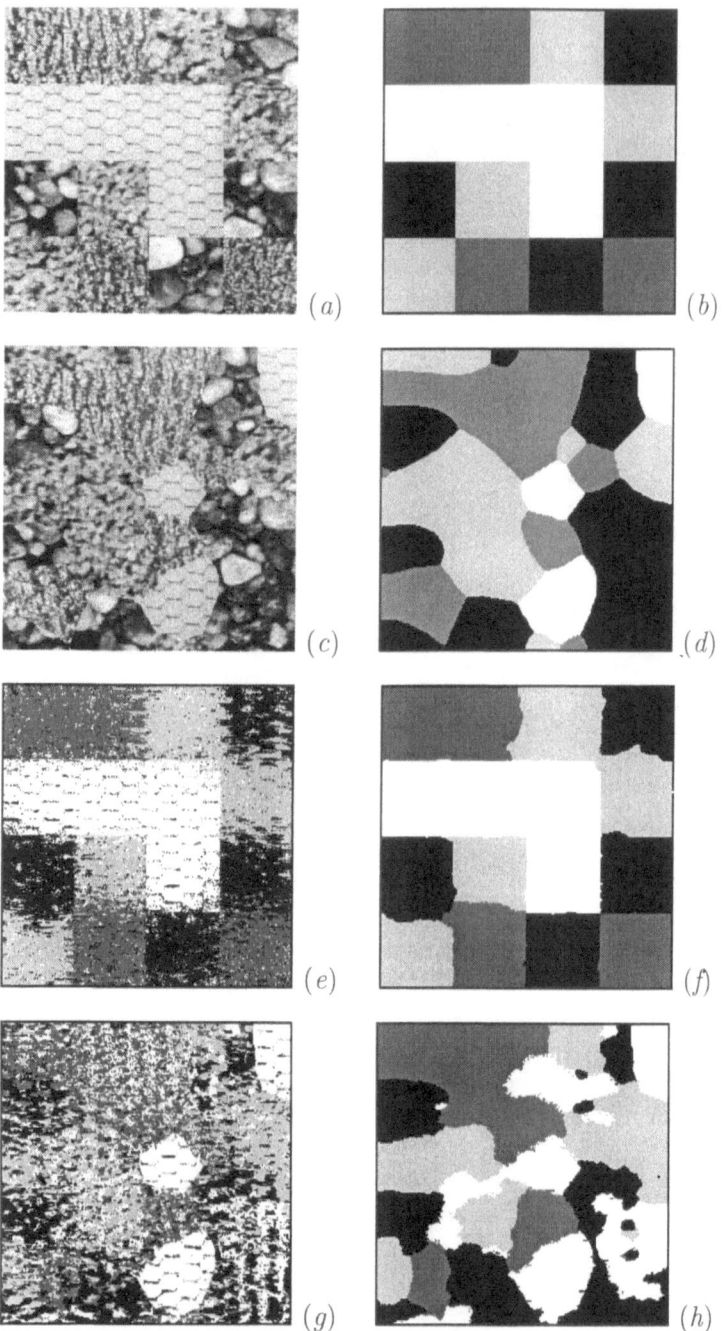

Figure 7.7. Segmentation of the four-region collage of the textures D23, D24, D29, D34 (the same notation as in Figure 7.5).

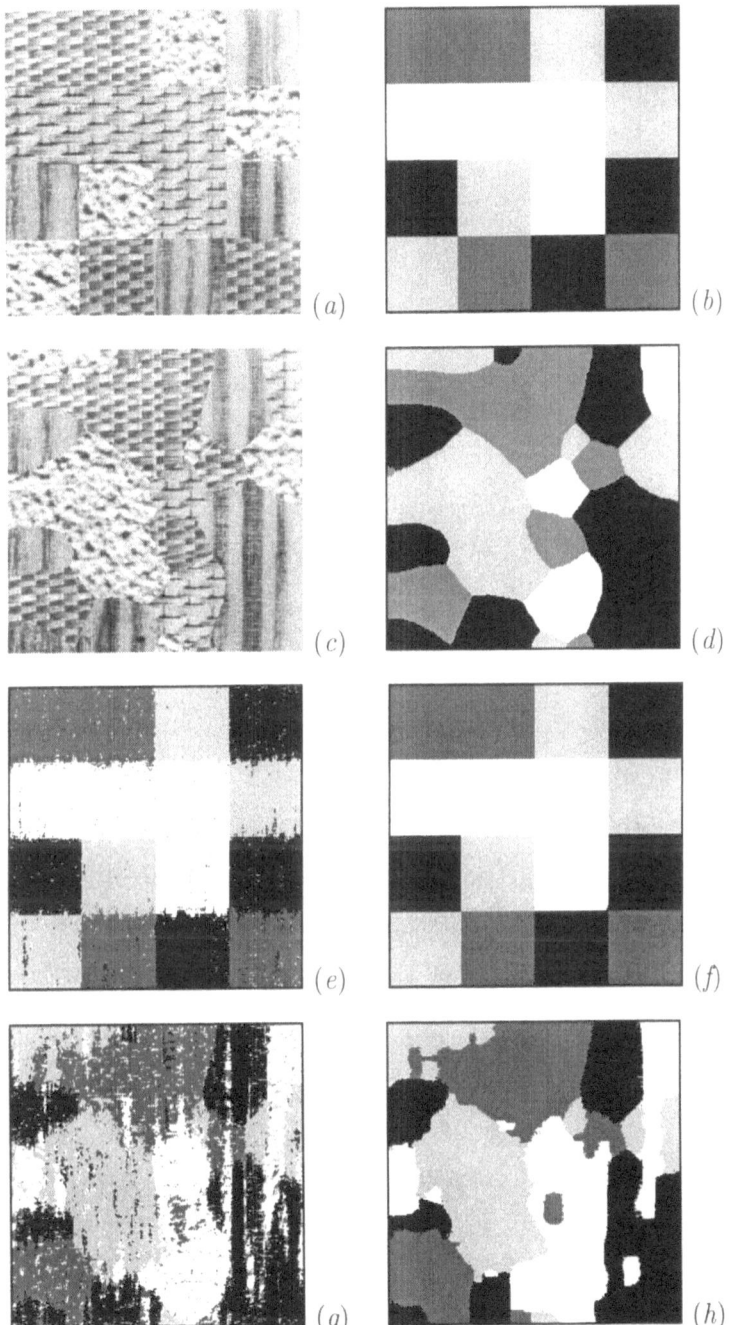

Figure 7.8.   Segmentation of the four-region collage of the textures D50, D55, D57, D65 (the same notation as in Figure 7.5).

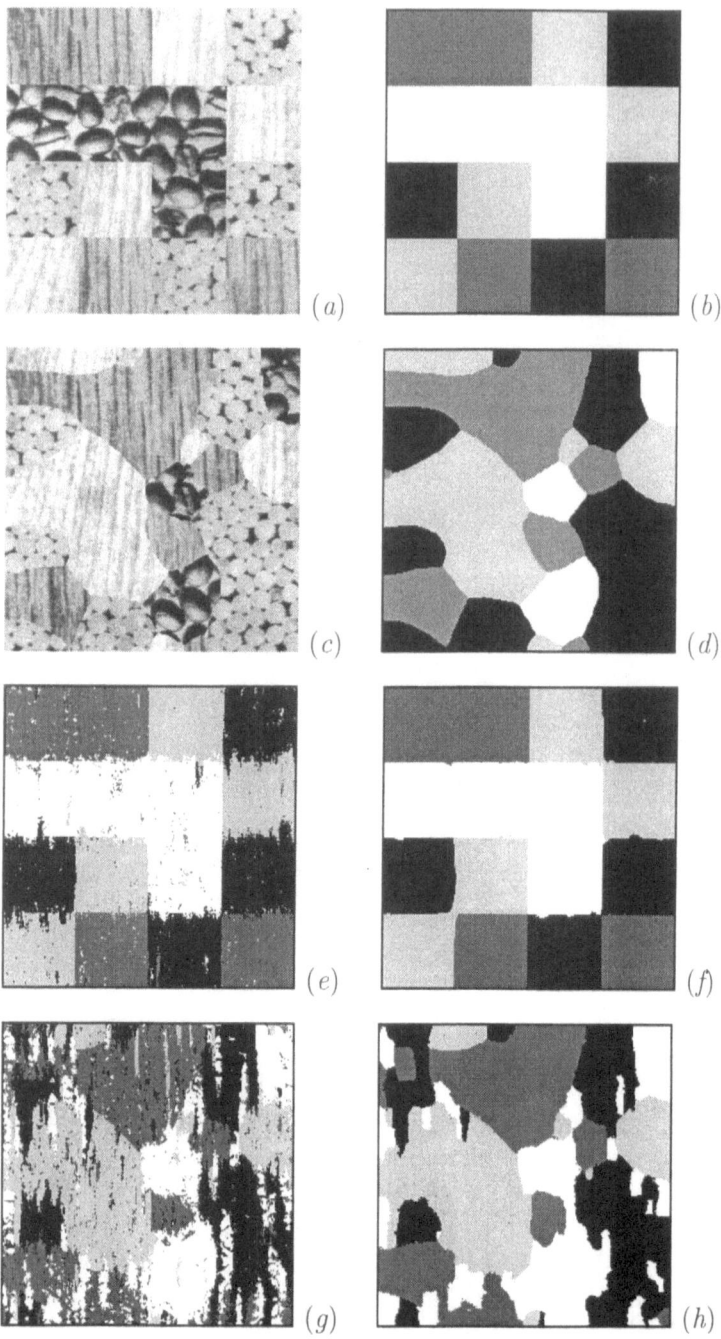

Figure 7.9. Segmentation of the four-region collage of the textures D66, D68, D69, D74 (the same notation as in Figure 7.5).

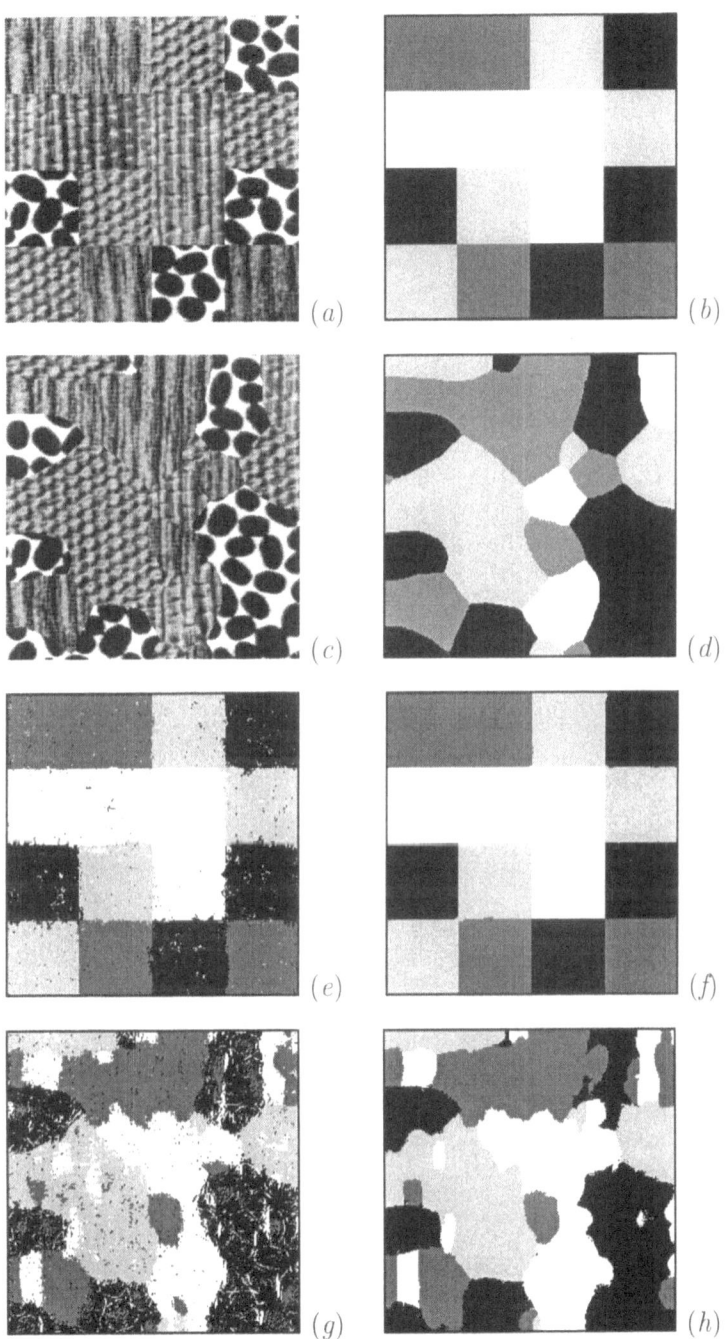

Figure 7.10.   Segmentation of the four-region collage of the textures D75, D76, D77, D79 (the same notation as in Figure 7.5).

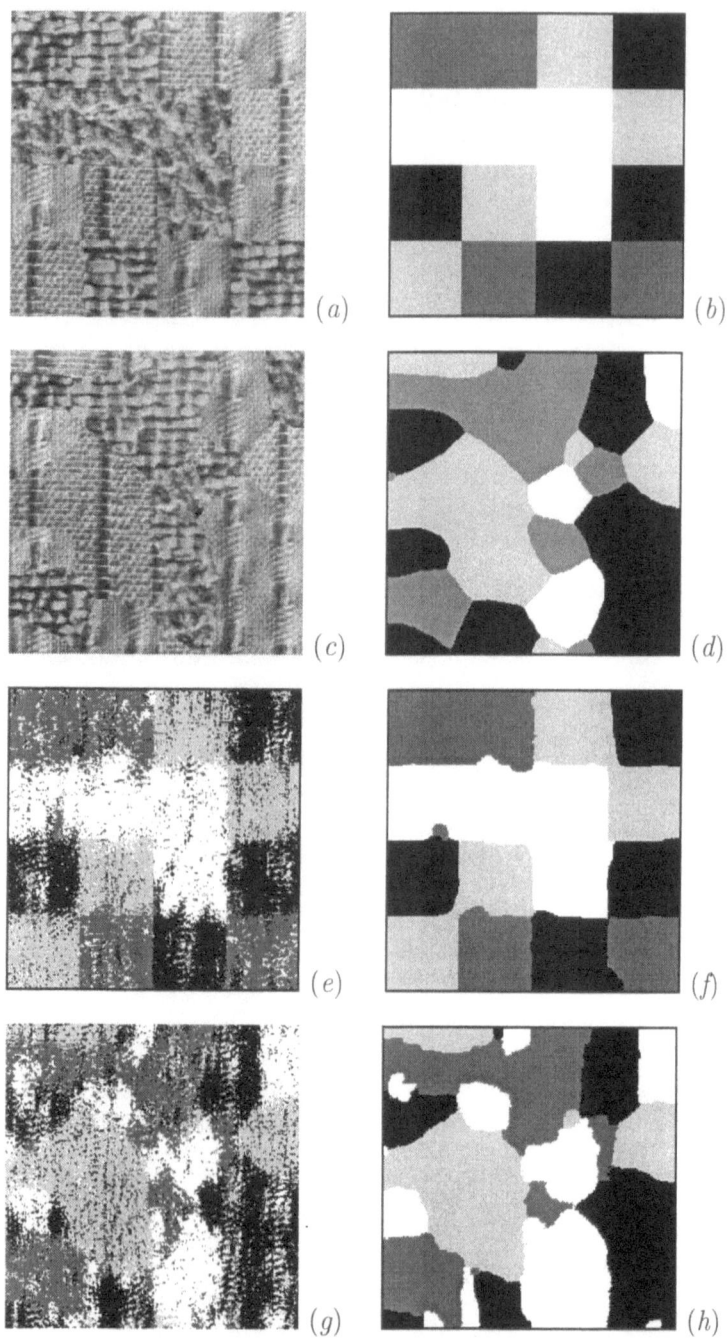

Figure 7.11. Segmentation of the four-region collage of the textures D83, D84, D85, D92 (the same notation as in Figure 7.5).

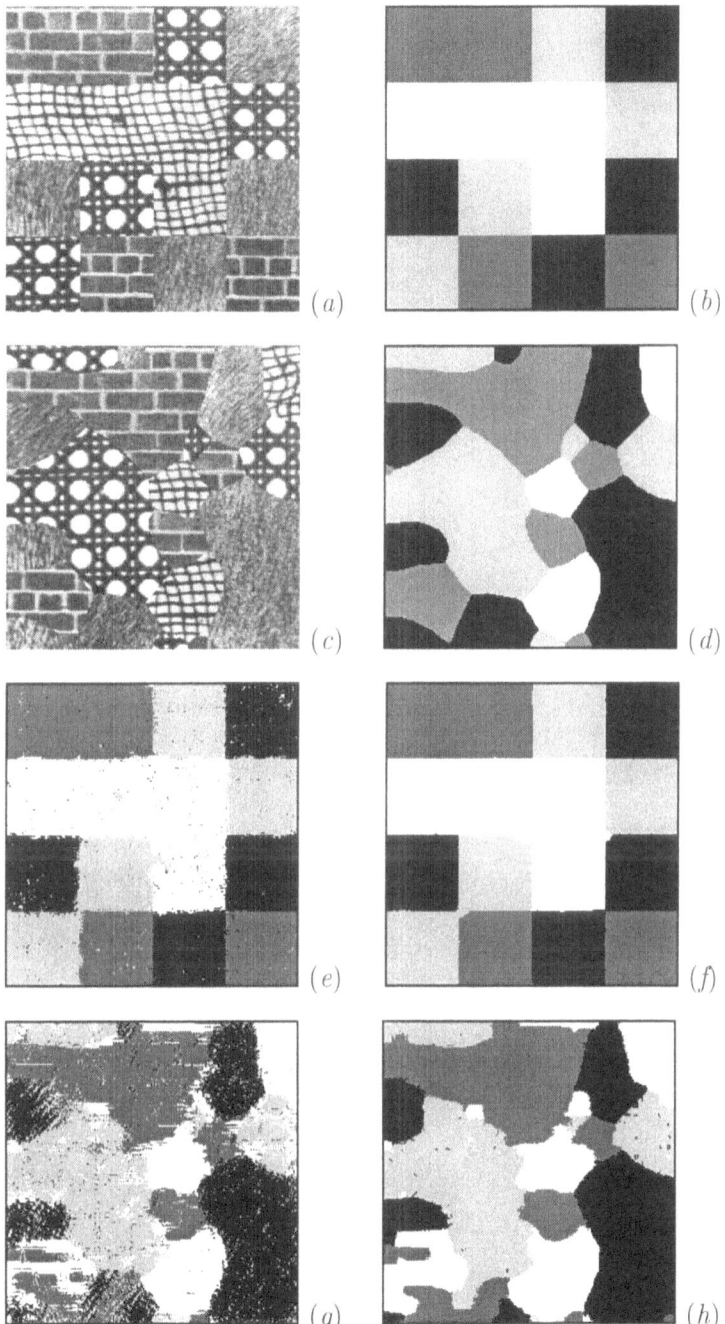

Figure 7.12.   Segmentation of the four-region collage of the textures D93, D95, D101, D103 (the same notation as in Figure 7.5).

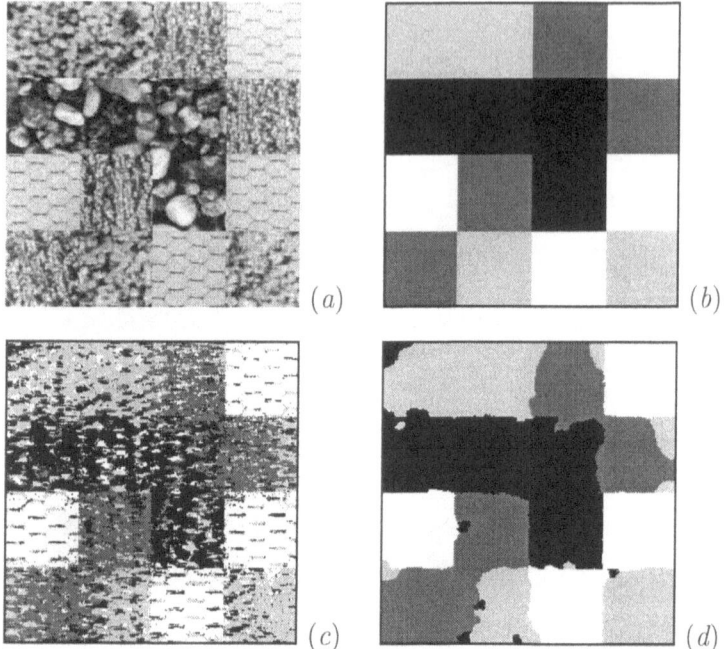

Figure 7.13. Segmentation of the four-region collage of the textures D23, D24, D29, D34 (the test collage (*a*), its ideal map (*b*), and initial (*c*) and final (*d*) segmentation maps; the training sample is the same as in Figure 7.7).

Inhomogeneous textures and relatively small regions of weakly homogeneous textures in the test images present the greatest difficulties in our supervised segmentation. In this case small patches of different texture types could be combined into arbitrary "weakly homogeneous" regions having incidentally the same pairwise statistics of gray levels and region labels which had been learnt from a given training sample. To illustrate such a possibility, the regions found by segmenting the above test collages are presented separately in Figures 7.15 – 7.18.

It is clear that sufficiently large patches of the homogeneous textures are recovered with a rather good precision whereas the small or insufficiently homogeneous patches result in the main segmentation errors.

7.2.3. COLLAGE OF 16 TEXTURES

Figures 7.19, 7.20, and 7.21 demonstrate the 16-region test collage 512×512 of various homogeneous, weakly homogeneous, and almost inhomogeneous textures from (Brodatz, 1966) and its ideal region map, the training pair

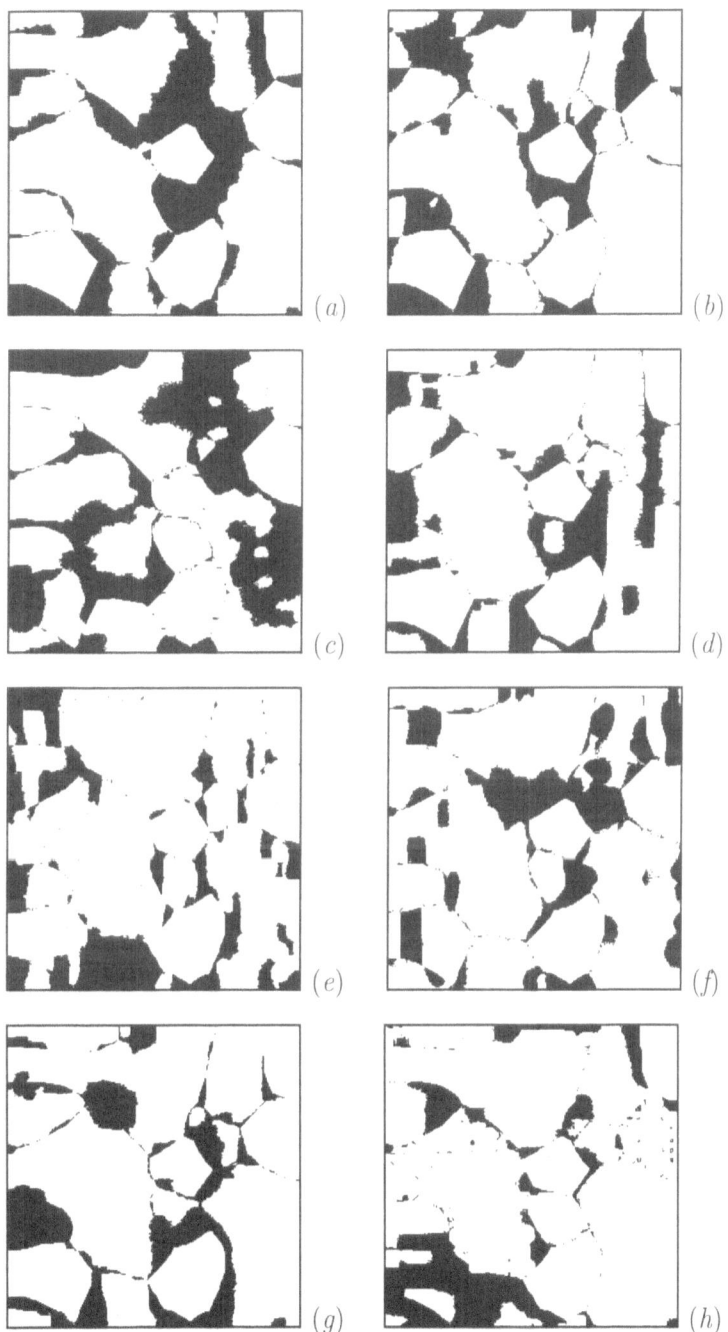

Figure 7.14. Differences between the ideal and segmentatin maps for the test collages in Figures 7.5 (*a*), 7.6 (*b*), 7.7 (*c*), 7.8 (*d*), 7.9 (*e*), 7.10 (*f*), 7.11 (*g*), 7.12 (*h*).

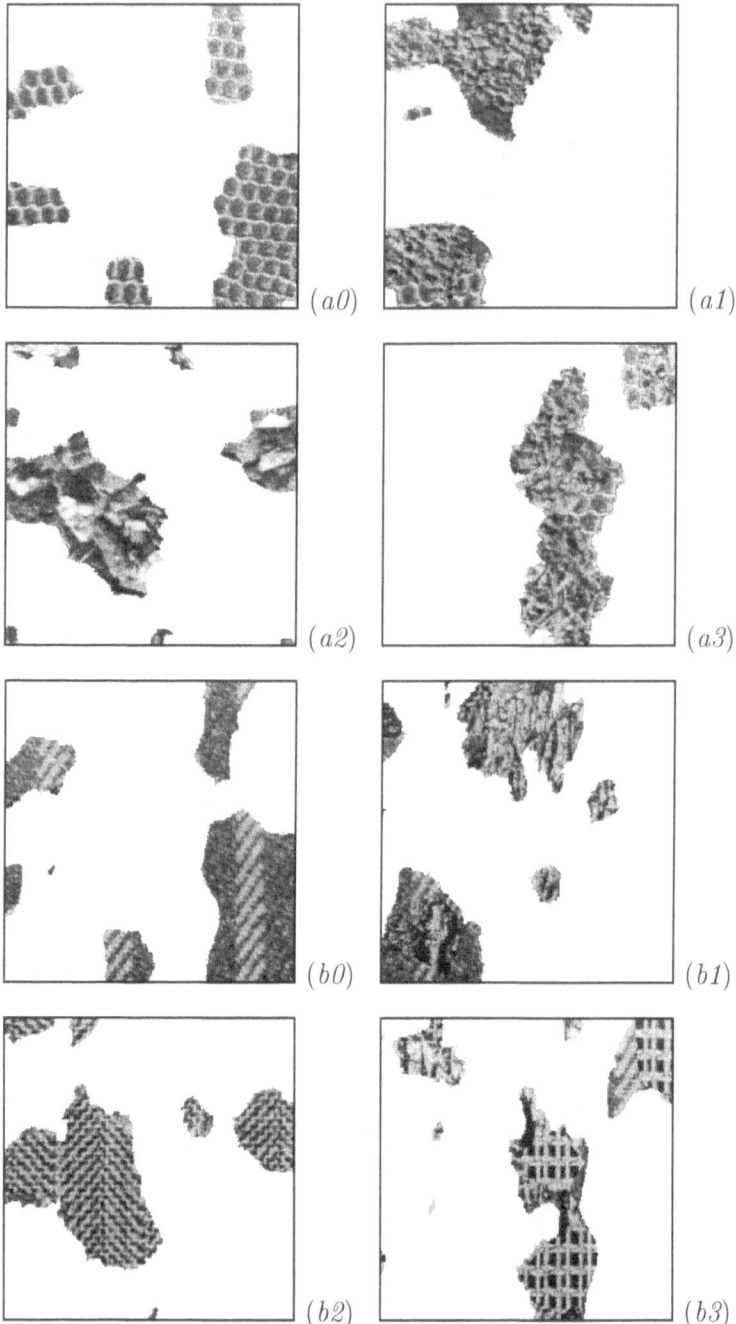

Figure 7.15. Texture regions *a0–a3* and *b0–b3* obtained by segmenting the test collages in Figures 7.5 and 7.6, respectively.

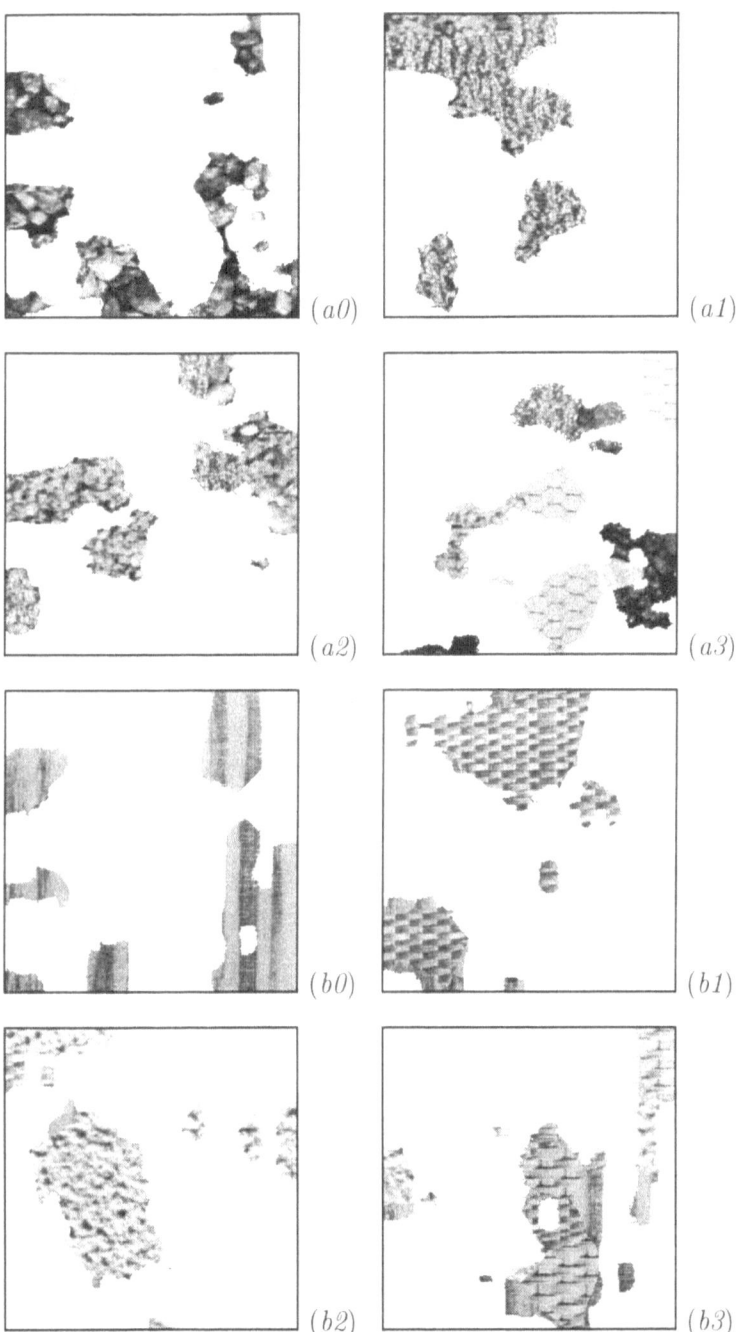

Figure 7.16. Texture regions *a0–a3* and *b0–b3* obtained by segmenting the test collages in Figures 7.7 and 7.8, respectively.

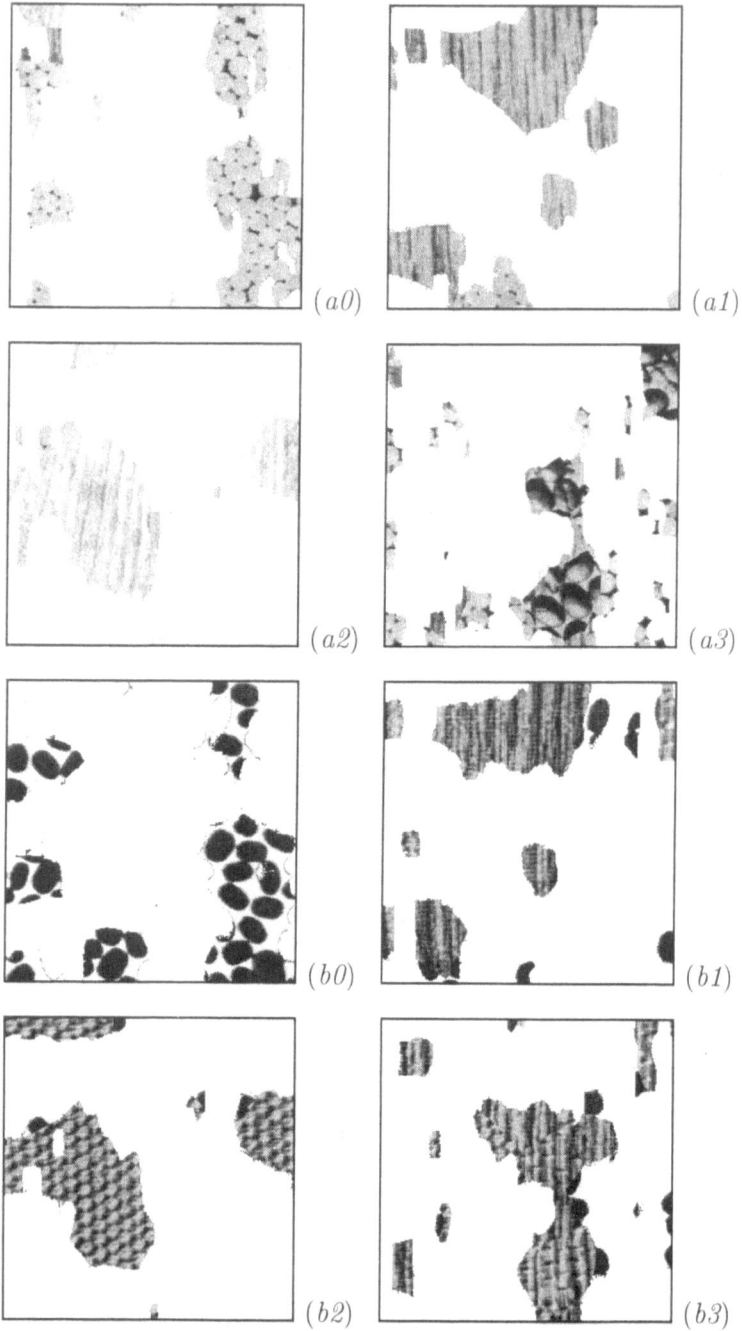

Figure 7.17. Texture regions *a0–a3* and *b0–b3* obtained by segmenting the test collages in Figures 7.9 and 7.10, respectively.

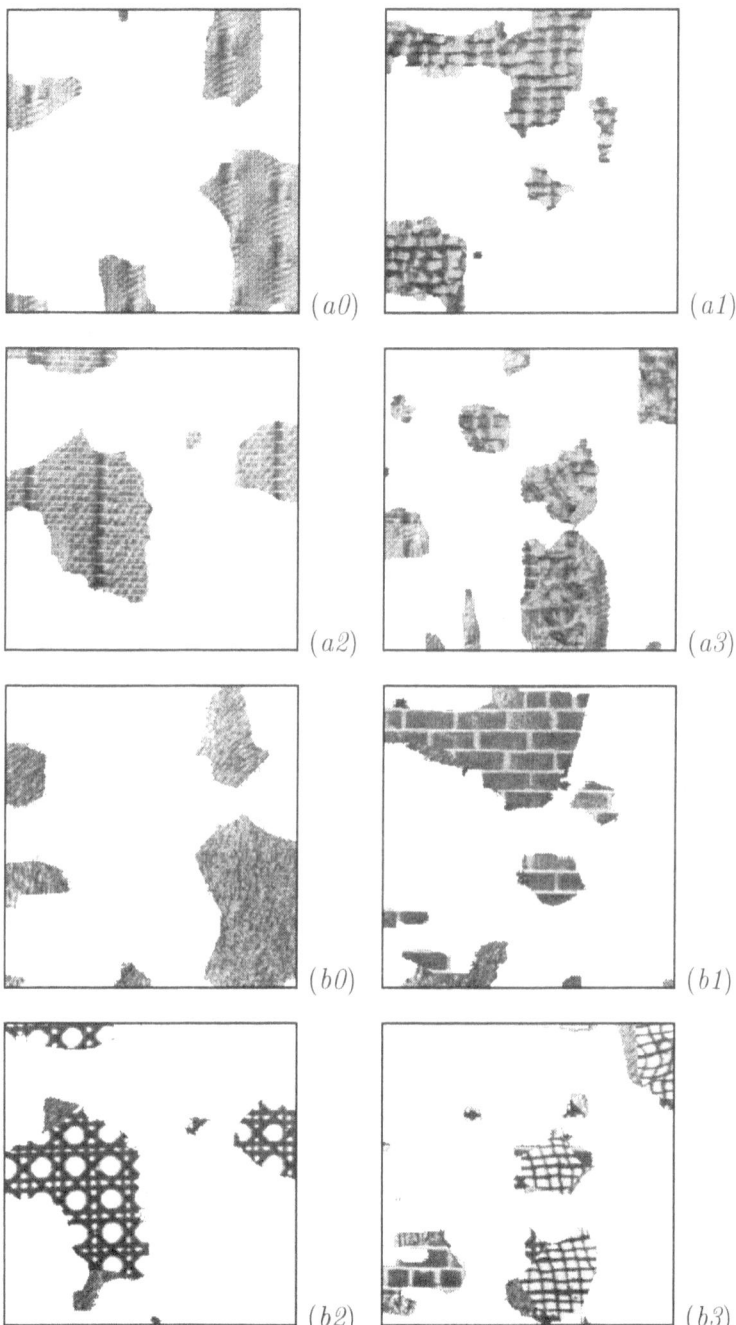

Figure 7.18. Texture regions *a0–a3* and *b0–b3* obtained by segmenting the test collages in Figures 7.11 and 7.12, respectively.

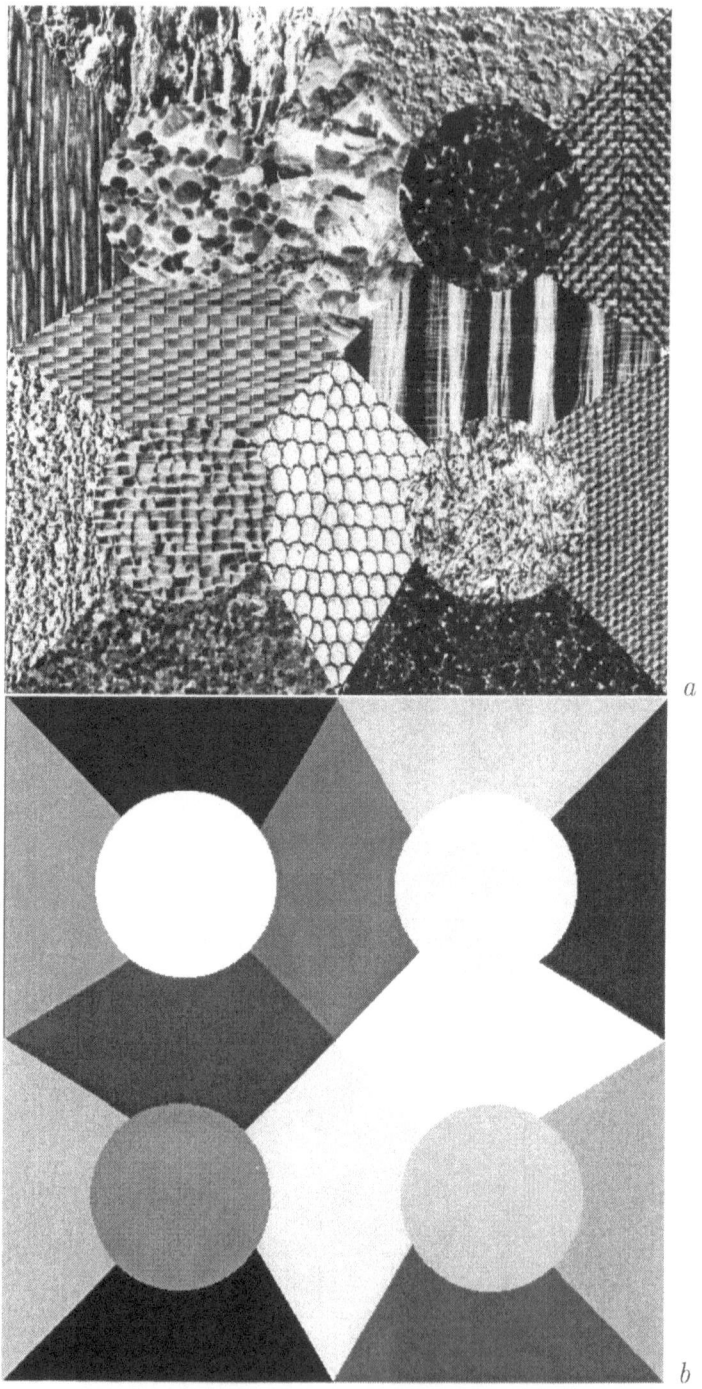

Figure 7.19.   16-region texture collage 512 × 512 (*a*) and its ideal region map (*b*).

TABLE 7.2. Absolute and relative errors in the final 16-region segmentation map.

Region	Size in the ideal map	Errors	In %%
0	15057	564	3.7
1	15241	3144	20.6
2	15321	634	4.1
3	19814	3227	16.3
4	15127	424	2.8
5	19503	7003	35.9
6	15440	1539	10.0
7	15311	716	4.7
8	14696	298	2.0
9	15007	1133	7.5
10	15440	1121	7.3
11	15374	1053	6.8
12	20281	4594	22.7
13	15440	810	5.2
14	19652	5682	28.9
15	15440	4297	27.8

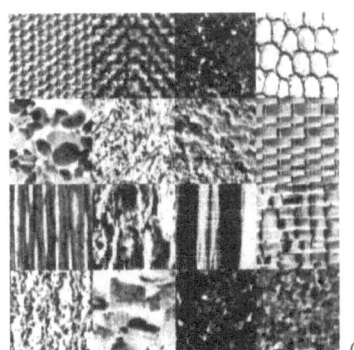 

Figure 7.20.   16-region training collage 256 × 256 (*a*) and its region map (*b*).

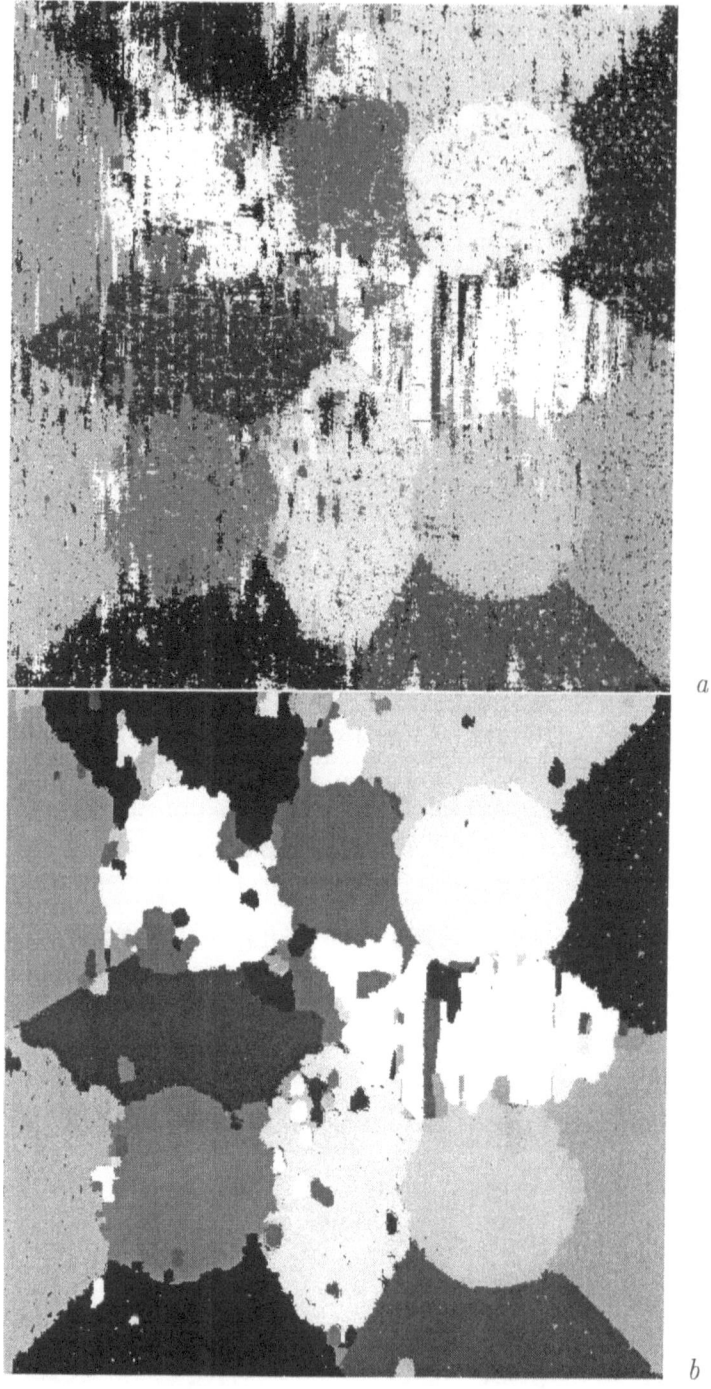

Figure 7.21.   Initial(*a*) and final (*b*) 16-region segmentation map.

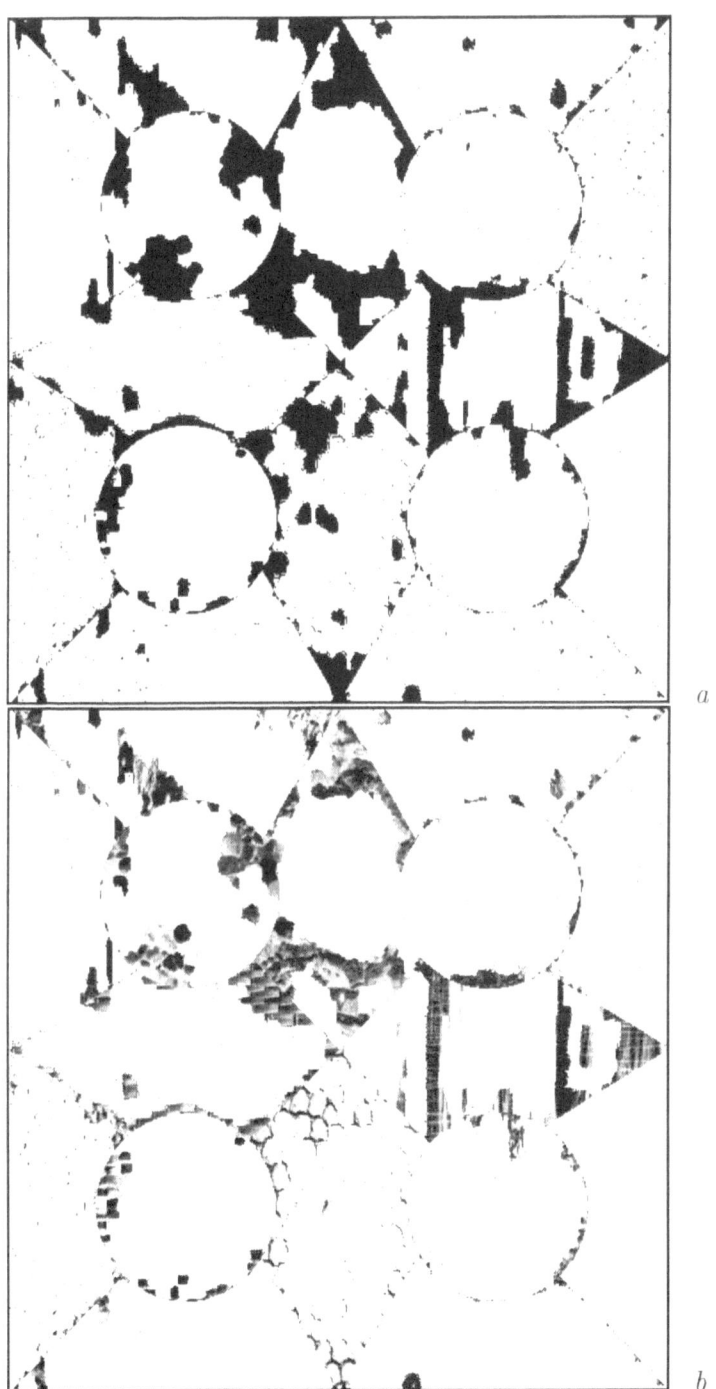

Figure 7.22.   Segmentation errors (*a*) and corresponding textures (*b*).

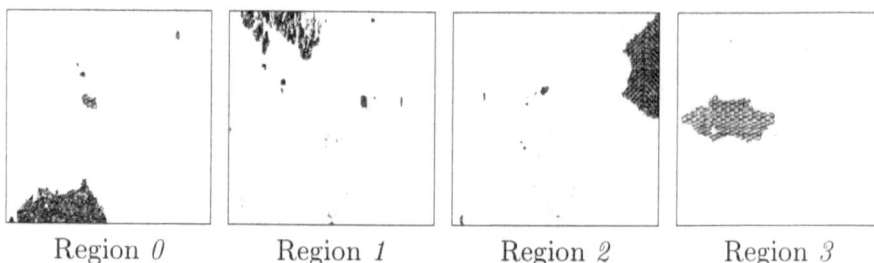

Region *0*   Region *1*   Region *2*   Region *3*

Figure 7.23.   Texture regions 0 . . . 3 after the segmentation (on a reduced scale).

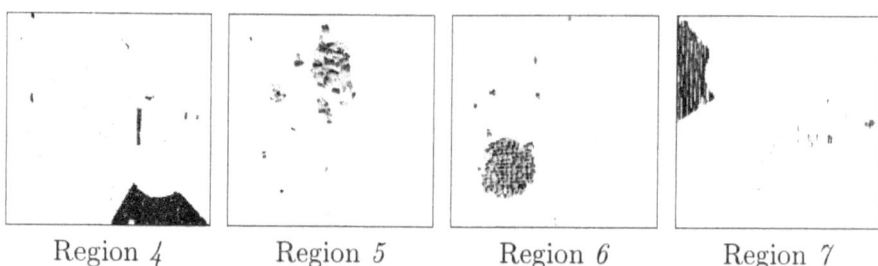

Region *4*   Region *5*   Region *6*   Region *7*

Figure 7.24.   Texture regions 4 . . . 7 after the segmentation (on a reduced scale).

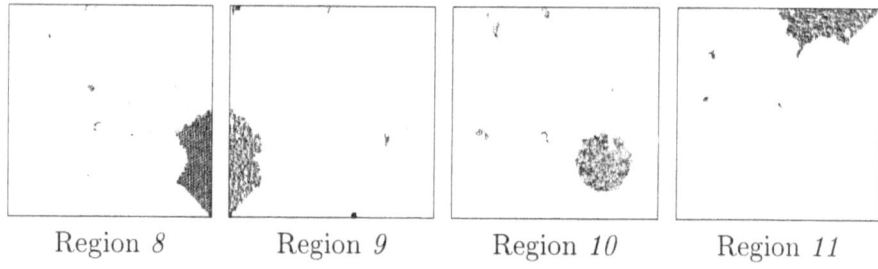

Region *8*   Region *9*   Region *10*   Region *11*

Figure 7.25.   Texture regions 8 . . . 11 after the segmentation (on a reduced scale).

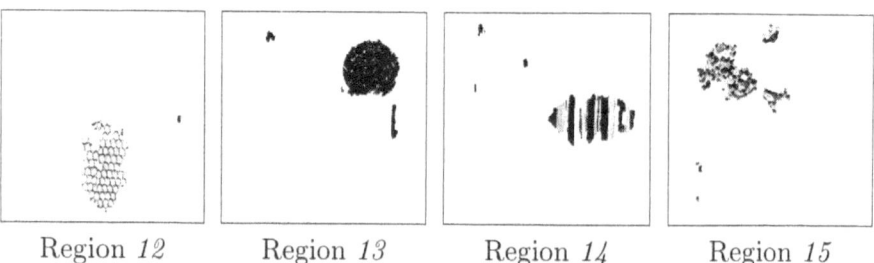

Region *12*   Region *13*   Region *14*   Region *15*

Figure 7.26.   Texture regions 12 . . . 15 after the segmentation (on a reduced scale).

256×256 with randomly arranged 16 patches 64×64 (one patch per texture type), and the resulting initial and final segmentation maps. The initial map in Figure 7.21, a is obtained by a two-stage CSA: (i) 1000 macrosteps with the control parameters $c_0 = 0.$, $c_1 = 10.$, $c_2 = 1.$ at the first stage and (ii) additional 300 CSA-macrosteps with the control parameters $c_0 = 0.$, $c_1 = 1.$, $c_2 = 0.001.$ at the second stage. The control parameters are changed at the second stage in order to accelerate the convergence of the joint GLD/RLCHs collected for the test image and current segmentation map to the desired training histograms. The total chi-square distance between these intra-region GLD/RLCHs is reduced, respectively, from $172,000$ to $32,200$ at the first stage and from $32,200$ to $12,400$ at the second stage. The final segmentation map is obtained by the similar two-stage CSA: (i) 300 macrosteps with the control parameters $c_0 = 0.$, $c_1 = 10.$, $c_2 = 1.$ to reduce the total chi-square distance between the intra- and inter-region GLD/RLCHs from $1,603,000$ to $67,400$ at the first stage and (ii) subsequent 300 CSA-macrosteps with the control parameters $c_0 = 0.$, $c_1 = 1.$, $c_2 = 0.001.$ to reduce this distance from $67,400$ to $52,100$ at the second stage.

Figures 7.23 – 7.26 shows 16 texture regions found by the final segmentation. The segmentation errors with respect to the ideal 16-region map , that is, the positions with different labels in the ideal region map and final segmentation map, are depicted by black pixels in Figure 7.22,a. Figure 7.22,b shows the textured areas in the test image which cause these segmentation errors. The total segmentation error (that is, difference between the obtained and ideal region maps) is about 13.82% (32239 pixels with different labeling among 262144 pixels in the lattice). Table 7.2 gives absolute and relative errors for each region in the test image. It is easily seen that most differences between the ideal and segmentation maps are due to similarities between subparts of different weakly homogeneous textures. The texture inhomogeneities result in significant distinctions between the inter- and intra-region statistics in the training and segmented images. But, visually the found textures are sufficiently homogeneous over the regions obtained by segmenting.

7.3. Natural piecewise-homogeneous images

7.3.1. DISCRIMINATION OF LANDFORMS

Human classification of complex topographic forms is usually based on salient textural attributes of the Earth's surface. Thus, as is shown in (Gimel'farb et al., 1999), the texture segmentation can be also used for discriminating different landform types if the digital elevation model (DEM) of the surface is represented by a range or slope image[1]. The Gibbs conditional model of Eq. (2.26) is able to roughly segment these images into some meaningful spatially homogeneous regions.

Here, the DEMs of north-central New Mexico, USA, in Figure 7.27 (Dikau et al., 1995) and of the Neckar catchment in southwest Germany in Figure 7.28 with spatial resolution of 200m are used as test areas.

Figure 7.27.   Test area in New Mexico (south-southeast view)

Figures 7.29 and 7.30 present both the range images of these DEMs (947 rows × 986 columns "New Mexico" and 654 × 478 "Neckar") and the corresponding landform segmentation maps produced by the automated Hammond–Dikau classification (Dikau et al., 1995). The classification is based on thresholding certain textural features computed for a moving window. The window is moving with no overlap over a given DEM, and three quantitative attributes measuring the slope, relief, and profile of a surface are calculated for each window position and compared to certain fixed thresholds. Here, the segmentation maps represent only three basic major landforms such as plains and tablelands (PTL), plains with hills and mountains (PHM), and open hills, hills, and mountains (OHM).

Parameters of the conditional Gibbs model of Eq. (2.26) are learnt from the training samples in Figures 7.31. Each sample contains a particular part

---

[1]The range and slope images of a DEM on a lattice **R** are obtained by grayscale coding of elevations or slope angles in the pixels, respectively.

Figure 7.28.   Test area in the Neckar catchment (north-northwest view)

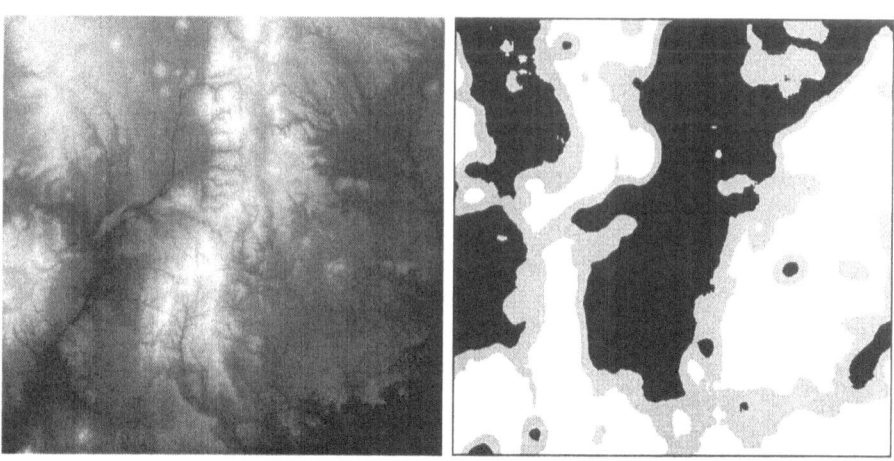

Figure 7.29.   Range image and Hammond–Dikau classification map of the DEM "New Mexico" (white PTL, gray PHM, and black OHM regions).

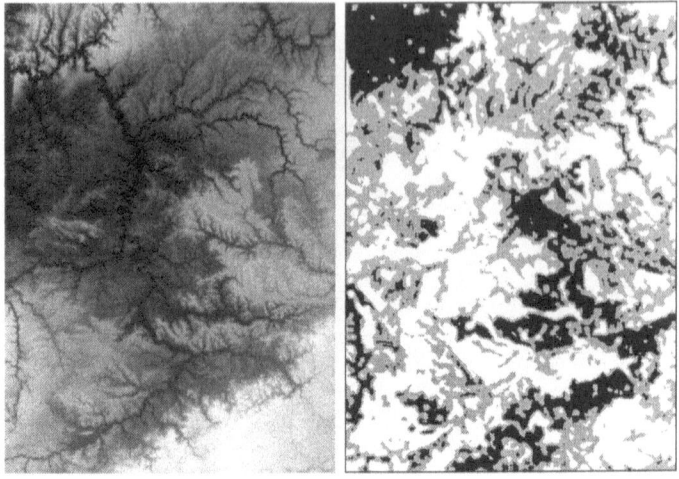

Figure 7.30. Range image and Hammond–Dikau classification map of the DEM "Neckar" (white PTL, gray PHM, and black OHM regions).

of the DEM (307 × 301 and 150 × 150 for "New Mexico" and "Neckar", respectively) and corresponding part of the landform map.

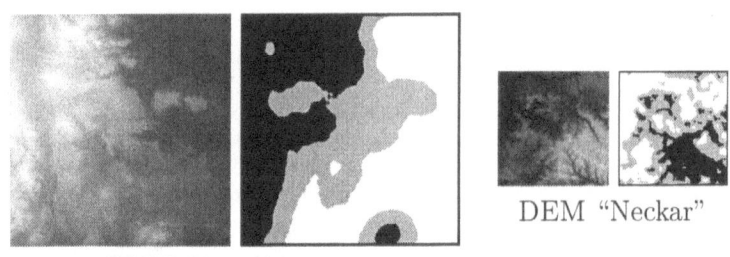

DEM "New Mexico"   DEM "Neckar"

Figure 7.31. Training samples of the DEM "New Mexico" and "Neckar".

Figure 7.32 shows the final region maps obtained by segmenting the training samples with the CSA under the learnt model parameters. The initial segmentation, starting from a sample of the IRF, exploits the simplified potential centering of Eq. (2.30). The obtained initial segmentation map, not shown here, is used as a starting point for the final segmentation based on the potential centering of Eq. (2.29). Both the initial and final segmentation perform 300 macrosteps of the CSA with the control parameters $c_0 = 0$, $c_1 = 1$, and $c_2 = 0.001$. We use the simple thresholding of Eq. (3.16) with the control parameter $c = 3$ for recovering the characteristic interaction structure. In both the cases we obtained 62–65 clique families

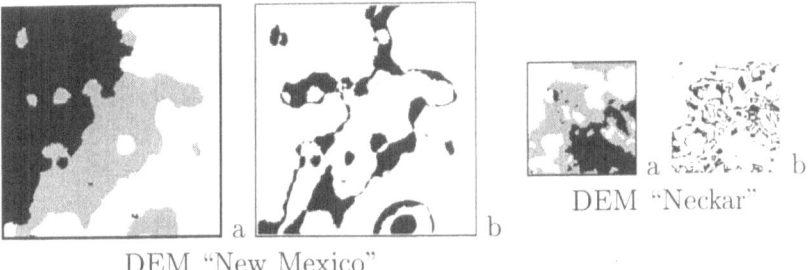

DEM "New Mexico"

DEM "Neckar"

Figure 7.32.   Segmentation map for the training sample (a) and its deviation from the training map (b). The coincident and non-coincident pixels are white and black, respectively.

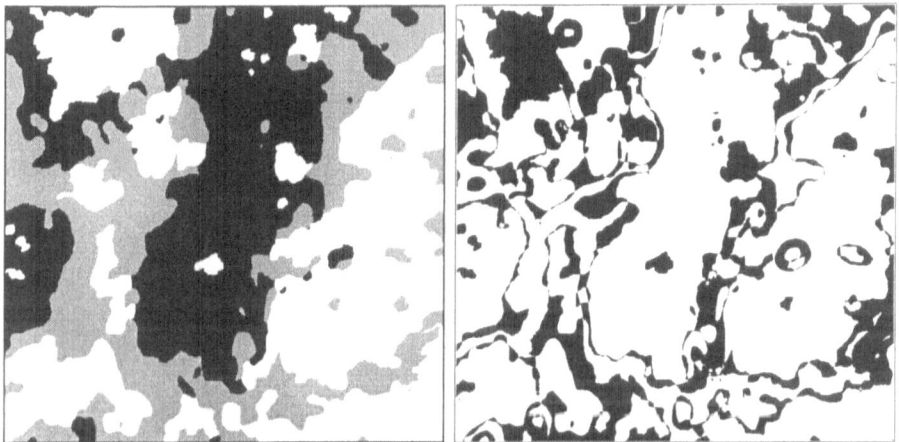

Figure 7.33.   Segmentation map for the DEM "New Mexico" and its deviation from the Hammond–Dikau classification. The coincident and non-coincident pixels are white and black, respectively.

for the initial and 45–50 clique families for the final segmentation, all the cliques being of a close-range type, namely, $|\mu_a| \leq 6$ and $|\nu_a| \leq 6$ for the search window with $\mu_{max} = 40$, $\nu_{max} = 40$.

The final segmentation maps give a reasonable fit to the goal training maps that validates the learnt model parameters: the overall difference between these maps is 15.8% and 18.0% for the DEM "New Mexico" and "Neckar", respectively. Table 7.3 presents the relative coincidences and differences between the OHM, PHM, and PTL landform regions in the training and segmentation maps.

As one might expect, there is a strong general correlation between the two maps, and the differences mostly reflect the fuzziness of the chosen

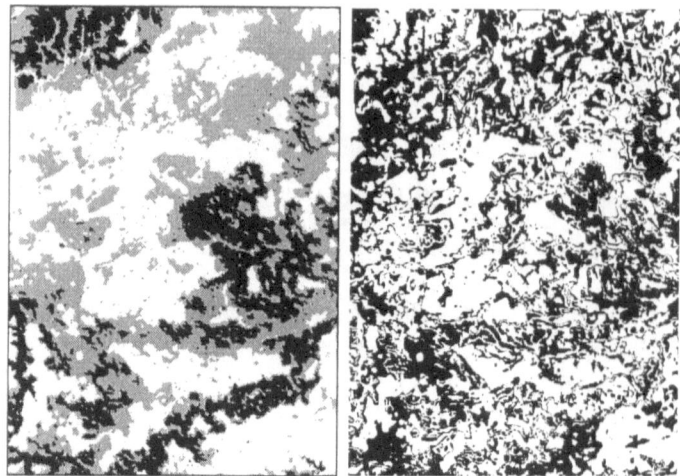

Figure 7.34.   Segmentation map for the DEM "Neckar" and its deviation from the Hammond–Dikau classification. The coincident and non-coincident pixels are white and black, respectively.

TABLE 7.3.   Confusion matrix for the landform classes in the Hammond–Dikau classification map with respect to segmentation of the training samples.

	DEM:	"New Mexico"			"Neckar"		
Segmentation map :		OHM	PHM	PTL	OHM	PHM	PTL
Hammond	OHM	91.7%	7.1%	1.2%	83.2%	16.5%	2.3%
–Dikau	PHM	13.8%	70.1%	16.1%	10.7%	76.0%	13.3%
classes	PTL	0.1%	12.6%	87.3%	0.0%	12.7%	87.3%

borders between the OHM and PHM or PHM and PTL landforms in the Hammond–Dikau classification algorithm.

The result of segmenting the range image "New Mexico" in Figure 7.29 using the same learnt parameters is displayed in Figure 7.33, and the corresponding grouped Hammond–Dikau classification map is shown in Figure 7.29. The overall difference between these maps is 27.9% (see Table 7.4). It is evident that the two maps possess rather similar macro-structures but differ in detail, especially, in the case of the intermediate PHM landform.

The similar results for the DEM "Neckar" with the overall difference of 41.0% are shown in Figure 7.34 and Table 7.4. It is clearly visible that they are somewhat less good than for the DEM "New Mexico". One reason might be that the landform characteristics in the DEM "Neckar" are more complex on the meso–scale than in the DEM "New Mexico" (see Figures 7.27 and 7.28). This results in more structured and complicated patterns of terrain segmentation.

TABLE 7.4.   Confusion matrix for the landform classes in the Hammond–Dikau classification map with respect to segmentation of the test samples.

DEM:		"New Mexico"			"Neckar"		
Segmentation map:		OHM	PHM	PTL	OHM	PHM	PTL
Hammond	OHM	76.6%	14.8%	8.6%	56.7%	39.3%	4.0%
–Dikau	PHM	17.3%	55.6%	27.1%	17.6%	47.6%	34.8%
classes	PTL	0.9%	20.3%	78.8%	3.9%	24.1%	72.0%

The slope data for the DEM in Figure 7.35 describes the landforms better than the height data. This gives a significant improvement of the obtained results, as can be seen in Figures 7.36 and 7.37. Here, our segmentation map is closer to the Hammond–Dikau classification (the overall differences of 20.1% and 32.9% for the training sample and whole slope image, respectively), and the differences are mainly in details corresponding to the intermediate cladd PHM as is indicated in Table 7.5.

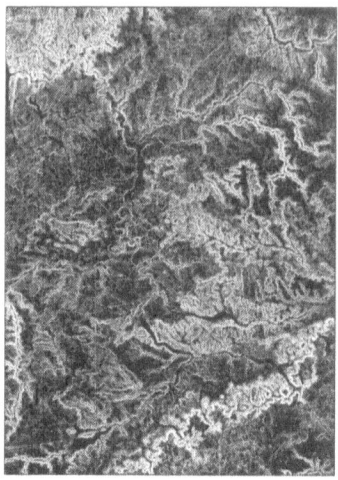

Figure 7.35.   Slope image of the DEM "Neckar".

Figure 7.36. Training sample of the slope image "Neckar" (a), segmentation map (b), and its deviation (c) from the Hammond–Dikau classification. The coincident and non-coincident pixels are white and black, respectively.

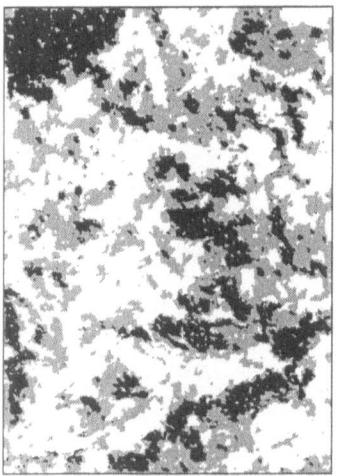 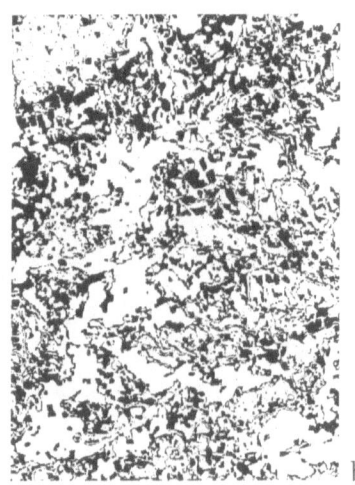

Figure 7.37. Segmentation map (a) of the slope image "Neckar" and its deviation (b) from the Hammond–Dikau classification. The coincident and non-coincident pixels are white and black, respectively.

Our segmentation uses only the height or slope data and no combination of specific geomorphometric derivates as the approach of Hammond–Dikau

TABLE 7.5. Confusion matrix for the landform classes in the Hammond–Dikau classification map with respect to segmentation of the slope image "Neckar".

Segmentation map:		Training sample in Figure 7.36			Total scene in Figure 7.37		
		OHM	PHM	PTL	OHM	PHM	PTL
Hammond	OHM	80.1%	19.8%	0.1%	65.3%	30.7%	4.0%
–Dikau	PHM	15.7%	72.0%	12.3%	14.4%	55.6%	30.0%
classes	PTL	0.1%	11.9%	88.0%	3.0%	17.0%	80.0%

does. Therefore, the supervised segmentation based on the Gibbs image model of Eq. (2.26) can be quoted as promising for discrimination between some meso-scale landforms. These and other experiments show that the terrain types sometimes really form spatially homogeneous textured regions, and spatial relationships between the elevations can be used as a geomorphometric feature for terrain classification. However, the efficacy of such tools for a more detailed landform discrimination remains to be investigated and further work has to be done to make the Gibbs model usable for geomorphometric applications.

7.3.2. GRAYSCALE IMAGES OF THE EARTH'S SURFACE

Aerial image. Figure 7.38 shows an aerial synthetic aperture radar (SAR) image of the Earth's surface with the four textured objects to be discriminated. The objects, chosen visually in the initial image, include radioshadows (the region 0 in the training map), vegetation areas and buildings (the region 1), grass fields (the region 2), and concrete ways (the region 3).

In this case we use the training map in Figure 7.38,b that contains mostly disjoint small patches of the desired four regions. Therefore, we cannot expect that the learnt model parameters will be representative of all the inter-region interactions. The training regions $0, \ldots, 3$ are coded by light gray, gray, dark gray, and black pixels, respectively. White pixels represent the region that takes no part in learning the model parameters.

Results of the initial and final segmentation are presented in Figure 7.40. Both the segmentation maps are obtained after 300 CSA macrosteps with the same control parameters as before. It is easily seen that the regions 1 and 2 with sufficiently different image textures are rather correct. But there are some apparent errors in discrimination between the radio-shadows and concrete ways because these objects are closely similar as regarding their GLHs and close-range GLDHs.

Space image. The gray-scale image in Figure 7.41 presents the resampled fragment 819×871 of a three-band SPOT image showing a part of the environs of Chernobyl (Ukraine). Gray values in each pixel are obtained by averaging the corresponding signals in the three spectral bands. The training sample 150×150 in Figure 7.42, a containing 36 square patches 25×25 represents the four visually chosen texture types, one patch 25×25 per texture type being placed in the 9 arbitrary positions within the sample. The visually chosen texture types are as follows: open water surfaces (the region 0 in the training map), forests (the region 1), open ground and sand (the region 2), and wetlands (the region 3).

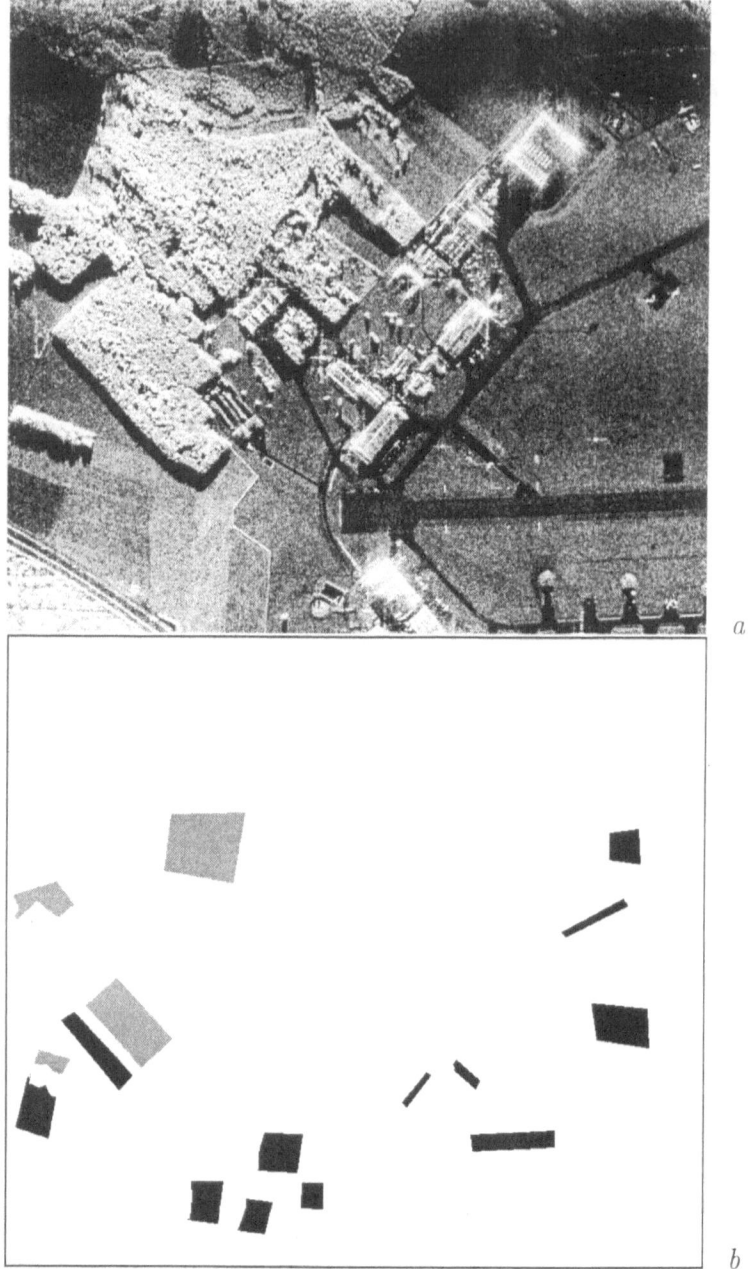

a

b

Figure 7.38. Initial SAR image (*a*) and the map of training patches (*b*).

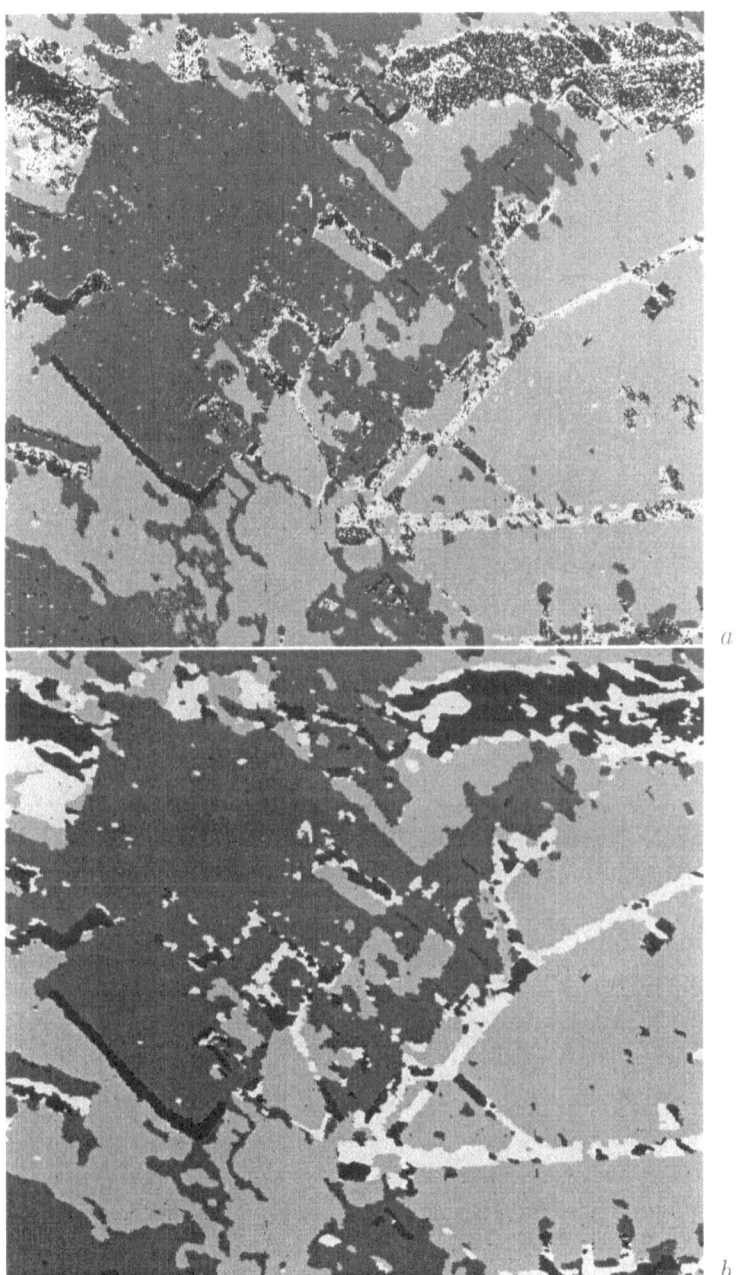

Figure 7.39.   Initial (*a*) and final (*b*) segmentation maps for the SAR image.

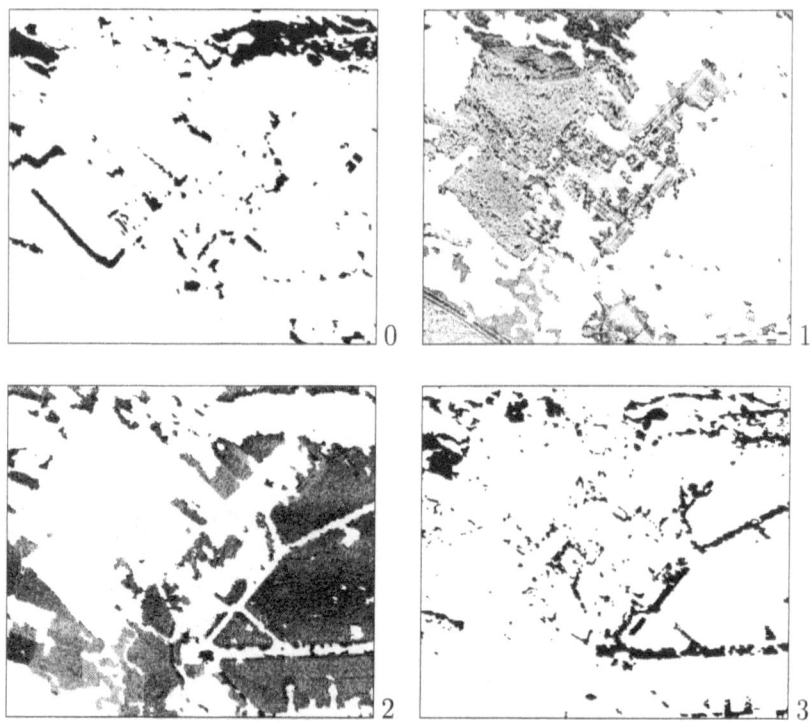

Figure 7.40.   Texture regions $0, \ldots, 3$ in the final segmentation map.

Figure 7.42 also shows the results of segmenting the training image by the CSA after learning the model parameters. Because the chosen training texture patches have such small sizes, the search window $\mathbf{W}$ is only 21×21, that is, $\mu_{\max} = \nu_{\max} = 10$. In such a case, our simple thresholding of Eq. (3.16) with the control parameter $c = 3$ or 4 does not work, and we set the thresholds for recovering the initial and final interaction structures in order to produce 8 and 2 characteristic clique families for the initial and final segmentation, respectively.

As one might expect, the segmentation of the training sample is almost ideal. The chi-square distance between the training and generated RLH and RLCHs is reduced from 3,236,227.6 for the starting IRF sample to 63.7 for the resulting segmentation map at the initial segmentation stage and from 368.8 for the initial segmentation map to 22.3 for the resulting one at the final stage. Both the stages use 300 CSA steps with the same control parameters as before.

But the learnt model parameters result in much less appropriate approximation of the training GLH and GLDHs by the generated signal statistics when the whole image in Figure 7.41 is segmented by the CSA. The ob-

Figure 7.41.   High-resolution space image of the Earth's surface.

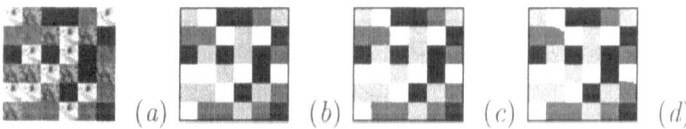

Figure 7.42.   Four-region training image (*a*) and map (*b*), and the initial (*c*) and final (*d*) segmentation maps for the training image. The regions $0, \ldots, 3$ are displayed by the black, dark gray, light gray, and white pixels, respectively.

tained initial and final segmentation maps are presented in Figures 7.43 and 7.44, respectively. In these cases the chi-square distance between the sample relative frequency distributions of region labels and region label coincidences in the training and generated samples is only reduced from 2,381,840.1 for the starting IRF sample to 408,623.8 for the resulting initial segmentation map and from 1,074,326.6 to 171,502.3 for the final segmentation map. Here, both the initial and final segmentation stages use 100 CSA steps with the control parameters $c_0 = 0$, $c_1 = 100$, and $c_2 = 1$. The initial interaction structure has the eight clique families with the following inter-pixel shifts: $(3,0)$, $(4,0)$, $(5,0)$, $(-5,1)$, $(-4,1)$, $(-5,2)$, $(-4,2)$, and $(-2,3)$. The final interaction structure contains the two close-range families $(1,0)$ and $(0,1)$.

The final segmentation map in Figure 7.44 gives a reasonably fair separation of the forest region from all the other ones because the training forest patch has the marked homogeneity and textural appearance. But there are errors in discriminating the almost uniform water surfaces from the uniform patches of wetland or open ground. These patches are assigned to the region 0 because our simplified Gibbs model of Eq. (2.26) does not discriminate between the pairwise interactions with the same gray level difference but with the different gray levels.

It is interesting to compare the above final segmentation map to Figure 7.45 that presents the segmentation by taking account of only the pixelwise interactions. In this case the gray levels are discriminated according to the GLHs for the training sample. Each gray level $q = g_i$ is assigned to the region k_q with the maximum conditional sample relative frequency in the training sample:

$$k_q = \arg \max_{k \in \mathbf{K}} F(k|q, \mathbf{g}^\circ, \mathbf{l}^\circ)$$

where

$$F(k|q, \mathbf{g}^\circ, \mathbf{l}^\circ) = \frac{F(q, k|\mathbf{g}^\circ, \mathbf{l}^\circ)}{\displaystyle\sum_{k \in \mathbf{K}} F(q, k|\mathbf{g}^\circ, \mathbf{l}^\circ)}.$$

It is seen that the regions obtained by our segmentation and displayed for convenience in Figure 7.46 describe the image under consideration much better than the regions in Figure 7.47 obtained by classification of the gray levels alone.

7.4. How to choose an interaction structure

7.4.1. THREE VARIANTS OF A CHOICE

In the above experiments the CSA approximates relative conditional sample frequencies of region labels for each gray level and of region label coincidences for each gray level difference in the clique families learnt for the training sample by the like frequencies for the test image and its segmentation map. We use the following two-stage CSA-based segmentation scheme.

- The initial stage starts from a random region map (a sample of the IRF) and takes into account only the intra-region interactions to form the initial segmentation map.
- The final stage starts from the initial region map and takes account of both the intra- and inter-region interactions to obtain the final segmentation map.

Each stage exploits its own characteristic interaction structure found by a particular criterion. In Chapter 3 we have already considered the simplest

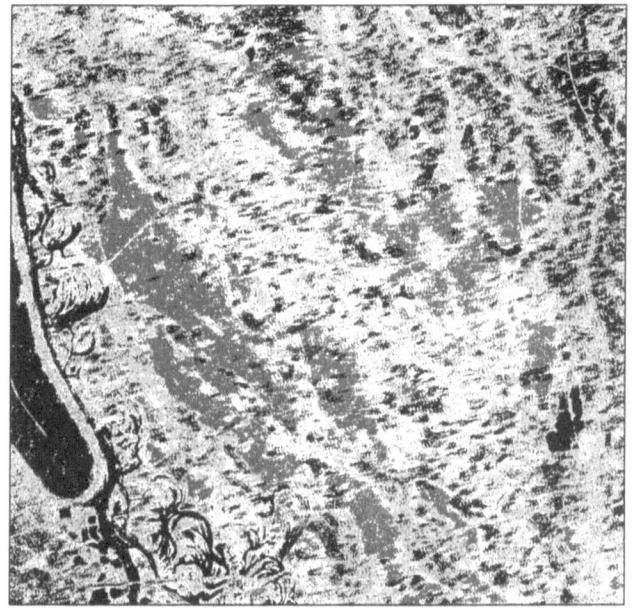

Figure 7.43.   Initial segmentation map.

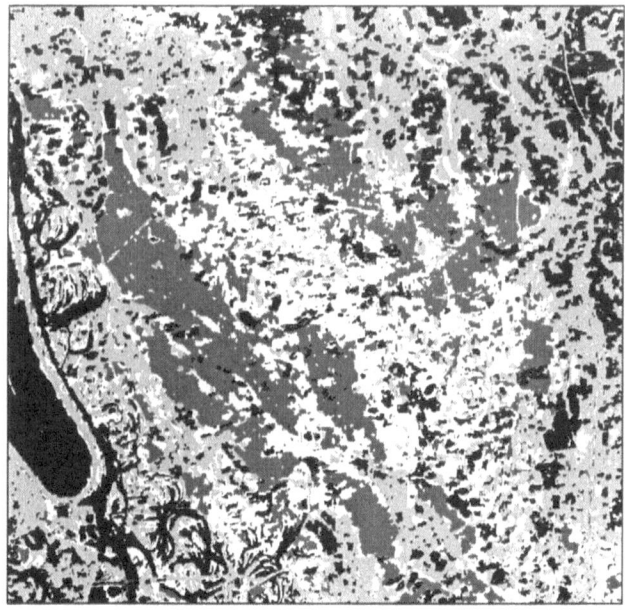

Figure 7.44.   Final segmentation map.

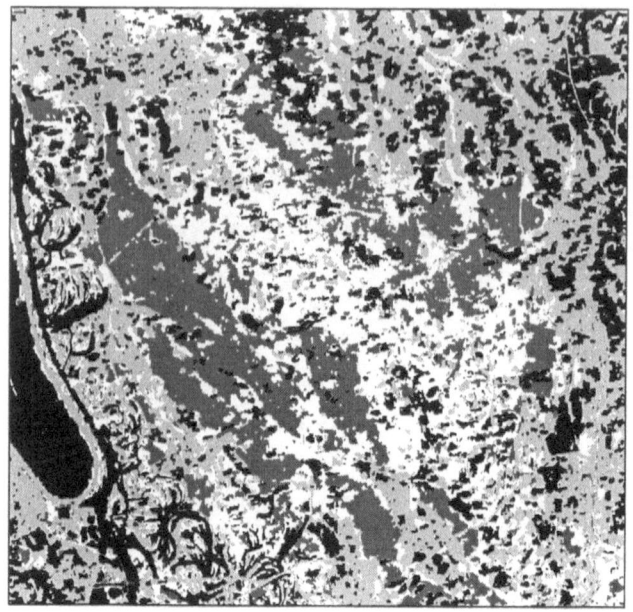

Figure 7.45.   Gray level classification using the GL/RLH for the training sample.

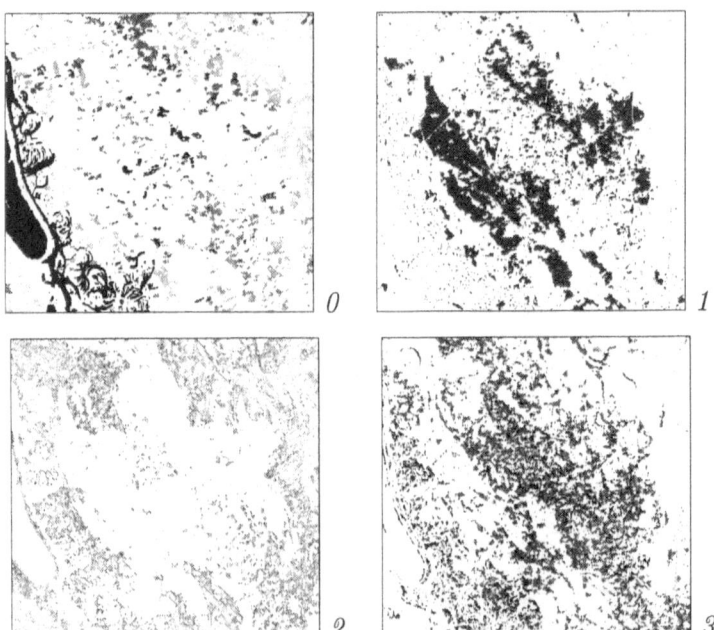

Figure 7.46.   Textured regions *0 – 3* for the final segmentation map in Figure 7.44 (on a reduced scale).

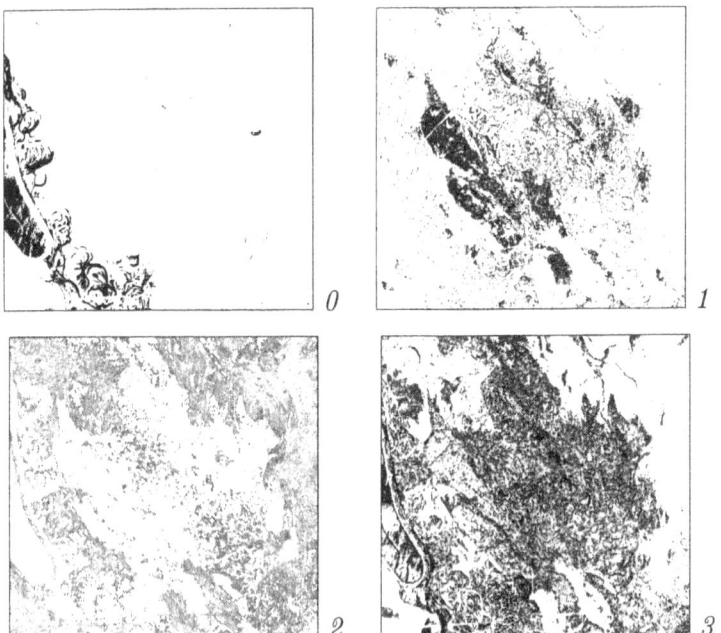

Figure 7.47.   Textured regions *0 – 3* for the segmentation map in Figure 7.45 (on a reduced scale).

thresholding of interaction maps based on Eq. (3.16). Here, it is compared to two more complicated choices of the interaction structure that is most characteristic for segmentation.

We consider the following three choices of the interaction structure for the conditional Gibbs model of Eq. (2.25) that take account of the GL/RLH and GLD/RLCHs in a given training sample $[\mathbf{g}^\circ, \mathbf{l}^\circ]$.

Choice 1:   Thresholding of the relative total energies $e_{a,[0]}(l^\circ|g^\circ)$ for the clique families in a search window $\mathbf{W}$ using the thresholds of Eqs. (3.16).

As indicated in Chapter 3, the first analytic approximation of the potentials for a particular clique family $\mathbf{C}_a$ is proportional to the centered conditional sample relative frequencies $F_{\mathrm{cn},a}(k,\alpha|d,l^\circ,g^\circ)$ of the region label coincidences in the cliques $\mathbf{c}_a \in \mathbf{C}_a$ that have a fixed gray level difference $d \in \mathbf{D}$. Thus the total relative energies are as follows:

$$e_{a,[0]}(l^\circ|g^\circ) = \rho_a \sum_{d\in\mathbf{D}} F_a(d|g^\circ) \sum_{k\in\mathbf{K}} \sum_{\alpha\in\{0,1\}} F_{\mathrm{cn},a}^2(k,\alpha|d,l^\circ,g^\circ) \qquad (7.1)$$

where $F_{\mathrm{cn},a}(k,\alpha|d,l^\circ,g^\circ) = F_a(k,\alpha|d,l^\circ,g^\circ) - \frac{1}{|\mathbf{K}|}$ and $F_a(d|g^\circ)$ is the sample relative frequency of gray level difference d obtained by normalizing the

GLDH for the training image. The conditional sample relative frequency of the region label coincidences $F_a(k, \alpha | d, l^\circ, g^\circ)$ is obtained from the corresponding joint GLD/RLCH collected over the training sample.

Choice 2: Sequential error minimization proposed by Zalesny (1996). Here, each successive new clique family which is added to a current interaction structure has to decrease at the most total segmentation error for the training sample. The total error is estimated by summing the partial errors in each pixel $i \in \mathbf{R}$. The partial error is estimated as follows. We place each region label $k \in \mathbf{K}$ in the pixel i and compute the sum of the potentials for all the cliques containing this pixel using the neighboring region labels taken from the training map. If the region label that yields the top sum differs from the true label in the training sample then the partial error is assumed to be found in this pixel.

Choice 3: Thresholding the relative total energies for the clique families separately in each region $k \in \mathbf{K}$ of the training sample. The relative total energy per region is as follows:

$$e_{a,[0]}(k | l^\circ, g^\circ) = \rho_a \sum_{d \in \mathbf{D}} F_a(d | g^\circ) \sum_{\alpha \in \{0,1\}} F_{cn,a}^2(k, \alpha | d, g^\circ, l^\circ). \qquad (7.2)$$

The overall volume of computations for the above Choices 1 and 3 is proportional to $|\mathbf{W}|$ and $|\mathbf{K}||\mathbf{W}|$, respectively. The Choice 2 results in the overall computational volume proportional to $|\mathbf{R}||\mathbf{W}|^2)$ which is too high to be practicable. It is simplified here by taking into account only a subset of the clique families $\mathbf{A_W}$ found by the Choice 1 with the low threshold of Eq. (3.16) with the control parameter $c = 0$. The Choice 2 may result in the minimum number of families due to taking account of each individual texture. The Choice 3 is introduced to allow for this feature, too, but with the lesser computational complexity. We use the simplified variant of the Choice 3 which ranks the families in the partial energies of Eq. (7.2) and chooses the same number $\tau = 4 \ldots 10$ of the top-rank families for each the region $k \in \mathbf{K}$.

7.4.2. IMPACT OF CHOSEN STRUCTURES ON SEGMENTATION

To compare the above Choices 1 – 3, experiments with several artificial collages 128×128 were conducted. The Choice 1 was tested using the threshold of Eq. (3.16) with the following two variants of the control parameter: (a) the value $c = 2$ and (b) the value c giving the same number of the cliques as in the Choice 2. The Choice 3 was used with the three thresholds (a) $\tau = 4$, (b) 8, and (c) 10.

For all the collages of the same size, the CSA started from the same random region map being a sample of the IRF. We use 200 macrosteps of the CSA both for the initial segmentation and for the final one with the same control parameters $c_0 = 0$, $c_1 = 1$, $c2 = 0.001$ at both the stages.

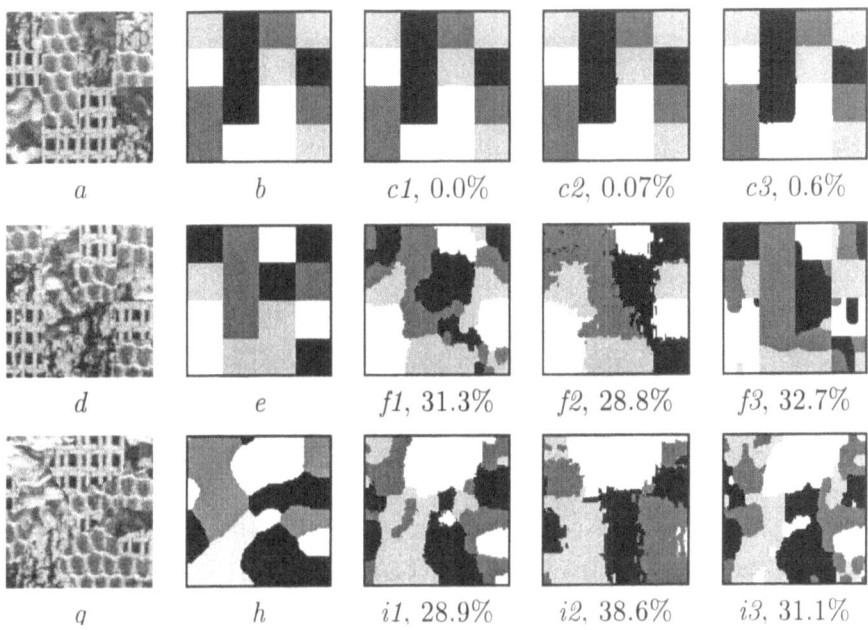

Figure 7.48.   Training collage (*a*) and test collages (*d*, *g*) of the Brodatz textures D3-D5-D12-D20, their true region maps (*b*, *e*, *h*), and segmentation maps (*c1–c3*, *e1–e3*, *h1–h3*), respectively; see also Table 7.6. The percentage value specifies the relative error of final segmentation. Numbers *1*, *2*, and *3* denote a particular Choice 1,*a*, 2, or 3, *c*, of the interaction structure, respectively).

Figures 7.48 and 7.49 present each training collage (*a*) with the same four-region map (*b*) in both cases, two test collages (*d*, *g*) with their true four-region maps (*e*, *h*), and results (*c*, *f*, *i*) of segmenting the training and test collages after learning the model parameters from the training sample. The collages in Figures 7.48 and 7.49 are formed from the Brodatz textures D3, D5, D12, D20 and D20, D65, D77, and D79, respectively. The true region maps have been used as masks to cut the corresponding patches from the individual image textures D3, ..., D79. Thus the training and the test images mostly contain different parts of the same textures. The maps *b* and *f* have four polygonal regions of size 4096 pixels. Each region is formed by four possibly disconnected squares 32×32. The map *h* has four arbitrary-shaped and possibly disconnected regions of size 5368, 5139, 2318, and 3559 pixels, respectively.

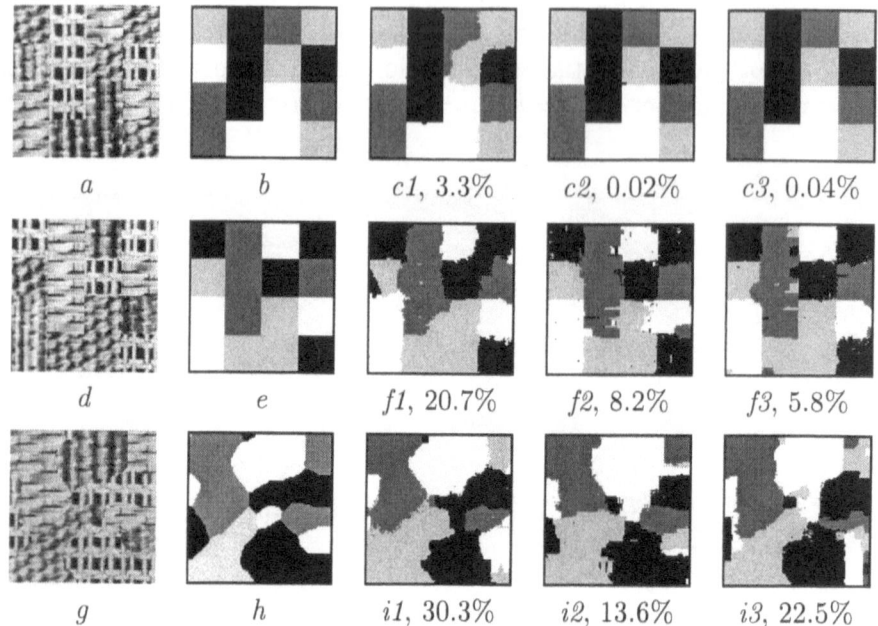

Figure 7.49.   Training collage (*a*) and test collages (*d*, *g*) of the Brodatz textures D20-D65-D77-D79, their true) region maps (*b*, *e*, *h*, and segmentation maps (*c1–c3*, *e1–e3*, *h1–h3*), respectively; see also Table 7.6. The percentage value specifies the relative error of final segmentation. Numbers *1*, *2*, and *3* denote a particular Choice 1,*a*, 2, or 3, *c*, of the interaction structure, respectively).

These collages show main difficulties of the supervised segmentation of the natual image textures, in particular, substantial differences between the training and test patches of the same texture type and due to arbitrary forms of the patches. In particular, the textures D3, D5 and D12 in Figure (7.48) have obvious inherent inhomogeneities (moreover, some parts of the texture D5 are very similar to D12). The test map *h* contains rather small subregions of each region where the texture patches are unlikely to exhibit the learnt long-range interactions.

Tables 7.6 and 7.7 show the numbers of the clique families found with different variants of the Choices 1-3 for the training samples in Figures 7.48 and 7.49, respectively, and the resulting relative errors for the initial and final segmentation of the training and test images.

Figures 7.50 and 7.51 present the the interaction maps and interaction structures found in the search window **W** by the three Choices 1,*a*, 2, and 3,*c*. The interaction maps and structures for the Choices 1,*a* and 3,*c* are displayed using gray-level coding of the total energies for the clique families (the blacker the pixel (μ, ν), the higher the energy).

TABLE 7.6. Numbers of the chosen clique families ($|\mathbf{A}|$, initial/final segmentation) and relative segmentation errors (ε_{ini}, ε_{fin}) for the initial and final segmentation of the training and test collages in Figure 7.48.

| Collage 128 × 128 | Choice 1 | | Choice 2 | Choice 3 | | |
	a	b		a	b	c
$\quad$ $|\mathbf{A}|$	42/34	20/11	20/11	16/6	32/12	40/13
a $\quad \varepsilon_{ini},\%$	14.4	24.3	20.9	19.0	13.9	12.4
$\quad \varepsilon_{fin},\%$	0.0	6.2	0.07	3.3	0.7	0.6
d $\quad \varepsilon_{ini},\%$	48.9	49.0	52.9	52.0	52.0	51.7
$\quad \varepsilon_{fin},\%$	31.3	36.6	28.8	38.2	35.9	32.7
g $\quad \varepsilon_{ini},\%$	46.5	47.7	51.4	49.8	48.5	46.8
$\quad \varepsilon_{fin},\%$	28.9	29.5	38.6	36.5	34.0	31.1

TABLE 7.7. Numbers of the chosen clique families ($|\mathbf{A}|$, initial/final segmentation) and relative segmentation errors (ε_{ini}, ε_{fin}) for the initial and final segmentation of the training and test collages in Figure 7.49.

| Collage 128 × 128 | Choice 1 | | Choice 2 | Choice 3 | | |
	a	b		a	b	c
$\quad$ $|\mathbf{A}|$	18/14	51/10	51/10	14/8	29/15	35/19
a $\quad \varepsilon_{ini},\%$	37.6	26.2	22.1	15.2	8.0	7.0
$\quad \varepsilon_{fin},\%$	3.3	2.0	0.2	1.2	0.2	0.04
d $\quad \varepsilon_{ini},\%$	45.3	32.2	31.8	33.3	34.3	34.2
$\quad \varepsilon_{fin},\%$	20.7	9.7	8.2	9.2	4.9	5.8
g $\quad \varepsilon_{ini},\%$	44.6	35.6	36.7	36.1	33.9	34.6
$\quad \varepsilon_{fin},\%$	30.3	17.0	13.6	23.0	21.2	22.5

The characteristic neighborhoods found by the Choice 2 reflect more features of the textures in a given training sample than the neighborhoods for two other Choices (especially, at the final segmentation stage). In turn, the Choice 3 results in more detailed neighborhoods than the Choice 1. The chosen structures themselves are quite different. In particular, the Choice 1 finds mostly the close-range clique families, the Choice 2 gives more long-range interactions and the Choice 3 stays in between them.

In spite of very different chosen structures, the results of segmenting

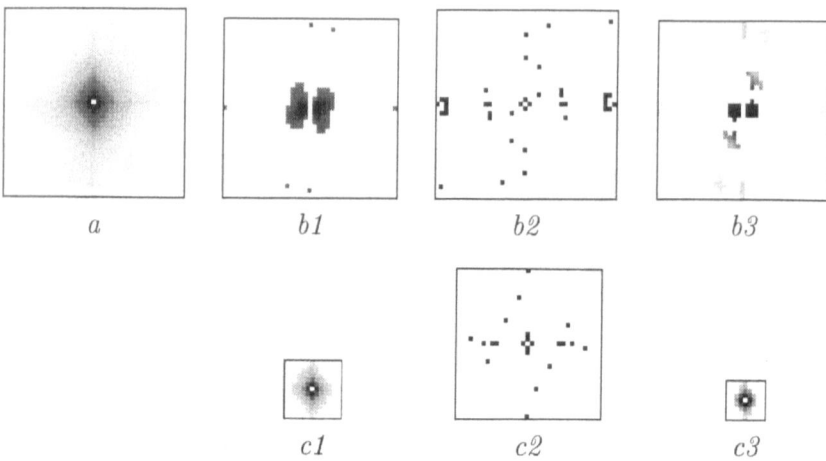

Figure 7.50. Interaction map (*a*) and interaction structures for the initial (*b*) and final (*c*) segmentation obtained for the training sample D3-D5-D12-D20 in Figure 7.48 by the Search 1,*a*, 2, and 3,*c*.

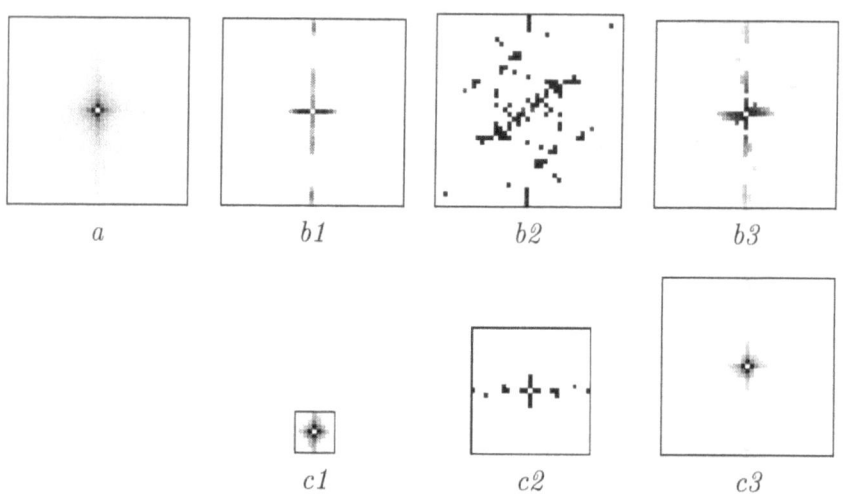

Figure 7.51. Interaction map (*a*) and interaction structures for the initial (*b*) and final (*c*) segmentation obtained for the training sample D20-D65-D77-D79 in Figure 7.49 by the Search 1,*a*, 2, and 3,*c*.

the training images are very similar: the error rates 0.0 and 3.3% for the Choice 1,*a*, 0.07 and 0.2% for the Choice 2, and 0.6 and 0.04% for the Choice 3,*c*. Curiously enough, both the structure sizes found by the Choice 2

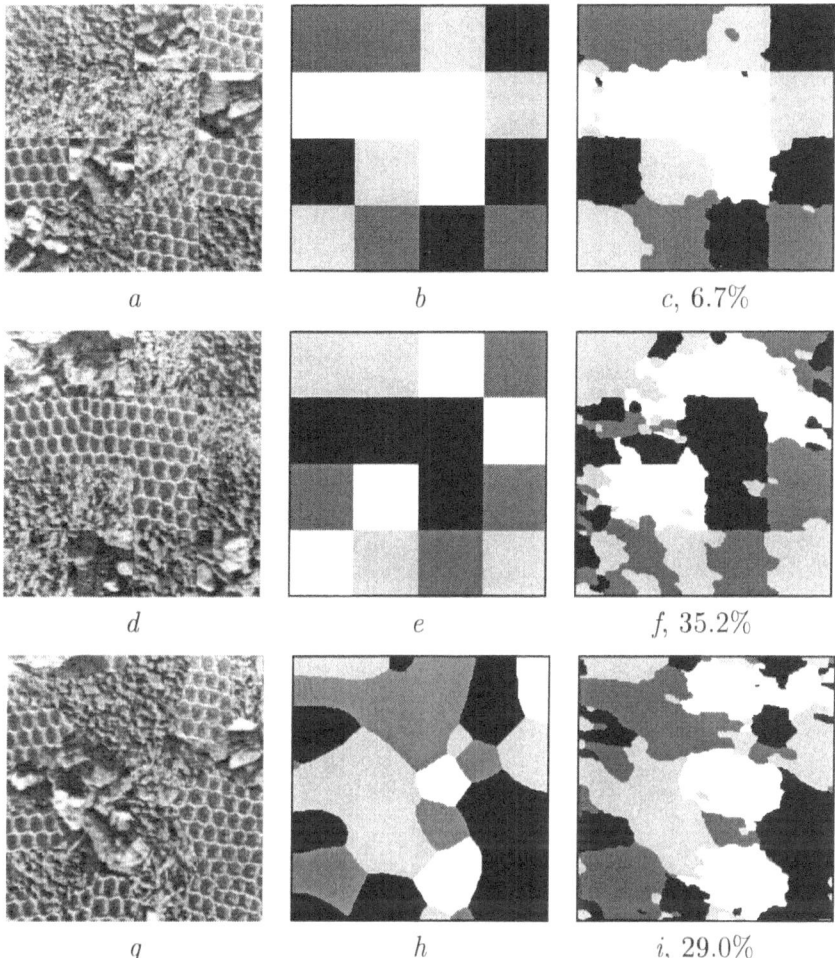

Figure 7.52.   Segmentation of the four-region collages of the textures D3, D4, D5, D9 (the training collage (*a*), test collages (*d, g*), their ideal region maps (*b, e, h*), and final segmentation maps (*c, f, i*); the percentage value specifies the total segmentation error).

and the resulting error rates are not always less than for the two other Choices. It is worth noting that the errors for the training samples are very low in spite of the inhomogeneities in and similarities between the different textures forming such a collage.

But, just due to these features, the test samples show the notably higher errors that do not differ much for all the Choices 1–3 if the chosen structures are about the same size. By ranking in these error rates, the Choices 2 and 3 outperform the Choice 1 (with a slight superiority of the Choice 2 in a number of experiments). But, the Choice 3 has much lower computational

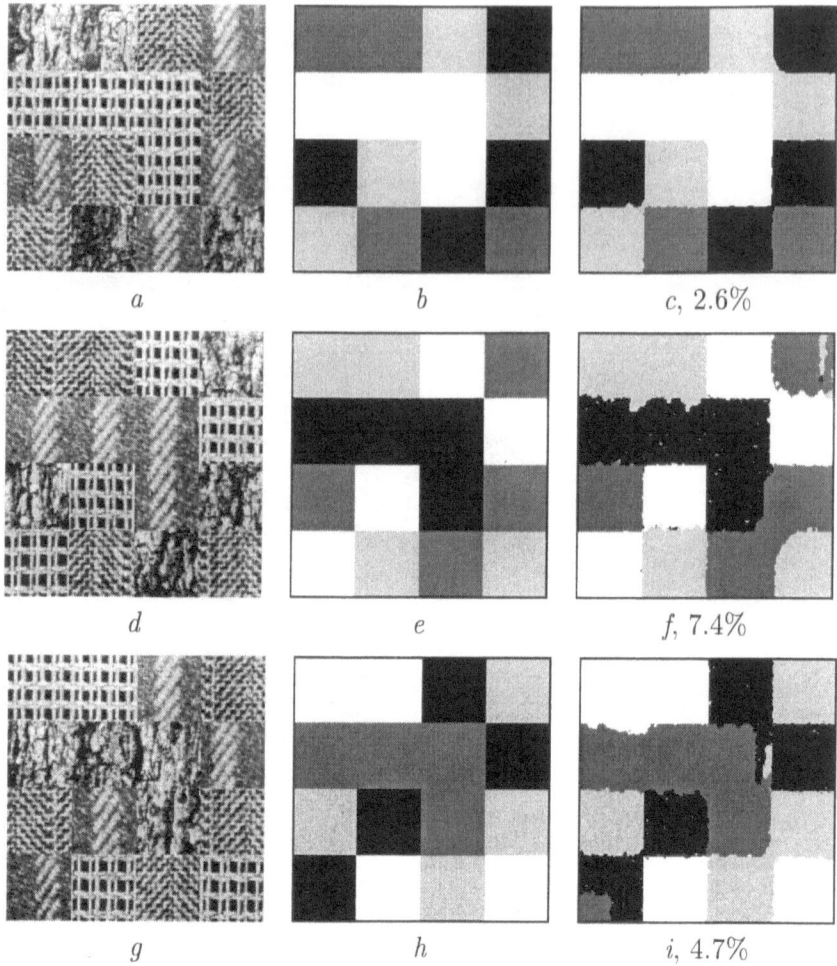

Figure 7.53. Segmentation of the collages from textures D11, D12, D17, D20 (the same notation as in Figure 7.52); to be continued in Figure 7.54.

complexity than the Choice 2 and preserves better the borders between the different regions as follows from Figures 7.48 and 7.49. Thus, in subsequent experiments with the collages 256 × 256 only the Choice 3,c with $\tau = 10$ is used. Some segmentation results are shown in Figures 7.52–7.57 where most of the individual textures, for instance, D3, D23, D50, D57, D66, D69, and D74, have notable spatial inhomogeneities. In all these figures, the percentage values specify the total segmentation error.

TABLE 7.8. Gibbs energies in the training sample of Figure 7.58 for choosing the initial and final interaction structures.

Region	D3	D4	D5	D9
Total size	23382	17502	17604	7048
Subregions	6	6	5	3
Initial segmentation				
Minimum own energy	42.0	5.4	19.9	-0.04
Maximum own energy	84.6	8.4	37.6	-0.02
Minimim total energy	214.6	99.2	206.8	7.7
Maximum total energy	528.9	370.7	528.9	8.2
Final segmentation				
Minimum own energy	111.8	53.0	78.7	3.0
Maximum own energy	237.5	117.0	167.0	7.8
Minimum total energy	242.4	225.0	234.4	242.4
Maximum total energy	528.9	528.9	528.9	528.9

TABLE 7.9. Gibbs energies in the training sample of Figure 7.59 for choosing the initial and final interaction structures.

Region	D9	D3	D5	D4
Total size	23382	17502	17604	7048
Subregions	6	6	5	3
Initial segmentation				
Minimum own energy	33.4	12.9	19.3	-0.04
Maximum own energy	65.7	15.7	40.9	-0.02
Minimim total energy	244.8	82.9	209.5	7.8
Maximum total energy	430.4	98.8	497.7	8.5
Final segmentation				
Minimum own energy	103.0	57.7	78.1	3.0
Maximum own energy	198.3	122.4	166.9	7.8
Minimum total energy	244.8	244.8	222.1	209.5
Maximum total energy	497.7	497.7	497.7	497.7

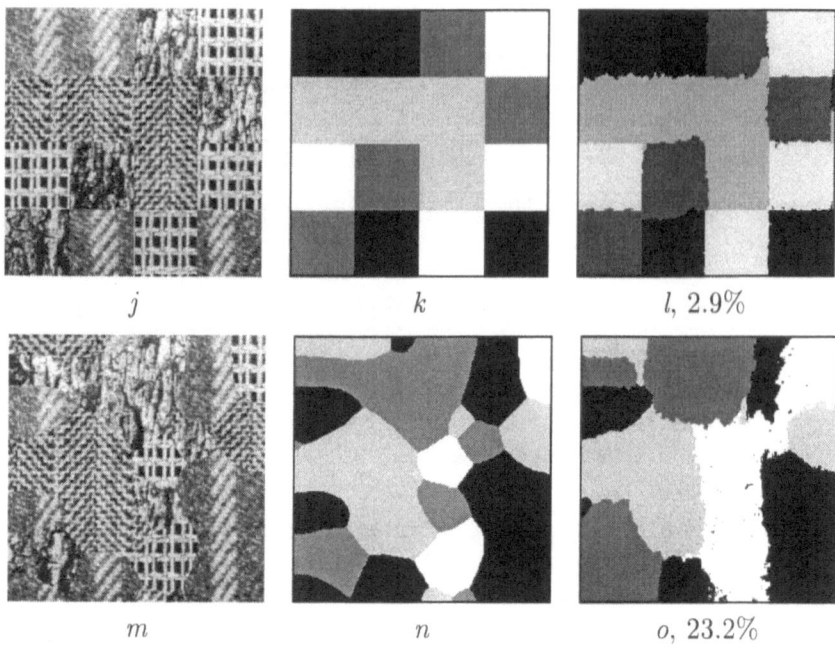

Figure 7.54.   Segmentation of the collages from textures D11, D12, D17, D20 (the continued Figure 7.53). Notice that most errors for the test images are due to the weakly homogeneous texture D12.

7.5. Do Gibbs models learn what we expect?

The above experiments provide the basis for discussing whether the learnt model parameters fit our expectations as regarding the correct discriminaton between the textures. The segmentation maps presented above show that our learning tends to adapt the conditional model of Eq. (2.26) more to peculiarities of the training sample caused by a particular arrangement and geometric form of the textured regions than to general discriminating features of the textures. It should be noted that the training sample has to contain sufficiently big patches of each texture type that allow for getting statistically valid estimates of the marginal signal probabilities from the collected GLD/RLCHs or GLC/RLCHs.

If some texture types in a training sample are represented by too small patches then the segmentation results may be significantly worse. In particular, due to the small size of the patches forming the region 3 of the training sample in Figure 7.58, the corresponding texture, D9, is represented by completely noncharacteristic clique families in the overall interaction structure. In this case the structure is recovered within the search window of size 81×81 by choosing for each region the ten top-rank clique families ordered

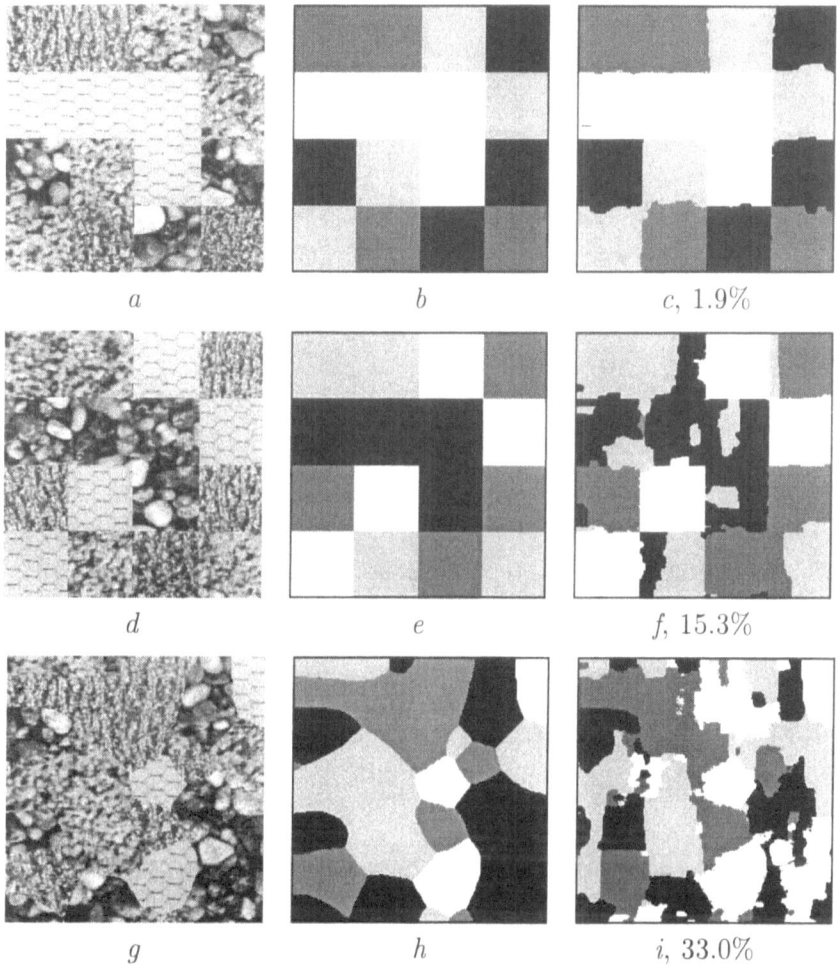

Figure 7.55. Segmentation of the collages from textures D23, D24, D29, D34 (the same notation as in Figure 7.52). Notice that most errors for the test images are due to the inhomogeneous texture D23 and weakly homogeneous texture D24.

by their energies of Eq. (7.2). The structures for the initial and final segmentation contain the 31 and 13 clique families, respectively, because many of them are shared by two or more regions. The relative total energies per region in Eq. (7.2) are called "own energies" in Table 7.8. These energies as well as the relative total energies of Eq. (7.1) are different for the training regions to an extent that the chosen structure is not really characteristic.

If we change the current positions of the textures, for instance, by placing the textures D9, D3, D5, and D4 to the regions 0 – 3, respectively, there are quite similar differences in the energies with respect to the region 3 (see

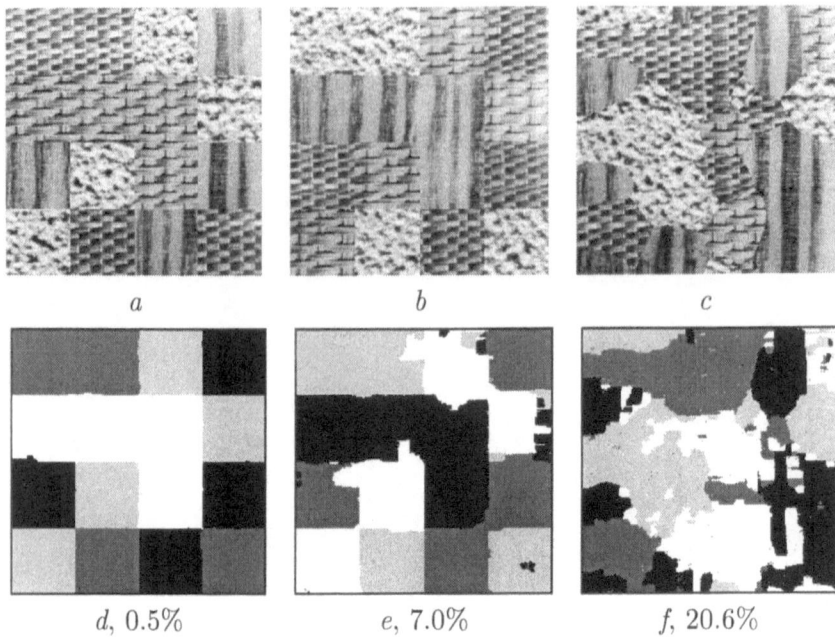

Figure 7.56.   Segmentation of the collages from textures D50, D55, D57, D65 (the training collage (*a*), test collages (*b*, *c*), and their final segmentation maps (*d*, *e*, *f*). The ideal training and test region maps are identical to those in Figure 7.52.

TABLE 7.10. Gibbs energies in the training sample of Figure 7.60 for choosing the initial and final interaction structures.

Region	D9	D3	D5	D4
Total size	23382	17502	17604	7048
Subregions	6	6	5	3
Initial segmentation				
Minimum own energy	1.0	9.3	6.9	8.4
Maximum own energy	1.2	12.2	18.6	8.6
Minimim total energy	79.2	89.6	171.0	10.1
Maximum total energy	91.5	102.1	435.7	18.6
Final segmentation				
Minimum own energy	44.3	51.4	61.4	49.9
Maximum own energy	86.8	108.2	135.3	105.5
Minimum total energy	200.7	180.2	196.4	184.0
Maximum total energy	435.7	435.7	435.7	435.7

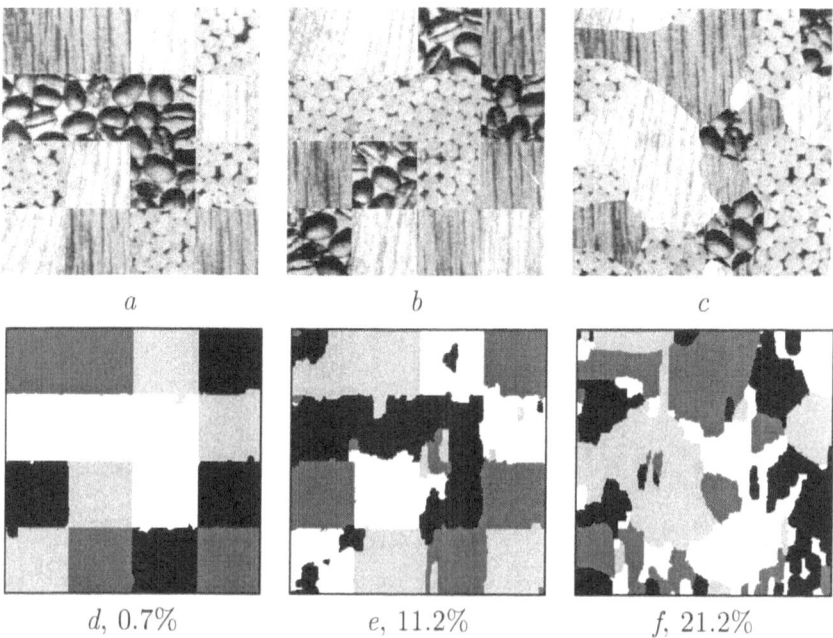

a b c

$d, 0.7\%$ $e, 11.2\%$ $f, 21.2\%$

Figure 7.57.   Segmentation of the collages from textures D66, D68, D69, D74 (the same notation as in Figure 7.56).

Tables 7.9 and 7.10). Figures 7.59 and 7.60 present the final segmentation maps obtained by interchanging the training and test collages formed by these textures and show the segmentation errors for these maps.

Ideally, we expect to obtain only slightly higher segmentation errors for arbitrary test images than for the training sample. The above and similar experiments with natural textures show that these expectations are in many cases unjustified in spite of very promising results obtained either for a training image or for a test image, large patches of which are used for learning. The low segmentation error rates obtained for these latter images may mislead in predicting the errors for the variety of test images. Also, texture inmohoheneities or different region statistics in the training and test images are outside the scope of the Gibbs models proposed in Chapter 2. The models presume both specific probability features of the regions in the piecewise-homogeneous image textures and translational homogeneity of the textures within each region. Thus, the images to be segmented need to meet particular constraints to ensure the admissible results of our CSA-based segmentation.

Due to estimating both the interaction structure and potentials, our models tend to learn more special than general texture features from a

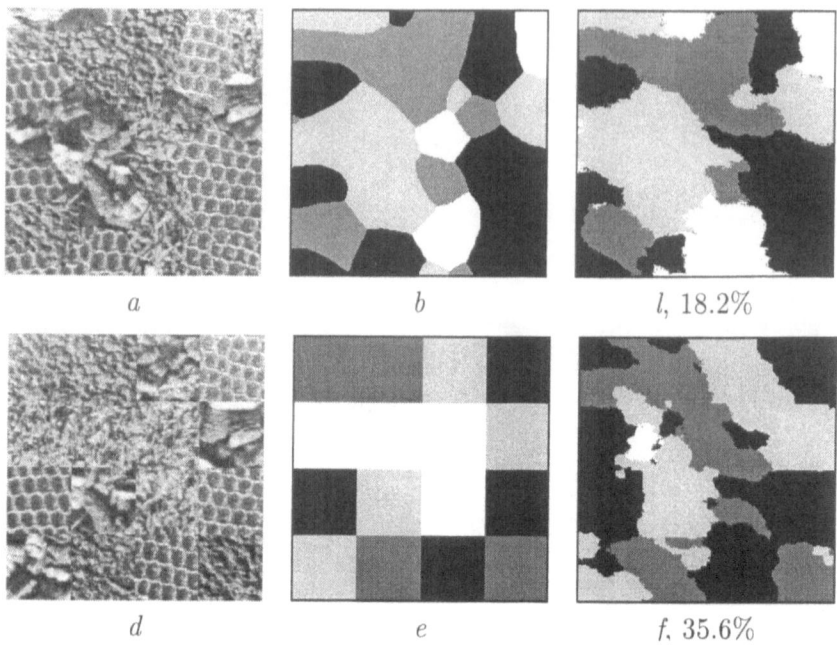

a b $l,\ 18.2\%$

d e $f,\ 35.6\%$

Figure 7.58. Segmentation of the collages from textures D3, D4, D5, D9 (the same notation as in Figure 7.52). The training (a) and test (d) collages are interchanged comparing
to those in Figure 7.52. Notice that most errors are due to the texture D9 represented
by too small patches in the training sample.

given training sample, especially, if sizes of textured regions are insufficient
to recover the most characteristic pairwise pixel interactions from the collected histograms. By changing the parameters and schedules of the CSA
and by varying the choices of characteristic interaction structure, slightly
better results for the individual training and test collages or a moderate
acceleration of the process may be obtained due to a smaller number of the
clique families or a smaller number of the CSA steps. But, it is unlikely
that such a tuning will enhance significally the segmentation results.

It is much more important for expecting the low segmentation errors,
that the image textures have to meet some specific constraints imposed
by our image models. The Gibbs models at hand presume the translation
invariance of the pixel interactions and expect at least a moderate homogeneity of the textures to be discriminated as well as correspondences
between the region sizes, forms, and relative arrangements of the regions
in the test and training samples.

Any natural texture demonstrates a big number of close-range and long-
range pixel interactions and the greater the range, the larger the training
regions have to be to obtain statistically consistent estimates of the model

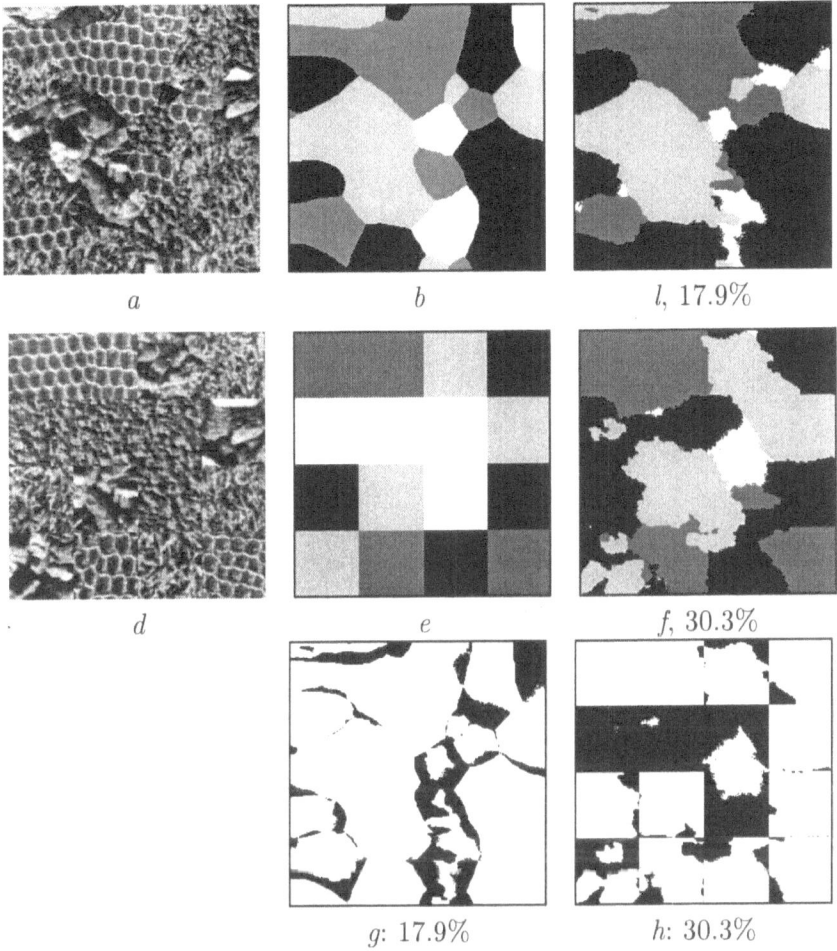

Figure 7.59.   Segmentation of the collages from textures D9, D3, D5, D4 (the same notation as in Figure 7.52). Black areas in the pictures (*g*) and (*h*) demonstrate the segmentation errors in the maps (*c*) and (*f*), respectively. Notice that most errors are now due to the texture D4 represented by too small patches in the training sample.

parameters. But, even in this case we can expect the high segmentation errors for the small patches of the textures in the test images where the model can mostly take only the inter-region interactions into account. It is obvious that the inter-region relations have much larger differences in most test images with respect to the training sample than the intra-region ones.

The MLE-based learning tends to closely adapt the model to a given training sample so that even very low errors for the training sample give no grounds for predicting the admissible separation of the same textures in another images. Also, if some sufficiently big patches of the textures are

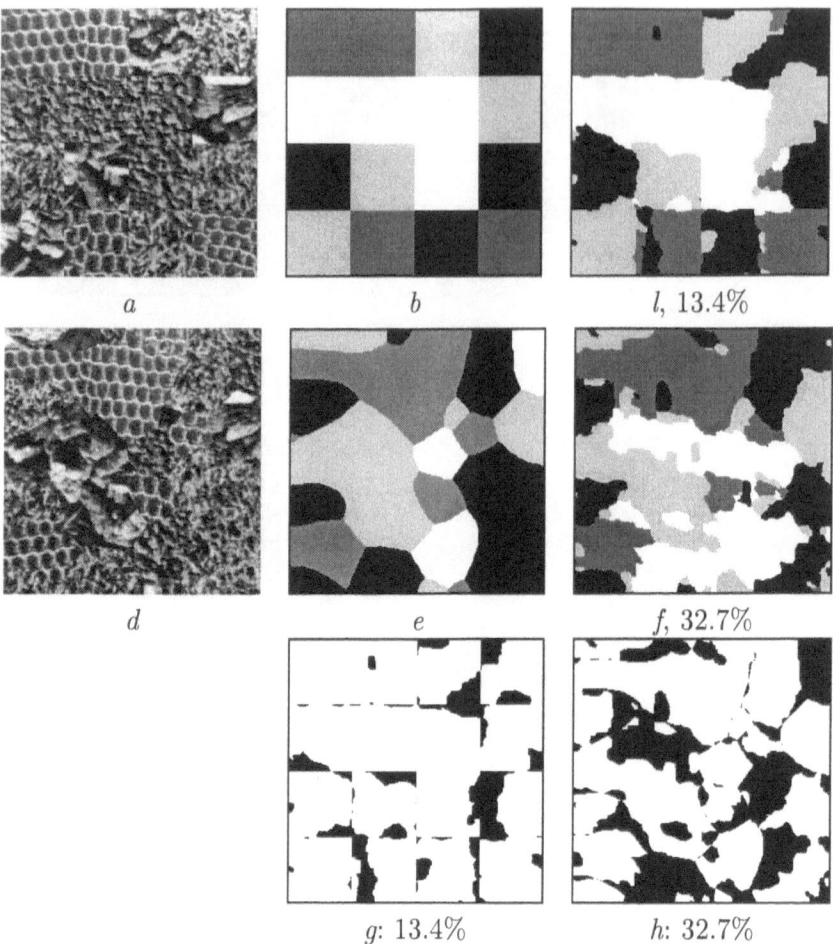

Figure 7.60.   Segmentation of the collages from textures D9, D3, D5, D4 (the same notation as in Figure 7.52). The training (*a*) and test (*d*) collages are interchanged comparing to those in Figure 7.59. Black areas in the pictures (*g*) and (*h*) demonstrate the segmentation errors in the maps (*c*) and (*f*), respectively.

cropped from the image to be segmented for forming the training sample, the segmentation errors can be expected to be small (especially, for and around the training patches). But, then it is hard to predict the same low errors for the like textures in the other images.

It seems that only by thorough testing the spatial homogeneity of the textures to be segmented and careful construction of the training samples we may bridge the gap between what is expected from the supervised segmentation and what is really learnt by the Gibbs models and displayed by the obtained segmentation maps.

Texture Modelling: Theory vs. Heuristics

Theoretical and experimental results presented in this book allow to conclude that the GRFs with multiple pairwise interactions hold much promise in simulating, retrieving, and segmenting various spatially homogeneous and piecewise-homogeneous image textures, DEMs, and other lattice-based spatial data. In principle, the models introduced in Chapter 2 can be supported by any arbitrary finite grid provided that the pairwise interactions between the grid sites form a uniform arc-colored neighborhood graph. But these GRFs are particularly attractive for modelling image textures because of the following features.

- In terms of these models, both the interaction structure and Gibbs potentials are explicitly related to particular signal co-occurrence histograms such as GLCH, GLDHs, GLC/RLCHs, and so on. Therefore we need not invent the potentials or assign the characteristic interactions on heuristic grounds and may anticipate them by analyzing the relevant histograms. Simultaneously, these models provide some grounding in theory for widely used heuristic GLDH-based features for texture recognition and allow us to gain a more penetrating insight into a physical meaning of pixel interactions in images.
- The feature of greatest practical utility is that the models permit us to compute the MLE of potentials by first analytic and then stochastic approximation and to use the initial estimates for recovering most characteristic interaction structure. This feature reduces considerably the heuristic part of modelling, too.
- The models allow to embed both texture simulation and supervised segmentation into the same computational framework. To consider segmentation as simulation of region maps described by a particular Gibbs model, we simply use a joint model of piecewise-homogeneous grayscale images and region maps that generalizes in a straightforward way the separate Gibbs models of the homogeneous images and region maps. These latter models take the form of particular cases of the joint model, and the conditional models of piecewise-homogeneous grayscale images, given a region map, and of region maps, given a grayscale image, are obtained easily from the joint model by fixing either a given region map or grayscale image.

239

– The models take into account that grayscale images can be subjected to simple gray range and limited scale/orientation transformations.

All the theoretical advantages over more traditional Gibbs image models such as auto-binomial or auto-normal one are obtained at the expense of a higher number of parameters to be learnt. But, the interrelations between Gibbs potentials and signal histograms and the proposed conditional MLE of potentials suggest well justified ways of how to describe the models with a lesser number of parameters.

Perhaps, the fact that the traditional scenario of Gibbs image modelling is replaced in this book by the alternative, CSA-based scenario is still a controversial subject. We argue that the traditional scenario based on the prior parameter estimation and subsequent image simulation by stochastic relaxation is not practicable because of an implicit goal, too slow convergence, and the absence of theoretically justified stopping rules. In the case of CSA, the simulation goal and stopping rules are defined in the explicit way because we have to closely approximate a chosen set of signal histograms for a training sample by similar histograms for simulated images, and this goal can be reached easily and reasonably fast by stochastic approximation.

In this book we assign the textures that can be described with a high degree of accuracy by our Gibbs models and modelling techniques to a particular class of stochastic textures. A large body of experiments show that many natural homogeneous and piecewise-homogeneous image textures belong to this class. To be certain, let us identify by eye which textures below are natural or simulated:

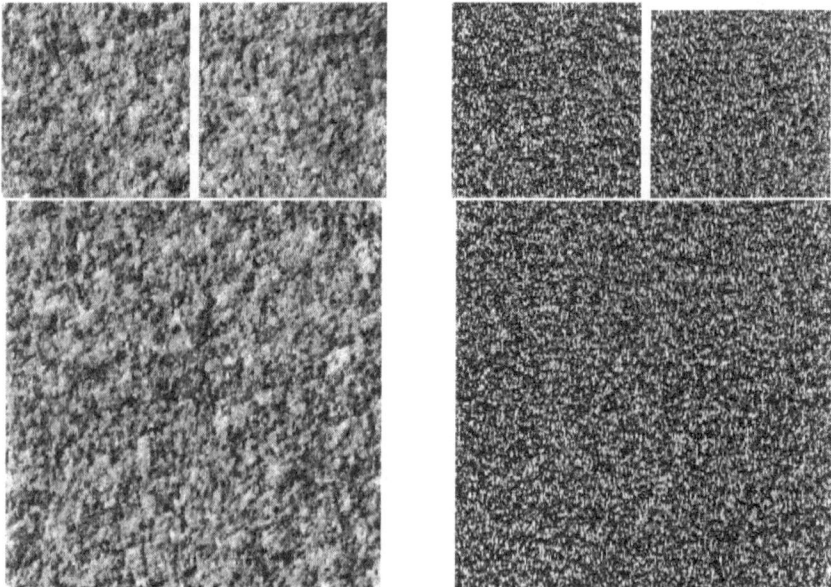

Textures "Fabrics15" and "Metal2" from the MIT "VisTex" database.

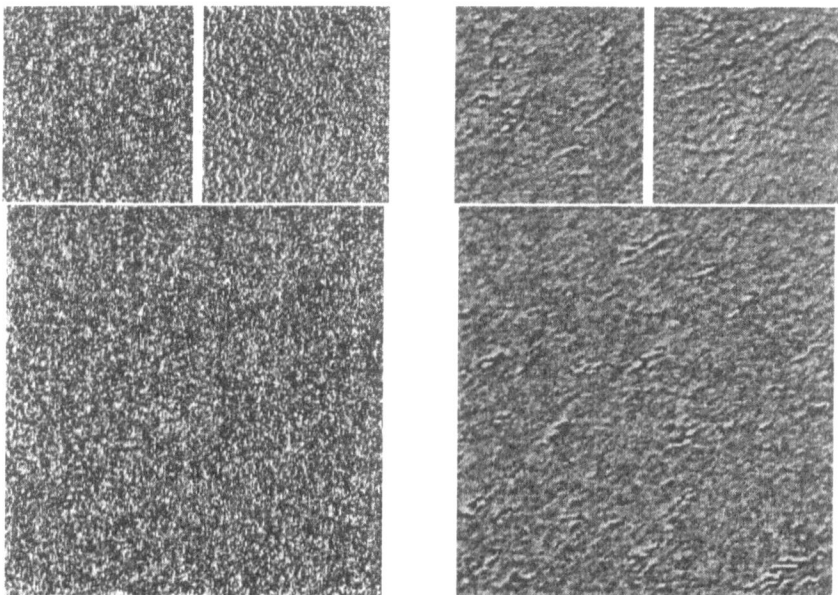

Textures "Metal3" and "Sand2" from the MIT "VisTex" database.

I do know that every upper right texture is natural and the two others are simulated, although you may agree that they are very similar. Of course, vastly more textures are outside the realm of this class of stochastic textures and must be modelled by other means.

But even for stochastic textures there still exist many open problems which are not solved in this book, in particular, an optimal search for most characteristic interaction structure, theoretically justified choices of control parameters and stopping rules for CSA, unsupervised segmentation by excluding or reducing the supervised learning of inter-region interactions which are most different in the training and test images, and so on.

I do hope that you, dear reader, will solve them soon or decide that they are not worth spending time on because you propose much better and easier ways for modelling these amazing and wonderful image textures...

References

K. Abend, "Compound decision procedures for pattern recognition", *Proc. National Electronic Conf.*, Volume 22, pp. 777–780, October 1966.

M. B. Averintsev, "On one method of describing random fields with discrete argument", *Problems of Information Transmission*, vol. 6, no. 2, pp. 100–108, 1970 [*In Russian*].

M. B. Averintsev, "Description of Markov random fields using Gibbs conditional probabilities". *Probability Theory and Its Applications*, vol. XVII, no. 1, pp. 21–35, 1972 [*In Russian*].

O. Barndorff-Nielsen, *Information and Exponential Families in Statistical Theory*. New York: John Wiley and Sons, 1978.

J. E. Besag, "Spatial interaction and the statistical analysis of lattice systems", *Journal of the Royal Statistical Society.*, vol. B36, pp. 192–236, 1974.

J. Besag, "On the statistical analysis of dirty pictures", *Journal of the Royal Statistical Society.*, vol. B48, pp. 259–302, 1986.

J. Besag, "Toward Bayesian image analysis", in *Advances in Applied Statistics: Statistics and Images: 1.* A Supplement to *Journal of Applied Statistics*, vol. 20, no. 5/6, pp. 107–119, 1993.

J. C. Bezdek and J. C. Dunn, "Optimal fuzzy partitions: a heuristic for estimating the parameters of a mixture of normal distributions", *IEEE Transactions on Computers*, vol. C-24, pp. 835–838, 1975.

A. T. Bharucha-Reid, *Elements of the Theory of Markov Processes and Their Applications"*. New York: McGraw-Hill, 1960.

P. Brodatz, *Textures: A Photographic Album for Artists an Designers*. New York: Dover Publications, 1966.

D. Brook, "On the distinction between the conditional probability and the joint probability approaches in the specification of nearest=neighbor systems", *Biometrika*, vol. 51, no. 3-4, pp. 481–483, 1964.

P. Carnevali, L. Coletti, and S. Partanello, "Image processing by simulated annealing", *IBM Journal of Research and Development*, vol. 29, no. 6, pp. 569–579, 1985.

S. Chatterjee, "Classification of natural textures using Gaussian Markov random field models", in *Markov Random Fields: Theory and Application* (R. Chellappa and A. Jain, eds), pp. 159–177. Boston: Academic Press, 1993.

R. Chellappa, "Two-dimensional discrete Gaussian Markov random field models for image processing", in: *Progress in Pattern Recognition 2* (L. N. Kanal and A. Rosenfeld, eds.), pp. 79–112. Amsterdam: Elsevier Science, North-Holland, 1985.

R. Chellappa and A. Jain (Eds), *Markov Random Fields: Theory and Application*. Boston: Academic Press, 1993.

R. Chellappa, R. L. Kashyap, and B. S. Manjunath, "Model-based texture segmentation and classification":, in *Handbook of Pattern Recognition and Computer Vision* (C. H. Chen, L. F. Pau, and P. S. P. Weng, eds), pp. 277–310. Singapore: World Scientific, 1993.

D. Chetverikov, "On some basic concepts of texture analysis", in *Proc. 2nd Int. Conf. on Computer Analysis of Images and Patterns*, (Wismar, GDR), pp. 196–201, September 1987.

D. Chetverikov, "GLDH based analysis of texture anisotropy and symmetry: an experimental study", in *Proceedings of the 12th IAPR International Conference on Pattern*

Recognition, Jerusalem, Israel, October 1994. Vol. I, pp. 444–448. Los Alamitos: IEEE Computer Society Press, 1994.

D. Chetverikov and R. M. Haralick, "Texture anisotropy, symmetry, regularity: recovering structure and orientation from interaction maps", in *Proceedings of the 6th British Machine Vision Conference*, (Birmingham, England), pp. 57–66, September 1995.

F. S. Cohen and D. B. Cooper, "Simple parallel hierarchical and relaxation algorithms for segmenting noncausal Markovian fields", *IEEE Transactions on Pattern Analysis and Machine Intelligence*, vol. 9, no. 2, pp. 195–219, 1987.

F. S. Cohen, Z. Fan, and S. Attali, "Automated inspection of textile fabrics using textural models", *IEEE Transactions on Pattern Analysis and Machine Intelligence*, vol. 13, no. 8, pp. 803–808, 1991.

F. S. Cohen and M. A. S. Patel, "Modeling and synthesis of images of 3D textured surfaces", *CVGIP: Graphical Models and Image Processing*, vol. 53, no. 6, pp. 501–510, 1991.

F. Comets, "On consistency of a class of estimators for exponential families of Markov random fields on the lattice", *Annals of Statistics*, vol. 20, pp. 455–486, 1992.

H. Cramer and M. R. Leadbetter, *Stationary and Related Stochastic Processes*. New York: John Wiley and Sons, 1967.

M. Creutz, *Quarks, Gluons and Lattices*. Cambridge: Cambridge University Press, 1983.

G. R. Cross and A. K. Jain, "Markov random field texture models", *IEEE Transactions on Pattern Analysis and Machine Intelligence*, vol. 5, no. 1, pp. 25–39, 1983.

H. Derin and W. S. Cole, "Segmentation of textured images using Gibbs random fields", *Computer Vision, Graphics, and Image Processing*, vol. 35, no. 1, pp. 72–98, 1986.

H. Derin and H. Elliot, "Modelling and segmentation of noisy and textured images using Gibbs random fields", *IEEE Transactions on Pattern Analysis and Machine Intelligence*, vol. 9, no. 1, pp. 39–55, 1987.

H. Derin, H. Elliot, R. Cristi, and D. Geman, "Bayes smoothing algorithm for segmentation of images modelled by Markov random fields", *IEEE Transactions on Pattern Analysis and Machine Intelligence*, vol. 6, no. 6, pp. 707–720, 1984.

H. Derin and P. A. Kelly, "Discrete-index Markov-type random processes", *Proceedings of the IEEE*, vol. 77, no. 10, pp. 1485–1510, 1989.

C. Derman, "A solution to a set of fundamental equations in Markov chains", *Proceedings of the American Mathematical Society*, vol. 5, pp. 332–334, 1954.

P. A. Devijver, "Hidden Markov mesh random field models in image analysis", *Advances in Applied Statistics: Statistics and Images: 1. A Supplement to J. of Applied Statistics*, vol. 20, no. 5/6, pp. 187–227, 1993.

R. Dikau, E. A. Brabb, R. K. Mark, and R. J. Pike, "Morphometric landform analysis of New Mexico", *Zeitschrift für Geomorphologie*, Suppl.–Bd. 101, pp. 109–126, 1995.

R. L. Dobrushin, "Gibbs random fields for the lattice systems with pairwise interaction", *Functional Analysis and Its Applications*, vol. 2, no. 4, pp. 31–43, 1968 [*In Russian*].

R. L. Dobrushin and S. A. Pigorov, "Theory of random fields", in *Proceedings of the 1975 IEEE-USSR Joint Workshop on Information Theory*, (Moscow, USSR), pp. 39–49, December 1975. New York: IEEE, 1976.

R. C. Dubes and A. K. Jain, "Random field models in image analysis", *Journal of Applied Statistics*, vol. 16, no. 2, pp. 131–164, 1989.

I. M. Elfadel and R. W. Picard, "Gibbs random fields, cooccurrences, and texture modelling", *IEEE Transactions on Pattern Analysis and Machine Intelligence*, vol. 16, no. 1, pp. 24–37, 1994.

C. Faloutsos, R. Barber, M. Flickner, J. Hafner, W. Niblack, and D. Petkovic, "Efficient and effective quering by image content", *Journal of Intelligent Information Systems*, vol. 3, pp. 231–262, 1994.

W. Feller, *An Introduction to Probability Theory and Its Applications*. Vol. 1. New York: John Wiley and Sons, 1970.

M. Flickner, H. Sawhney, W. Niblack, J. Ashey, Q. Huang, B. Dom, M. Gorkani, J. Hafner, D. Lee, D. Petkovic, D. Steele, and P. Yanker, "Query by image and video content: the QBIC system", *IEEE Computer*, vol. 28, pp. 23–32, 1995.

S. B. Gelfand and S. K. Mitter, "On sampling methods and annealing algorithms", in *Markov Random Fields: Theory and Applications* (R. Chellappa and A. Jain, eds), pp. 499–515. Boston: Academic Press, 1993.

D. Geman, S. Geman, C. Graffigne, and P. Dong, "Boundary detection by constrained optimization", *IEEE Transactions on Pattern Analysis and Machine Intelligence*, vol. 12, no. 7, pp. 609–628, 1990.

S. Geman and D. Geman, "Stochastic relaxation, Gibbs distributions, and the Bayesian restoration of images", *IEEE Transactions on Pattern Analysis and Machine Intelligence*, vol. 6, no. 6, pp. 721–741, 1984.

D. Geman, G. Reynolds, and C. Yang, "Stochastic algorithms for restricted image spaces and experiments in deblurring", in *Markov Random Fields: Theory and Applications* (R. Chellappa and A. Jain, eds), pp. 39–68. Boston: Academic Press, 1993.

B. Gidas, "Nonstationary Markov chains and convergence of the annealing algorithm", *Journal of Statistical Physics*, vol. 39, no. 1-2, pp. 73–131, 1985.

B. Gidas, "Parameter estimation for Gibbs distributions from fully observed data", in *Markov Random Fields: Theory and Applications* (R. Chellappa and A. Jain, eds), pp. 471–483. Boston: Academic Press, 1993.

G. L. Gimel'farb, "Gibbs random fields and compound Bayesian decisions at the lower level of digital image processing", *Pattern Recognition and Image Analysis: Advances in Mathematical Theory and Applications in the USSR*, vol. 1, no. 1, pp. 39–49, 1991.

G. L. Gimel'farb, "Texture modeling by multiple pairwise pixel interactions", *IEEE Trans. Pattern Analysis and Machine Intell.*, vol. 18, no. 11, pp. 1110–1114, 1996.

G. L. Gimel'farb, "Non-Markov Gibbs texture model with multiple pairwise pixel interactions", in *Proceedings of the 13th IAPR International Conference on Pattern Recognition*, Vienna, Austria, August 1996. Vol. II, pp. 591-595. Los Alamitos: IEEE Computer Society Press, 1996.

G. L. Gimel'farb, "Gibbs models for Bayesian simulation and segmentation of piecewise-uniform textures", *Ibid.*, pp. 760–764.

G. L. Gimel'farb, "Probabilistic models in computer vision: possibilities and limitations", *Vistas in Astronomy*, vol. 40, no. 4, pp. 487–494, 1996.

G. L. Gimel'farb, *Gibbs Fields with Multiple Pairwise Pixel Interactions for Texture Simulation and Segmentation*. Research Report no. 3202, 68 p. INRIA Sophia Antipolis, July 1997.

G. L. Gimel'farb, "Analytic approximation of Gibbs potentials to model stochastic textures", in *Proceedings of the First Joint Australia & New Zealand Biennial Conference on Digital Image & Vision Computing: Techniques and Applications*, Auckland, New Zealand, December 1997, pp. 153–158. Palmerston North: Production Technology, 1997.

G. L. Gimel'farb, J. Schmidt, and A. Braunmandl, "Gibbs fields with multiple pairwise interactions as a tool for modelling grid-based data", in *Process Modelling and Landform Evolution* (S. Hergarten, H. J. Neugebauer, eds.) *Lecture Notes in Earth Sciences 78*. Berlin e.a.: Springer, pp. 47–74, 1999.

G. L. Gimel'farb and A. K. Jain, "On retrieving textured images from an image data base", *Pattern Recognition*, vol. 29, no. 9, pp. 1461–1483, 1996.

G. L. Gimel'farb and A. V. Zalesny, "Models of Markov random fields in the problems of simulating and segmenting textured images", in *Means to Intellectualize Cybernetic Systems*, pp. 27–36. Kiev: V. M. Glushkov Inst. of Cybernetics, 1989 [*In Russian*].

G. L. Gimel'farb and A. V. Zalesny, "Low-level Bayesian segmentation of piecewise-homogeneous noisy and textured images", *International Journal of Imaging Systems and Technology*, vol. 3, no. 3, pp. 227–243, 1991.

G. L. Gimel'farb and A. V. Zalesny, "Probabilistic models of digital region maps based on Markov random fields with short- and long-range interaction", *Pattern Recognition Letters*, vol. 14, no. 5, pp. 789–797, 1993.

G. L. Gimel'farb and A. V. Zalesny, "Markov random fields with short- and long-range interaction for modelling gray-scale textured images", in *Proceedings of the Fifth International Conference on Computer Analysis of Images and Patterns* (D. Chetverikov and W. Kropatsch, eds), Budapest, Hungary, September 1993, pp. 275–282. *Lecture Notes in Computer Science 719*, Berlin: Springer, 1993.

G. R. Grimmett, "A theorem about random fields", *Bulletin of the London Mathematical Society*, vol. 5, pp. 81–84, 1973.

J. M. Hammersley and D. C. Handscomb, *Monte-Carlo Methods*, London: Methuen, 1964.

R. M. Haralick, "Statistical and structural approaches to textures", *Proceedings of the IEEE*, vol. 67, no. 8, pp. 786–804, 1979.

R. M. Haralick and L. G. Shapiro, *Computer and Robot Vision*, Vol. 1. Reading: Addison-Wesley, 1992.

M. Hassner and J. Sklansky, Markov random fields as models of digitized image texture, in: *Proceedings of the IEEE Computer Society Conference on Pattern Recognition and Image Processing*, Chicago, USA, May-June 1978, pp. 346–351. IEEE Computer Society Press, 1978.

M. Hassner and J. Sklansky, "The use of Markov random fields as models of textures", *Computer Graphics and Image Processing*, vol. 12, no. 4, pp. 357–370, 1980.

B. Holt and L. Hartwick, "Visual image retrieval for application in art and art history", *Proceedings of the SPIE*, vol. 2185, pp. 70–81, February 1994.

A. Isihara, *Statistical Physics*. Orlando: Academic Press, 1971.

M. Jacobsen, "Existence and unicity of MLE in discrete exponential family distributions", *Scandinavian Journal of Statistics*, vol. 16, pp. 335–349, 1989.

A. K. Jain and G. Gimel'farb, "Retrieving textured images from an image data base", in *Proceedings of the 9th Scandinavian Conference on Image Analysis*, Uppsala, Sweden, June 1995. Vol. 1, pp. 441–448. Uppsala: SSAIA, 1995.

B. Julesz, "Textons, the elements of texture perception, and their interactions", *Nature*, no. 290, pp. 91–97, 1981.

B. Julesz and J. R. Bergen, "Textons: the fundamental elements in preattentive vision and perception of textures", *Bell Systems Technical Journal*, vol. 62, no. 6, pp. 1619–1645, 1983.

N. Karssmeijer, "A relaxation method for image segmentation using a spatially dependent stochastic model", *Pattern Recognition Letters*, vol. 11, no. 1, pp. 13–23, 1990.

R. L. Kashyap, Random field models on torous lattices for finite images, in *Proceedings of the IEEE 5th International Conference on Pattern Recognition*, Miami Beach, USA, December 1980, pp. 1103-1105. IEEE Computer Society Press, 1980.

R. L. Kashyap, Analysis and synthesis of image patterns by spatial interaction models, in *Progress in Pattern Recognition 1* (L. N. Kanal and A. Rosenfeld, eds.), pp. 149–186. Amsterdam: North-Holland, 1981.

R. L. Kashyap, "Image models", in *Handbook on Pattern Recognition and Image Processing* (T. Y. Young and K.-S. Fu, eds), pp. 247–279. Orlando: Academic Press, 1986.

R. L. Kashyap and R. Chellappa, "Estimation and choice of neighbors in spatial-interaction models of images", *IEEE Transactions on Information Theory*, vol. 29, no. 1, pp. 60–72, 1983.

M. G. Kendall and A. Stuart, *The Advanced Theory of Statistics*. Vol. 3. London: Charles Griffin, 1966.

J. G. Kemeney and J. L. Snell, *Finite Markov Chains*. Princeton: van Nostrand Co., 1960.

S. Kirkpatrick, C. D. Gelett, and M. P. Wecchi, "Optimization by simulatewd annealing", *Science*, no. 220, pp. 671–680, 1983.

S. Kirkpatrick and R. H. Swendsen, "Statistical mechanics and disordered systems", *Communications of the ACM*, vol. 28, no. 4, pp. 363–373, 1985.

P. J. M. van Laarhoven, *Theoretical and Computational Aspects of Simulated Annealing*, Amsterdam: Stichting Mathematisch Centrum, 1988.

D. S. Lebedev, A. A. Bezruk, and V. M. Novikov, *Markov Probabilistic Model of Image and Picture*. Preprint: Institute of Information Transmission Problems, Academy of Sciences of the USSR. Moscow: VINITI, 1983 [*In Russian*].

D. S. Lebedev, *Statistical Theory of Video Data Processing*, Moscow: Moscow Inst. of Physics and Technology (MFTI), 1988. [*In Russian*].

E. Levitan, M. Chang, and G. Herman, "Image-modeling Gibbs priors", *Graphical Models and Image Processing*, vol. 57, no. 2, pp. 117–130, 1995.

S. Z. Li, *Markov Random Field Modeling in Computer Vision*. Tokyo: Springer, 1995.

T. M. Liggett, *Interacting Particle Systems*. New York: Springer, 1985.

E. Lloyd (Ed.), *Handbook of Applicable Mathematics*. Vol. VI: Statistics. Part A. Chichester: John Wiley and Sons, 1984.

G. Lohmann, "Co-occurrence-based analysis and synthesis of textures, in *Proceedings of the 12th IAPR International Conference on Pattern Recognition*, Jerusalem, Israel, October 1994. Vol. 1, pp. 449–453. Los Alamitos: IEEE Computer Society Press, 1994.

D. Marr, *Vision*. San Francisco: Freeman, 1982.

J. L. Marroquin, "Deterministic Bayesian estimation of Markov random fields with applications to computational vision", in *Proceedings of the First International Conference on Computer Vision*, London, England, June 1987, pp. 597–601.

J. Marroquin, "Deterministic interaction particle models for image processing and computer graphics", *CVGIP: Graphical Models and Image Processing*, vol. 55, no. 5, pp. 408–417, 1993.

J. Marroquin, S. Mitter, and T. Poggio, "Probabilistic solution of ill-posed problems in computational vision", *Journal of the American Statistical Association*, vol. 82, no. 397, pp. 76–89, 1987.

N. Metropolis, A. W. Rosenbluth, M. N. Rosenbluth, A. H. Teller, and E. Teller, "Equations of state calculations by fast computing machines", *Journal of Chemical Physics*, vol. 21, pp. 1087–1091, 1953.

J. Moussouris, "Gibbs and Markov random systems with constraints", *Journal of Statistical Physics*, vol. 10, pp. 11–33, 1974.

M. B. Nevel'son and R. Z. Has'minskiĭ, *Stochastic Approximation and Recursive Estimation*. Providence: AMS, 1973 [*Translated from Russian*].

R. W. Picard and I. M. Elfadel, "Structure of Aura and Cooccurrence Matrices for the Gibbs Texture Model", *Journal of Mathematical Imaging and Vision*, vol. 2, no. 1, pp. 5–25, 1992.

R. Pickard, C. Graszyk, S. Mann, J. Wachman, L. Pickard, and L. Campbell, *VisTex Database*. Cambridge, Mass.: MIT Media Laboratory, 1995.

G. X. Ritter, J. N. Wilson, and J. L. Davidson, "Image algebra: an overview", *Computer Vision, Graphics, and Image Processing*, vol. 49, no. 3, pp. 297–331, 1990.

The Oxford English Dictionary (the compact edition). "Structure", "Texture", Vol. II, p. 1165, p. 3274. Oxford: Oxford University Press, 1971.

The Oxford English Dictionary (2nd edition). "Structure", Vol. XVI, pp. 959–960; "Texture", Vol. XVII, pp. 854–855. Oxford: Clarendon Press – Oxford University Press, 1989.

The World Book Dictionary (C. L. Barnhart and R. K. Barnhart, eds). "Structure", "Texture", Vol. 2, p. 2077, p. 2170. Chicago: World Book, 1990.

M. Tuceryan and A. K. Jain, "Texture analysis", in *Handbook on Pattern Recognition and Computer Vision* (C. H. Chen, L. F. Pau, and P. S. P. Weng, eds), pp. 235–276. Singapore: World Scientific, 1993.

M. Wasan, *Stochastic Approximation*. Cambridge: Cambridge University Press, 1969.

Webster's New International Dictionary of English Language. "Structure", "Texture", p. 2501, p. 2614. Spingfield: Merriam, 1959.

Webster's Third New International Dictionary of English Language. "Structure", "Texture", p. 2267, p. 2366. Spingfield: Merriam-Webster, 1986.

G. Winkler, *Image Analysis, Random Fields and Dynamic Monte Carlo Methods*. Berlin: Springer, 1995.

L. Younes, "Estimation and annealing for Gibbsian fields", *Annales de l'Institut Henri Poincaré - Probabilités et Statistiques*, vol. 24, no. 2, pp. 269–294, 1988.

A. V. Zalesny, "Homogeneity & texture. General approach", in *Proceedings of the 12th IAPR International Conference on Pattern Recognition*, Jerusalem, Israel, October 1994, Vol. 1, pp. 592–594. Los Alamitos: IEEE Computer Society Press, 1994.

A. V. Zalesny, *Personal communication*, 1996.

Index

Computational Imaging and Vision

Kluwer Academic Publishers – Dordrecht / Boston / London